The Handbook
of Environmental Chemistry

Editor-in-Chief: O. Hutzinger

Volume 2 Reactions and Processes
Part N

Advisory Board:
D. Barceló · P. Fabian · H. Fiedler · H. Frank · J. P. Giesy · R. A. Hites
T. A. Kassim · M. A. K. Khalil · D. Mackay · A. H. Neilson
J. Paasivirta · H. Parlar · S. H. Safe · P. J. Wangersky

The Handbook of Environmental Chemistry
Recently Published and Forthcoming Volumes

Environmental Specimen Banking
Volume Editors: S. A. Wise and P. P. R. Becker
Vol. 3/S, 2006

Polymers: Chances and Risks
Volume Editors: P. Eyerer, M. Weller
and C. Hübner
Vol. 3/V, 2006

The Rhine
Volume Editor: T. P. Knepper
Vol. 5/L, 03.2006

Persistent Organic Pollutants
in the Great Lakes
Volume Editor: R. A. Hites
Vol. 5/N, 2006

Antifouling Paint Biocides
Volume Editor: I. Konstantinou
Vol. 5/O, 2006

Estuaries
Volume Editor: P. J. Wangersky
Vol. 5/H, 2006

The Caspian Sea
Volume Editors: A. Kostianoy and A. Kosarev
Vol. 5/P, 2005

Marine Organic Matter: Biomarkers,
Isotopes and DNA
Volume Editor: J. K. Volkman
Vol. 2/N, 2005

Environmental Photochemistry Part II
Volume Editors: P. Boule, D. Bahnemann
and P. Robertson
Vol. 2/M, 2005

Air Quality in Airplane Cabins
and Similar Enclosed Spaces
Volume Editor: M. B. Hocking
Vol. 4/H, 2005

Environmental Effects
of Marine Finfish Aquaculture
Volume Editor: B. T. Hargrave
Vol. 5/M, 2005

The Mediterranean Sea
Volume Editor: A. Saliot
Vol. 5/K, 2005

Environmental Impact Assessment of Recycled
Wastes on Surface and Ground Waters
Engineering Modeling and Sustainability
Volume Editor: T. A. Kassim
Vol. 5/F (3 Vols.), 2005

Oxidants and Antioxidant Defense Systems
Volume Editor: T. Grune
Vol. 2/O, 2005

Marine Organic Matter: Biomarkers, Isotopes and DNA

Volume Editor:
John K. Volkman

With contributions by
J. Albaigés · M. A. Altabet · J. M. Bayona · E. A. Canuel
C. Corinaldesi · R. Danovaro · A. Dell'Anno · H. R. Harvey
S. W. Jeffrey · G. M. Luna · S. Schouten · B. R. T. Simoneit
J. S. Sinninghe Damsté · M. Pagani · R. D. Pancost
J. K. Volkman · S. G. Wakeham · S. W. Wright

 Springer

Environmental chemistry is a rather young and interdisciplinary field of science. Its aim is a complete description of the environment and of transformations occurring on a local or global scale. Environmental chemistry also gives an account of the impact of man's activities on the natural environment by describing observed changes.

The Handbook of Environmental Chemistry provides the compilation of today's knowledge. Contributions are written by leading experts with practical experience in their fields. The Handbook will grow with the increase in our scientific understanding and should provide a valuable source not only for scientists, but also for environmental managers and decision-makers.

The Handbook of Environmental Chemistry is published in a series of five volumes:

Volume 1: The Natural Environment and the Biogeochemical Cycles
Volume 2: Reactions and Processes
Volume 3: Anthropogenic Compounds
Volume 4: Air Pollution
Volume 5: Water Pollution

The series Volume 1 The Natural Environment and the Biogeochemical Cycles describes the natural environment and gives an account of the global cycles for elements and classes of natural compounds.
The series Volume 2 Reactions and Processes is an account of physical transport, and chemical and biological transformations of chemicals in the environment.
The series Volume 3 Anthropogenic Compounds describes synthetic compounds, and compound classes as well as elements and naturally occurring chemical entities which are mobilized by man's activities.
The series Volume 4 Air Pollution and Volume 5 Water Pollution deal with the description of civilization's effects on the atmosphere and hydrosphere.

Within the individual series articles do not appear in a predetermined sequence. Instead, we invite contributors as our knowledge matures enough to warrant a handbook article.
Suggestions for new topics from the scientific community to members of the Advisory Board or to the Publisher are very welcome.

Library of Congress Control Number: 2005930943

ISSN 1433-6839
ISBN-10 3-540-28401-X Springer Berlin Heidelberg New York
ISBN-13 978-3-540-28401-7 Springer Berlin Heidelberg New York
DOI 10.1007/b11682

This work is subject to copyright. All rights are reserved, whether the whole or part of the material is concerned, specifically the rights of translation, reprinting, reuse of illustrations, recitation, broadcasting, reproduction on microfilm or in any other way, and storage in data banks. Duplication of this publication or parts thereof is permitted only under the provisions of the German Copyright Law of September 9, 1965, in its current version, and permission for use must always be obtained from Springer. Violations are liable for prosecution under the German Copyright Law.

Springer is a part of Springer Science+Business Media

springer.com

© Springer-Verlag Berlin Heidelberg 2006
Printed in Germany

The use of registered names, trademarks, etc. in this publication does not imply, even in the absence of a specific statement, that such names are exempt from the relevant protective laws and regulations and therefore free for general use.

Cover design: E. Kirchner, Springer-Verlag
Typesetting and Production: LE-TEX Jelonek, Schmidt & Vöckler GbR, Leipzig

Printed on acid-free paper   02/3141 YL – 5 4 3 2 1 0

Editor-in-Chief

Prof. em. Dr. Otto Hutzinger
Universität Bayreuth
c/o Bad Ischl Office
Grenzweg 22
5351 Aigen-Vogelhub, Austria
hutzinger-univ-bayreuth@aon.at

Volume Editor

Dr. John K. Volkman
CSIRO Marine and Atmospheric
Research Laboratories
Castray Esplanade,
7000 Hobart, Australia
john.volkman@csiro.au

Advisory Board

Prof. Dr. D. Barceló
Dept. of Environmental Chemistry
IIQAB-CSIC
JordiGirona, 18–26
08034 Barcelona, Spain
dbcqam@cid.csic.es

Prof. Dr. P. Fabian
Lehrstuhl für Bioklimatologie
und Immissionsforschung
der Universität München
Hohenbachernstraße 22
85354 Freising-Weihenstephan, Germany

Dr. H. Fiedler
Scientific Affairs Office
UNEP Chemicals
11–13, chemin des Anémones
1219 Châteleine (GE), Switzerland
hfiedler@unep.ch

Prof. Dr. H. Frank
Lehrstuhl für Umwelttechnik
und Ökotoxikologie
Universität Bayreuth
Postfach 10 12 51
95440 Bayreuth, Germany

Prof. Dr. J. P. Giesy
Department of Zoology
Michigan State University
East Lansing, MI 48824-1115, USA
Jgiesy@aol.com

Prof. Dr. R. A. Hites
Indiana University
School of Public
and Environmental Affairs
Bloomington, IN 47405, USA
hitesr@indiana.edu

Dr. T. A. Kassim
Department of Civil
and Environmental Engineering
College of Science and Engineering
Seattle University
901 12th Avenue
Seattle, WA 98122-1090, USA
kassimt@seattleu.edu

Prof. Dr. M. A. K. Khalil
Department of Physics
Portland State University
Science Building II, Room 410
P.O. Box 751
Portland, OR 97207-0751, USA
aslam@global.phy.pdx.edu

Prof. Dr. D. Mackay
Department of Chemical Engineering
and Applied Chemistry
University of Toronto
Toronto, ON, Canada M5S 1A4

Prof. Dr. A. H. Neilson
Swedish Environmental Research Institute
P.O. Box 21060
10031 Stockholm, Sweden
ahsdair@ivl.se

Prof. Dr. J. Paasivirta
Department of Chemistry
University of Jyväskylä
Survontie 9
P.O. Box 35
40351 Jyväskylä, Finland

Prof. Dr. Dr. H. Parlar
Institut für Lebensmitteltechnologie
und Analytische Chemie
Technische Universität München
85350 Freising-Weihenstephan, Germany

Prof. Dr. S. H. Safe
Department of Veterinary
Physiology and Pharmacology
College of Veterinary Medicine
Texas A & M University
College Station, TX 77843-4466, USA
ssafe@cvm.tamu.edu

Prof. P. J. Wangersky
University of Victoria
Centre for Earth and Ocean Research
P.O. Box 1700
Victoria, BC, V8W 3P6, Canada
wangers@telus.net

The Handbook of Environmental Chemistry
Also Available Electronically

For all customers who have a standing order to The Handbook of Environmental Chemistry, we offer the electronic version via SpringerLink free of charge. Please contact your librarian who can receive a password or free access to the full articles by registering at:

springerlink.com

If you do not have a subscription, you can still view the tables of contents of the volumes and the abstract of each article by going to the SpringerLink Homepage, clicking on "Browse by Online Libraries", then "Chemical Sciences", and finally choose The Handbook of Environmental Chemistry.

You will find information about the

– Editorial Board
– Aims and Scope
– Instructions for Authors
– Sample Contribution

at springer.com using the search function.

Preface

Environmental Chemistry is a relatively young science. Interest in this subject, however, is growing very rapidly and, although no agreement has been reached as yet about the exact content and limits of this interdisciplinary discipline, there appears to be increasing interest in seeing environmental topics which are based on chemistry embodied in this subject. One of the first objectives of Environmental Chemistry must be the study of the environment and of natural chemical processes which occur in the environment. A major purpose of this series on Environmental Chemistry, therefore, is to present a reasonably uniform view of various aspects of the chemistry of the environment and chemical reactions occurring in the environment.

The industrial activities of man have given a new dimension to Environmental Chemistry. We have now synthesized and described over five million chemical compounds and chemical industry produces about hundred and fifty million tons of synthetic chemicals annually. We ship billions of tons of oil per year and through mining operations and other geophysical modifications, large quantities of inorganic and organic materials are released from their natural deposits. Cities and metropolitan areas of up to 15 million inhabitants produce large quantities of waste in relatively small and confined areas. Much of the chemical products and waste products of modern society are released into the environment either during production, storage, transport, use or ultimate disposal. These released materials participate in natural cycles and reactions and frequently lead to interference and disturbance of natural systems.

Environmental Chemistry is concerned with reactions in the environment. It is about distribution and equilibria between environmental compartments. It is about reactions, pathways, thermodynamics and kinetics. An important purpose of this Handbook, is to aid understanding of the basic distribution and chemical reaction processes which occur in the environment.

Laws regulating toxic substances in various countries are designed to assess and control risk of chemicals to man and his environment. Science can contribute in two areas to this assessment; firstly in the area of toxicology and secondly in the area of chemical exposure. The available concentration ("environmental exposure concentration") depends on the fate of chemical compounds in the environment and thus their distribution and reaction behaviour in the environment. One very important contribution of Environmental Chemistry to the above mentioned toxic substances laws is to develop laboratory test methods, or mathematical correlations and models that predict the environmental fate of new chemical compounds. The third purpose of this Handbook is to help in the basic understanding and development of such test methods and models.

The last explicit purpose of the Handbook is to present, in concise form, the most important properties relating to environmental chemistry and hazard assessment for the most important series of chemical compounds.

At the moment three volumes of the Handbook are planned. Volume 1 deals with the natural environment and the biogeochemical cycles therein, including some background information such as energetics and ecology. Volume 2 is concerned with reactions and processes in the environment and deals with physical factors such as transport and adsorption, and chemical, photochemical and biochemical reactions in the environment, as well as some aspects of pharmacokinetics and metabolism within organisms. Volume 3 deals with anthropogenic compounds, their chemical backgrounds, production methods and information about their use, their environmental behaviour, analytical methodology and some important aspects of their toxic effects. The material for volume 1, 2 and 3 was each more than could easily be fitted into a single volume, and for this reason, as well as for the purpose of rapid publication of available manuscripts, all three volumes were divided in the parts A and B. Part A of all three volumes is now being published and the second part of each of these volumes should appear about six months thereafter. Publisher and editor hope to keep materials of the volumes one to three up to date and to extend coverage in the subject areas by publishing further parts in the future. Plans also exist for volumes dealing with different subject matter such as analysis, chemical technology and toxicology, and readers are encouraged to offer suggestions and advice as to future editions of "The Handbook of Environmental Chemistry".

Most chapters in the Handbook are written to a fairly advanced level and should be of interest to the graduate student and practising scientist. I also hope that the subject matter treated will be of interest to people outside chemistry and to scientists in industry as well as government and regulatory bodies. It would be very satisfying for me to see the books used as a basis for developing graduate courses in Environmental Chemistry.

Due to the breadth of the subject matter, it was not easy to edit this Handbook. Specialists had to be found in quite different areas of science who were willing to contribute a chapter within the prescribed schedule. It is with great satisfaction that I thank all 52 authors from 8 countries for their understanding and for devoting their time to this effort. Special thanks are due to Dr. F. Boschke of Springer for his advice and discussions throughout all stages of preparation of the Handbook. Mrs. A. Heinrich of Springer has significantly contributed to the technical development of the book through her conscientious and efficient work. Finally I like to thank my family, students and colleagues for being so patient with me during several critical phases of preparation for the Handbook, and to some colleagues and the secretaries for technical help.

I consider it a privilege to see my chosen subject grow. My interest in Environmental Chemistry dates back to my early college days in Vienna. I received significant impulses during my postdoctoral period at the University of California and my interest slowly developed during my time with the National Research Council of Canada, before I could devote my full time of Environmental Chemistry, here in Amsterdam. I hope this Handbook may help deepen the interest of other scientists in this subject.

Amsterdam, May 1980   *O. Hutzinger*

Twenty-one years have now passed since the appearance of the first volumes of the Handbook. Although the basic concept has remained the same changes and adjustments were necessary.

Some years ago publishers and editors agreed to expand the Handbook by two new open-end volume series: Air Pollution and Water Pollution. These broad topics could not be fitted easily into the headings of the first three volumes. All five volume series are integrated through the choice of topics and by a system of cross referencing.

The outline of the Handbook is thus as follows:

1. The Natural Environment and the Biochemical Cycles,
2. Reaction and Processes,
3. Anthropogenic Compounds,
4. Air Pollution,
5. Water Pollution.

Rapid developments in Environmental Chemistry and the increasing breadth of the subject matter covered made it necessary to establish volume-editors. Each subject is now supervised by specialists in their respective fields.

A recent development is the accessibility of all new volumes of the Handbook from 1990 onwards, available via the Springer Homepage springeronline.com or springerlink.com.

During the last 5 to 10 years there was a growing tendency to include subject matters of societal relevance into a broad view of Environmental Chemistry. Topics include LCA (Life Cycle Analysis), Environmental Management, Sustainable Development and others. Whilst these topics are of great importance for the development and acceptance of Environmental Chemistry Publishers and Editors have decided to keep the Handbook essentially a source of information on "hard sciences".

With books in press and in preparation we have now well over 40 volumes available. Authors, volume-editors and editor-in-chief are rewarded by the broad acceptance of the "Handbook" in the scientific community.

Bayreuth, July 2001   *Otto Hutzinger*

Contents

Sources and Cycling of Organic Matter
in the Marine Water Column
H. R. Harvey   1

Lipid Markers for Marine Organic Matter
J. K. Volkman 27

Pigment Markers for Phytoplankton Production
S. W. Wright · S. W. Jeffrey 71

Molecular Tools for the Analysis of DNA in Marine Environments
R. Danovaro · C. Corinaldesi · G. M. Luna · A. Dell'Anno 105

Biological Markers for Anoxia
in the Photic Zone of the Water Column
J. S. Sinninghe Damsté · S. Schouten 127

Atmospheric Transport
of Terrestrial Organic Matter to the Sea
B. R. T. Simoneit 165

Controls on the Carbon Isotopic Compositions
of Lipids in Marine Environments
R. D. Pancost · M. Pagani 209

Isotopic Tracers of the Marine Nitrogen Cycle: Present and Past
M. A. Altabet 251

**Degradation and Preservation
of Organic Matter in Marine Sediments**
S. G. Wakeham · E. A. Canuel 295

Sources and Fate of Organic Contaminants in the Marine Environment
J. M. Bayona · J. Albaigés 323

Subject Index 371

Foreword

The oceans play a vital role in moderating the Earth's climate and in providing food for the Earth's human inhabitants and yet many of the processes of carbon and nutrient cycling are still not well understood. Modern advances in molecular biology are revealing a myriad of uncultured organisms in marine ecosystems, many having unknown ecology and function. These organisms have a rich variety of unusual genes and biochemistries which produce a diverse array of organic compounds ranging from colourful carotenoids and chlorophylls to lipids with structures ranging from the simple to the complex.

This book brings together 10 chapters on the use of lipid biomarkers, pigments, isotopes and molecular biology to ascertain the sources and fate of organic matter (both natural and pollutant) in the sea and underlying sediments. The authors are expert in their field and they have been able to bring their broader knowledge of marine processes to provide both an overview of the state-of-the-art and knowledge gaps with sufficient detail to satisfy the needs of specialists and non-specialists alike. All are very busy researchers at the leading edge of their science and I am grateful that they were able to find the time to write these reviews.

A characteristic feature of today's marine science is the need for multidisciplinary approaches. Thus the skills and knowledge of the chemist, biologist, physical oceanographer and modeller are needed to unravel the interactions between organisms in marine food-webs and the cycling of the major elements. A multi-marker approach is also desirable – an approach that makes use of biomarkers, isotopes and DNA which might be thought of as the ultimate biomarker. Advances in methodology have played a major role with a range of highly sensitive "hyphenated techniques" now available including gas chromatography and high performance liquid chromatography linked to mass spectrometry systems (GC-MS and HPLC-MS) for compound identification. Continuous flow GC-irm-MS systems can now provide stable isotope values for compounds separated by GC. Methods are also now available to measure the ^{14}C-content of individual compounds and thus estimate their age which has revealed that some of the more refractory compounds in the sea may have been synthesized many hundreds (or in some cases thousands) of years previously.

My intention as editor has been to include detailed information of practical use to new researchers. I hope that the book can provide a roadmap for the analysis of the different organic compounds found in the sea, atmosphere and sediments. In addition, detailed information is provided on the fundamental concepts underlying the use of isotopes, lipids and pigments for studying organic matter cycling. The book opens with a broad overview of the carbon cycle in the sea followed by chapters on lipid, pigment and DNA biomarkers for studying its sources and sinks. Much of this organic matter is remineralised (i.e. becomes food for consumers), but a small proportion sinks to the depths and an even smaller proportion becomes incorporated into the sedimentary record either as the original biochemicals or as diagenetically altered forms.

Distributions of biomarkers in sediments can provide a great deal of information about the type of environment present at the time of deposition. Specific environmental types can be recognized such as the example discussed here of photic zone anoxia. Biomarkers are used by petroleum companies to identify the likely sources for petroleum based on the fingerprint of molecules preserved in the oil. These same molecules can be used to identify pollution of the oceans together with the many hundreds of manufactured compounds that are unfortunately found throughout the marine realm. Biomarker distributions can be used to decipher the many environmental changes that have occurred in the Earth's past. Such information can greatly assist our understanding of the effects of climate change, so it is vital to ascertain how the organic matter preserved in sediments relates to water column processes.

I hope that you find this book interesting, useful and enjoyable.

Hobart, Tasmania, June 2005   John K. Volkman

Sources and Cycling of Organic Matter in the Marine Water Column

H. Rodger Harvey

University of Maryland Center for Environmental Science, Chesapeake Biological Laboratory, PO Box 38, Solomons, MD 20688, USA
harvey@cbl.umces.edu

1	Introduction	1
2	Global Reservoirs of Organic Carbon	3
3	Defining the Compartments – The Size Continuum	4
4	The Flux of Organic Carbon in the Ocean	5
5	The Importance of DOM	8
6	Kinetics of Organic Matter Recycling	10
7	Organic Matter Composition During Decay	12
8	Pathways for Preservation	16
9	The Role of Microbes in Organic Matter Cycling	18
10	Concluding Remarks	20
	References	21

Abstract The organic carbon cycle operates on multiple time scales with a only small fraction of the global reservoir actively exchanged. For the marine system, the sources are principally recently synthesized material from autotrophic production which annually contribute 44–50 Pg/year of new organic carbon. This is supplemented by terrestrial carbon arriving from rivers, erosion and the atmosphere which contribute to the complex mixture present on oceanic waters. The focus of this review is to highlight the major sources or organic carbon and describe how the interaction of biological, chemical and physical processes provides an efficient mechanism for its eventual recycling.

Keywords Carbon reservoirs · Diagenesis · DOM · Global carbon cycle · Microbial loop · Particles · POC

1
Introduction

The cycling of organic carbon in the marine environment is a key process in the global carbon cycle. Marine systems are roughly equal to the terrestrial system as a source of new organic carbon to the biosphere, contributing an estimated 44–50 Pg/year of new production [1]. Over 80% of this amount is in the open ocean [2]. Yet only a small fraction (< 1%) of this material escapes recycling in the water column or active sediments to be ultimately buried and preserved in the sedimentary record [3, 4]. The interaction of biological, chemical and physical processes in oceanic systems thus provides an efficient mechanism for the production of new organic carbon as well as its eventual recycling as part of the global carbon cycle.

The sources of organic matter in the oceans are myriad, and dependent upon the intensity of the autochthonous signal and the proximity and magnitude of inputs from rivers, coastal erosion, and the atmosphere (Fig. 1). Although organic carbon is ultimately a product of biological synthesis, its sources are often viewed as a dichotomy between terrestrial inputs of particles and dissolved fractions, and primary production by phytoplankton in the water column. Primary production by algae is the larger of these two sources to the marine system, but terrestrial material eroded from rivers has received heightened interest in recent years as a recorder of changing coastal systems and increased sea level. The balance between these two end members is highly variable in differing ocean regions, ranging from systems such as the Arctic which receive large freshwater and erosional inputs [5] to the pelagic

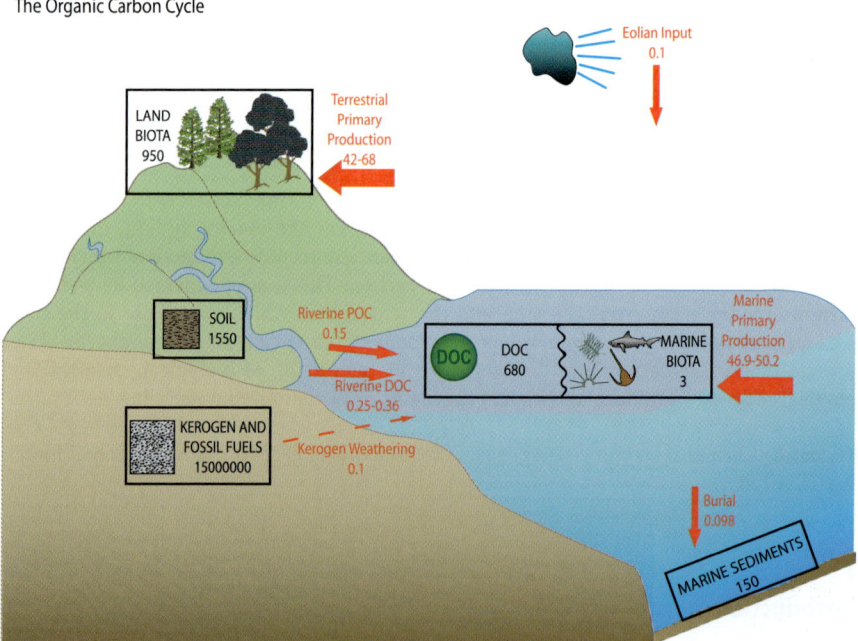

Fig. 1 The global organic carbon cycle. The major reservoirs (10^{15} G C) are shown as boxes with arrows depicting fluxes (10^{15} G C year^{-1}) of the cycle

Pacific, which is dominated by marine-derived material. Atmospheric input is a quantitatively minor fraction from the perspective of total organic input, but has indirect importance for transport of essential trace metals needed for phytoplankton growth [6, 7]. Atmospheric transport may also be of unique consequence, since the organic materials deposited range from soil-derived particles to highly labile dissolved forms of local and remote origins [8, 9].

The intention of this review is not to provide a comprehensive discussion of the processes that alter the organic matter signature, but instead focus on the major sources and how biological processing in the marine water column alters the amount and composition of organic matter in marine systems. Recent reviews of the literature which detail processes and organic character are emphasized. The active carbon cycle is a dynamic environment where single measures of organic carbon content integrate complex mixtures; mixtures that arise from the combined effects of multiple sources and varied reactivity.

2
Global Reservoirs of Organic Carbon

An examination of organic matter cycling in marine systems must begin with the realization that the vast majority of organic carbon does not actively participate in the global carbon cycle, but is retained as finely distributed material in sedimentary rocks (Table 1). Fossil fuel combustion has returned a measurable, albeit minor, fraction of this material back to the active carbon cycle in recent years [10, 11], largely as CO_2. Of the global total, only about 0.1% of the organic reservoir actually cycles through the active pool. Within this active cycle, soils which represent the largest pool, with decreasing amounts of organic matter contained in land biota, dissolved organic matter in seawater, and surficial marine sediments. The smallest fraction includes marine biota and particulate pools, encompassing highly variable

Table 1 Major reservoirs of organic carbon on Earth

Reservoir	Size (Pg C)	References
Kerogen and fossil fuels	15 000 000	Berner, 1989 (3) [115]
Soil	1550	Lal, 2003 [116]
Land biota	950	Olson et al., 1985 [117]
Ocean DOC	680	Hansell and Carlson, 1998 [118]
Marine surface sediments	150	Emerson and Hedges, 1988 [119]
Marine biota	3	Siegenthaler and Sarmiento, 1993 [120]

mixtures ranging from recently synthesized material as intact living cells to heavily degraded detrital substances with little resemblance to their original precursor. Although the particulate organic carbon (POC) reservoir is small, it undergoes rapid exchange and plays a central role in both amount and composition of organic mater which reaches underlying sediments. Over long time scales, the small fraction of organic matter remaining after extensive exposure to degradative processes is transferred to the geological reservoir.

A complication in describing each organic reservoir is that they comprise complex fractions having multiple origins and different turnover times. A recent example is black carbon, which represents a refractory and chemically complex product of incomplete combustion. It includes both ancient fossil fuels and modern biomass, including vegetation burns and forest fires. Operationally defined, the presence of black carbon in particles from the atmosphere, ice, rivers, soils and marine sediments suggests that this material is ubiquitous in the environment [12–14]. Black carbon accumulates in sediments and thus appears refractory, comprising 10–50% of sedimentary organic carbon [15] and having much older ages than other organic fractions [16]. Recent evidence suggests that black carbon also comprises a significant fraction of marine DOM in coastal zones [17]. The widespread presence of this organic component suggests that it represents an important fraction of the ocean's carbon cycle, yet its poorly defined structure and multiple origins complicates interpretation of its cycling and transfer from the active carbon cycle.

3
Defining the Compartments – The Size Continuum

The physical size (or more appropriately the density) of the organic fraction is an important control over where recycling occurs. Given the operational definitions inherent in the collection of samples prior to analysis of organic matter composition, the size distribution from dissolved molecules to large particles is an important influence over the fraction which is sampled and subsequently measured. The distribution of organic matter in the ocean is continuous yet variable, with the overall total abundance decreasing as size increases (Fig. 2). Although particles represent a quantitatively small fraction of the total organic carbon present in marine waters, they have historically attracted much attention, largely due to the ability of oceanographers and geochemists to collect them in traps or filter material from seawater in adequate amounts for chemical characterization.

Traditional collections have used filters having a variety of pore sizes or mesh supports, generally from 0.2 to 1.0 µm which operationally define the particulate fraction before analysis. Particles for organic analysis are often collected on glass fiber filters (e.g. GF/F nominally 0.7 µm pore size) which can be made organic-free through combustion. Depending on the definition

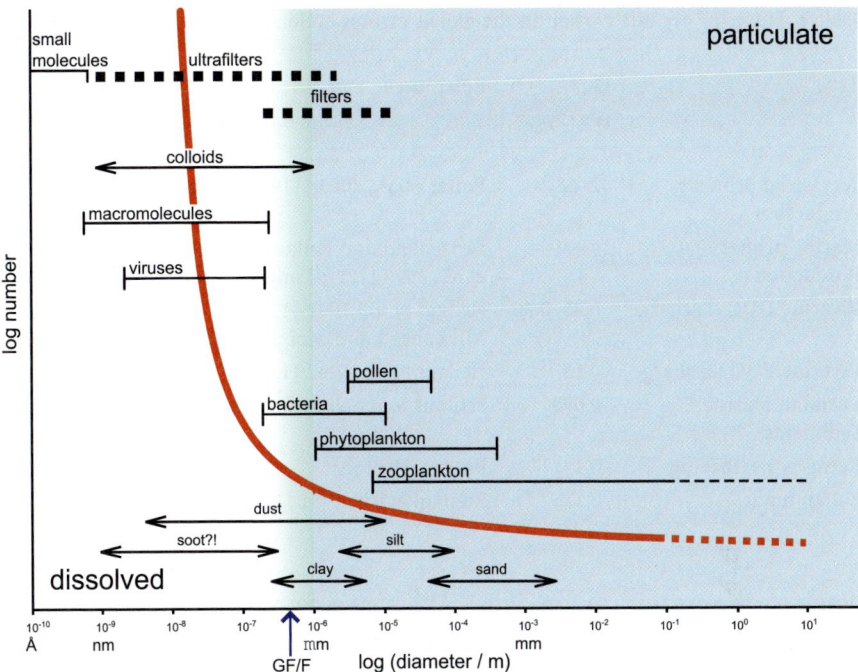

Fig. 2 The log abundance of particles versus log diameter in aquatic environments together with major components and collection ranges. Ranges among varying compartments are shown as well as arrows of major inorganic and soot components. The *vertical shading* shows the major cutoff for commonly used glass fiber filters (GF/F)

of what constitutes a particulate fraction [18], such filters might be considered either quantitative or highly selective (Fig. 2). Comparative measures of the organic composition of differing size fractions have shown important differences suggesting that the context of collection is required to fully interpret the organic signatures observed.

4
The Flux of Organic Carbon in the Ocean

The movement of organic carbon between compartments and its eventual recycling to inorganic phases are illustrated in Fig. 1 and summarized in Table 2. In the ocean the autotrophic production by phytoplankton represents the major source of organic carbon [19], supplemented by terrestrial material supplied by rivers. Most particulate forms of terrestrial matter, however, are rapidly deposited in coastal shelf and slope environments [20], with the general character of particles as seen in molecular biomarkers and isotopic values shifting to one where marine phytoplankton in surface waters

Table 2 Fluxes of organic carbon in the global carbon cycle

Type	Flux (Pg C year^{-1})	References
Terrestrial primary Production	42–68	Potter et al., 2003 [119]; Schimel et al., 2001 [120]
Marine primary Production	44–50	Behrenfeld and Falkowski, 1997 [121]; Antoine et al., 1996 [122]; Longhurst et al., 1995 [27]
Riverine DOC discharge	0.25–0.36	Hedges et al., 1997 [124]; Aitkenhead and McDowell, 2000 [125]
Riverine POC discharge	0.15	Hedges et al., 1997 [126]
Burial in marine Sediments	0.098	Schlunz and Schneider, 2000 [36]
Kerogen weathering	0.1	Berner, 1989 [3]
Eolian input	0.1	Romankevich, 1984 [127]

dominate the organic carbon signal. Although autotrophic production occurs in the lighted surface waters, sinking provides the major pathway for transport of particulate organic carbon (POC) from surface waters to the ocean depths and sediments. Estimates of the transfer of material and losses during sinking have often relied on data from particle (i.e. sediment) traps [21] which have shown that larger particle settling accounts for the majority of the flux, but also show an exponential decrease of surface productivity flux with depth [22, 23]. Such estimates come with the realization that the efficiency of such traps are affected by particle sinking rates, hydrodynamics at the opening, trap design and the nature of the particles themselves [24, 25]. All suggest, however, that in oxic waters most (> 80%) of the particulate organic material originating in surface waters is recycled at depths < 1000 m.

To understand the movement of POC, an extensive comparison of organic carbon flux estimates was conducted by Lampitt and Antia [26], who examined a total of 68 data years of trap deployments to provide a global picture of carbon flux to the deep (> 2000 m) ocean and its seasonal variability. Calculations included estimates of total annual primary production derived from long-term satellite observations at the same sites [27]. The annual range was large, with organic carbon flux varying by a factor of 375 when extreme values seen in high latitude environments are included (Table 3). Excluding high latitudes where episodic primary production is common and variable; however, a much narrower range was evident, with organic carbon flux varying by a factor of 11. The estimated range was similar to that estimated for primary production (factor of 5) for the same stations. In comparing the relationship between primary production and flux, they also found organic carbon reaching deep waters to comprise from 0.4 to 2.7% of annual primary production

Table 3 Particle flux and composition compiled from 68 data years of deep (> 1000 m) trap deployments in all major ocean basins by Lampitt and Anita (1997) [26]. Maximum and minimum flux in all ocean basin are shown. Columns include all except polar stations which show large variability. Rates in g m^{-2} year^{-1}

	All ocean basins collected			Sites excluding polar oceans		
	Max	Min	Median+SD	Max	Min	Median+SD
Dry weight	147.88	0.259	22.3 ± 22.0	66.26	7.77	22.89 ± 13.66
Organic carbon	5.24	0.014	1.00 ± 0.94	3.07	0.26	1.02 ± 0.74
C_{org} 2000	5.94	0.007	1.37 ± 1.27	4.24	0.38	1.50 ± 1.08
Inorganic carbon	3.64	0.001	1.40 ± 0.90	3.64	0.60	1.68 ± 0.83
Opaline silica	8.92	0.10	1.60 ± 2.02	8.92	0.37	1.91 ± 1.94

(Fig. 3). This suggests that for many ocean basins where primary production is not episodic (i.e. polar oceans) that there is a large scale balance in the fraction of new primary production which is exported from upper ocean waters over annual cycles despite known seasonal variability [28, 29]. Recent models of particulate flux have explored the complex interactions which occur during sedimentation [30, 31] and suggested that mesozooplankton are more important in decreasing particle fluxes than macrozooplankton, particularly in midwater zones where much POC is remineralized. In the context of organic matter cycling, it reinforces the long held belief that the vast majority of organic matter produced in oceanic surface waters as particles are recycled during descent, never to be incorporated into oceanic sediments.

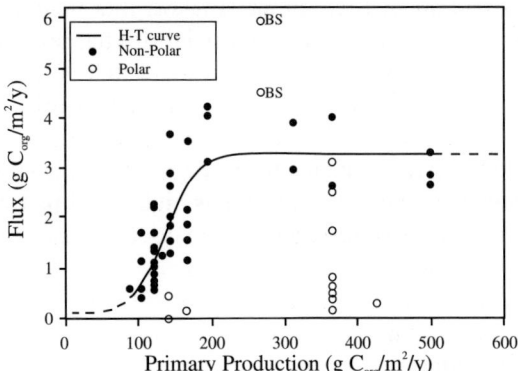

Fig. 3 Relationship between annual primary production and flux of organic carbon at 2000 m depth in the oceans. The line represents hyperbolic tangent fit with polar environments (*open circles*) excluded. Redrawn from Lampitt and Antia (1997) [26] with modifications. BS represents sites in the Bering Sea excluded from the line fit

Recent estimates and modeling have shown that at least part of the variability observed in the flux of organic carbon might also be due to the fraction of mineral ballast associated with sinking particles [32]. The presence of mineral matrices affects the time particles spent in the water column, with organic materials associated with denser minerals having more rapid transit to the ocean floor. In addition, mineral matrices have been suggested to provide direct physical protection of organic material through either adsorptive processes or perhaps as binding agents [33, 34], thus influencing the amount and composition of organic matter that survives descent and is incorporated into sediments.

Among the varied sources of organic carbon to marine systems, terrestrial organic matter is an important component, yet its fate in the ocean is not clear [35]. Much arrives through river transport, with estimates of the flux of organic carbon to the sea ranging from 0.25–0.36 Pg C year for dissolved OC and less for particles (Table 1). The range encompasses much variability, due in part to the lack in uniformity in the estimates themselves. Some of the issues which affect the accuracy of estimates have been discussed by Schlünz and Schneider (2000) [36] in their compilation of published estimates of terrestrial transport by rivers. They noted a lack of uniformity in approaches and assumptions, particularly for flux estimates where data may not include seasonal trends in discharge or measures of both particulate and dissolved components. This appears particularly true for Asian rivers, which account for 40% of the total annual sediment discharge but are poorly documented.

Despite these gaps, it is apparent that terrestrial organic matter represents a large source of reduced organic carbon to marine systems which principally arrives in dissolved form. Much of this terrestrial export by rivers appears to be derived from soils [37] and includes the highly degraded remnants of vascular plants which have been used to provide a detailed suite of molecular structures as tracers of their input (Ittekot, this volume). The primary drainage sources which account for terrestrial discharge are varied, but the majority has been estimated to be from forested catchments, with decreasing contributions from other forests, cultivated lands, wetlands, grasslands, tundra and deserts. Eolian input of terrestrial carbon to the ocean surface has been difficult to quantify, partly due to the highly variable and complex wind patterns. Estimates for total carbon range as high as 0.1 pg C year [38] and is particularly important for terrestrial input to open ocean areas [39, 40].

5
The Importance of DOM

Although the dissolved organic phases of carbon which pass through various filters (Fig. 2) have been long studied (see [41]), intense interest did not developed until the late 1980s. In several papers describing new approaches using high temperature combustion as well as oceanographic surveys of surface and deep waters, Suzuki et al. [42] and Sugimura and Suzuki [43] challenged early observations, stating that previous wet chemical measures of DOC concentration in ocean waters were substantial underestimates. Although this work has subsequently been discounted [44, 45], the initial reports led to a revolution in interest to understand dissolved organic matter (DOM) in aquatic systems and a variety of new analytical approaches were developed to examine both their concentration and chemical character. The outpouring of research on the dynamics and cycling of DOM has led to a much better understanding of its composition and cycling and a greater appreciation of its important role in the global carbon cycle. A number of recent reviews have discussed the chemical composition and cycling of DOM [46, 47] and a comprehensive presentation of sources, character and cycling of marine DOM is now available, reflecting the rapid progress in the field [48]. Its total contribution to the organic carbon pool places it as an essential component of the global cycle (Fig. 1) and a crossroads for many components of organic carbon during recycling.

In the context of global carbon estimates, Del Giorgio and Duarte [49] have argued that present estimates of DOC may not reflect its important role. They noted that DOC also represents a substantial fraction of total primary production which is not captured in satellite estimates of chlorophyll or standard ^{14}C incorporation measures used to quantify particles. By including estimated values for oceanic algal respiration and DOC production together with measures of primary production as seen in particles, they calculated that estimates of gross primary production would be enhanced by up to 48%. Such a correction would elevate the values seen in Table 2 for primary production to 69.4 to 72.3×10^{15} g C year^{-1}. The inclusion of DOC dynamics has the potential to substantially increase the total amounts of new production and export in the open ocean.

Both the chemical character and general distribution of DOM show parallels with that seen for particles. DOM has consistently been found to show highest concentration in surface waters, and compositional analysis suggests that most is derived from biological production [50]. Direct sources are varied, but direct inputs from phytoplankton [51] and sloppy feeding by macrozooplankton are significant sources as well as organic material leached from soils [52]. As with particles, the organic composition of DOM includes a significant portion which cannot be characterized at the molecular level [53, 54]. Much of the DOM as defined by ultrafiltration is low molecular weight (< 1000 Da) [55, 56] and resistant to biological recycling.

The high abundance and refractory nature of this low molecular weight dissolved organic material in the ocean might seem at odds with observation that its major recycler are bacteria which rapidly take up low molecular weight compounds. Amon and Benner [57], proposed that low molecular weight does not equate with lability. They postulated that DOM exists in a "size-reactivity continuum", suggesting that particulate organic material might follow a transition through dissolved materials, with bioreactivity decreasing in concert with molecular size along the path:

POM → High molecular weight DOM → Low molecular weight DOM

Each size fraction comprises a continuum of organic compositions and reactivities in multiple states of decay. This reactivity continuum would also help explain the relatively old age of deep-water DOM in several ocean basins, with an apparent age of 400–600 years [58], yet relatively young DOM is seen in coastal environments since this is where most appears to be produced [59, 60]. This might also explain recent observations that the fraction of DOM which cannot be easily characterized at the molecular level increases with decreasing molecular weight [61, 62]. In the context of organic matter cycling in the water column, the similarity in many of the processes that affect particles and dissolved fractions reinforces the need for integrated information and detailed composition on multiple organic matter pools to understand the pathways for cycling. A significant avenue for removal of DOM in surface waters is also photooxidation, with exposure leading to significant losses seen for chromophoric dissolved organic matter, and specific molecular markers for vascular plants such as lignins [63] and lipids of phytoplankton [64, 65].

6
Kinetics of Organic Matter Recycling

The majority of organic matter produced in surface waters by autotrophic organisms is not incorporated into surface sediments, but is recycled in the water column or at the sediment-water interface. The same is also true for terrestrial material carried in rivers or deposited across the water-atmosphere interface, although the efficiency of these recycling terms are more poorly constrained. The changes that these mixtures of organic materials undergo are both complex and selective, with the general observation of decreasing concentration with increasing water depth and increasing recalcitrance whether as particles or in dissolved phases. There are notable exceptions, including the rapid deposition of algal blooms to the sea floor [66, 67], or water column discontinuities which impede sedimentation (e.g. Black Sea), but for oxic water columns, the majority of labile compounds are degraded during decent. The fact that a variable, but ultimately very small, fraction of organic matter present in surface waters is ultimately incorporated in sediments illustrates the efficiency at which heterotrophic processes act on material prior to sediment incorporation.

Given the importance of phytoplankton as a dominant source of particles, there has been much effort to understand the liability and turnover of algal material during water column decent and at the sediment interface. These studies range from unialgal cultures in static incubations to field programs following bloom dynamics. Early work on algal carbon dynamics [68] first suggested that algal carbon might be variable in its degradation. This suggested that while algal carbon measured as POC was often considered as a single compound class during recycling, observed rates of carbon loss represented a composite of rates among the various biochemical classes. Based upon changes in POC seen in algal carbon degradation experiments, Westrich and Berner [69] developed multi-first-order rate equations (the multi-G model) to describe the utilization of multiple pools of algal carbon. Organic carbon loss could be described by a series of exponential decreases in specific components, with a first order rate used to describe the overall decrease observed.

A number of studies have examined the fate of algae in the water column, yielding a range of turnover times of total organic carbon under oxic and anoxic conditions from 3.7 to 256 days ([70] and references therein). As these and other authors have noted, the wide range of reported degradation rates is a likely consequence of both the differing reactivity among specific biochemical pools in concert with differences in the duration of experiments and their environmental conditions. A study by Harvey et al. [71] reported results on the degradation sequence for major biochemical classes (protein, lipid, and carbohydrate) in two diverse marine phytoplankters (a diatom and cyanobacterium). The major biochemical fractions of organic carbon were found to degrade at significantly different rates in both algae, with kinetics following multiple first order kinetics. In these microbial dominated experiments, carbohydrates were most rapidly utilized followed by protein and then lipid (Table 4).

Parallel incubations with oxygen as the major variable showed that substantially lower rates occur when oxygen was absent, even though levels of microbial activity were equal or greater than under oxic conditions. Subsequent work by Nguyen and Harvey [72] observed that dinoflagellates showed similar kinetics of carbon turnover. Perhaps most important for understanding organic carbon cycling is the observation that degradation rates of major biochemical fractions differed by a maximum of 4-fold for all algae despite differences in size, cellular organization and chemical composition (Table 4). The reactivity of algal derived material under oscillating redox conditions [73] and estimated removal near the sediment water interface [74] have often shown rates intermediate between these purely oxic and anoxic laboratory conditions. Although such differences are important for tracing heterotrophic processes and organic matter utilization, it illustrates the rapidity at which most algal POC is removed before sediment incorporation.

Table 4 First order decay constants (k years^{-1}) and corresponding turnover time (τ days^{-1}) for algal cells and various biochemical components during water column degradation of phytoplankton by microbes. Additional rates for individual organic classes are included where available for comparison

Biochemical fraction	Oxic		Anoxic	
	k (year^{-1})	τ	k (year^{-1})	τ
Diatom[a]				
POC	13	28.5	2.9	125
Total lipid	8.3	43.7	2.7	135
Protein	21	17.1	15	24
Carbohydrate	34	10.8	7.2	50
Fatty acids				
Polyunsaturated	34	10.7	13.6	26.8
Monounsaturated	37.7	9.8	9.3	39.5
Saturated	25.7	14.1	8.2	44.4
Sterols	33.4	10.9	2.6	142
Phytol	15–31*	11–29*	3–3.4*	114–120*
Cyanobacteria[b]				
POC	15	23.8	2.5	146
Total lipid	8.2	44.3	2.3	160
Protein	22	16.6	6.2	58
Carbohydrate	34	10.7	4.1	88
Dinoflagellate				
POC	24	15	5.6	65
Protein	34	11	5.2	70
THAA-C	18	21	3.8	96

* Range of rates seen for diatom and cyanobacterium incubations. Algal rates based on Harvey et al., 1995 [93]; Nguyen and Harvey, 1997 [128]. See Canuel and Martens (1996) [129] sediments.

7
Organic Matter Composition During Decay

A difficulty in understanding the sources and processing of organic matter in the water column is its dynamic state, with multiple compartments in various stages of biosynthesis, metabolism and decomposition. Kinetic information obtained from controlled degradation experiments can be used to describe the compositional changes that autotrophic material undergoes during degradation. This approach has been used to illustrate the potential impact of variable degradation rates of multi-component biomarker mixtures during diagenesis [74].

We can use measured rates for POC and several biochemical compartments to examine how water column degradation alters organic composition even when starting from well characterized material. Just as the overall rate of POC recycling is a composite of many degradative rates, each of these compartments in turn contain a large suite of individual molecules. This undoubtedly is a simplistic explanation of a much more complex process, but can serve to illustrate the compositional changes in POC over time which impacts interpretation.

To represent a typical phytoplankton we first approximate the distribution of major biochemical groups of a phytoplankton cell. Absolute amount are highly variable [75, 76], but a composite value among biochemical classes suggests a composition of 35% protein, 16% lipid and 40% carbohydrate. These values sum to 90% of the total organic matter observed. The final 10% represents material contained within the POC, but which cannot be classified into one of the three biochemical pools. This would include nucleic acids or perhaps mineral-bound material that is not extractable [77]. Using this composite as a "typical cell", we can apply measured removal rates to follow the changing composition of POC during early diagenesis as particulate material sinks through an oxic water column. For this exercise, the three major biochemical components are matched to their respective first order decay constants (Table 4), which vary only slightly among phytoplankton, but do differ between oxic and anoxic environments.

The calculated losses among the major biochemical classes and POC are shown in Fig. 4a. Following their prescribed first-order rate constants, all three classes decrease quickly over the 100 day time frame shown. POC which was quantified independently, follows a first order rate as well, with a loss of > 94% over the period. What is quickly evident is that while both overall POC and individual biochemical groups are lost as carbon is efficiently recycled, the major biochemical classes are lost more rapidly than the total POC. The unidentified material originally presents as a small component of carbon in algal cells rapidly accounts for the majority of organic matter remaining. As a result, the composition of particles evolves from living cellular material where most components can be assigned to one of three major biochemical groups to one which the bulk of organic material cannot be identified to biochemical class (Fig. 4a). Although the amount of total organic carbon remaining is small given the efficiency of mineralization, the organic composition is less clear than the cellular material from which it originated. These results parallel observations seen for many environments where much of the total organic pool is not amenable to characterization at the molecular level [78].

While these results illustrate that the overall character of organic matter can show rapid changes in composition during its recycling, such measures

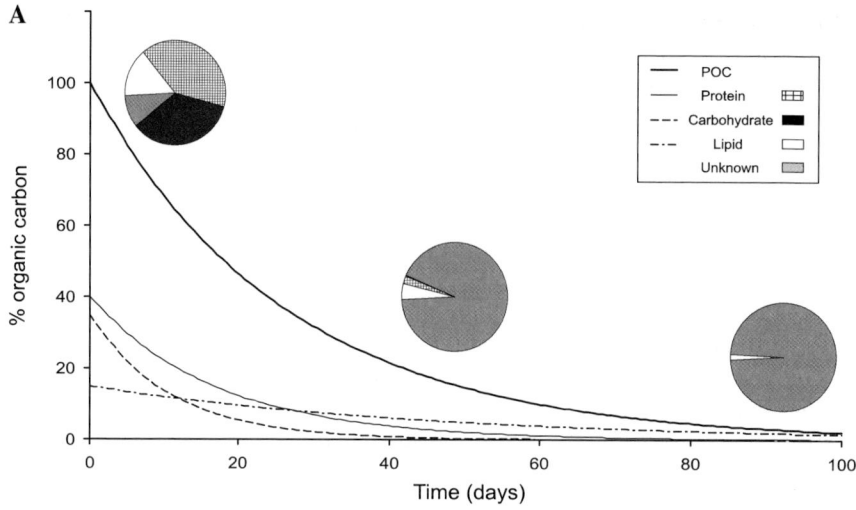
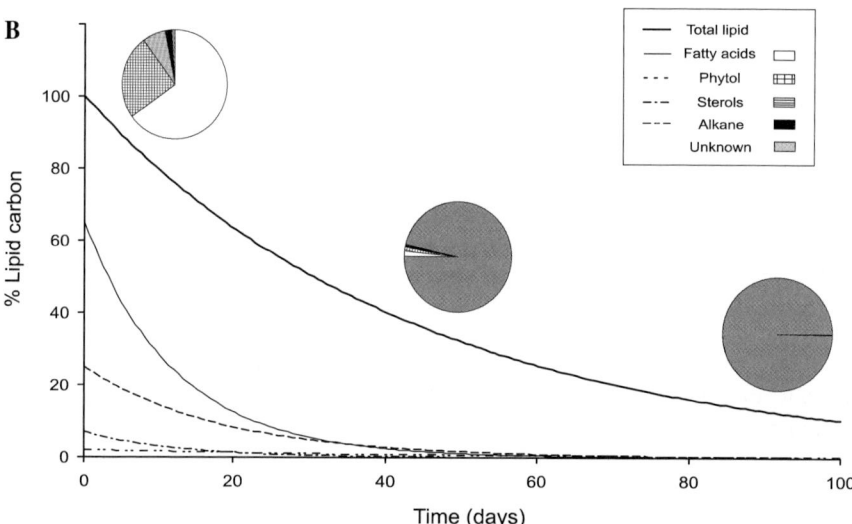

Fig. 4 The changes in amount and distribution of organic carbon and major biochemical components during degradation of "typical" algal material. *Panel A* shows changes in major biochemical components and POC over a 100 day decomposition sequence in oxic waters. *Panel B* shows the lipid fraction of the same algal material and changes in lipid composition over the same time frame. Although most organic matter is efficiently recycled, both major biochemical groups and specific fractions reveal an increasing fraction of unidentified composition over time

are not the norm. More often, either total POC or individual chemical classes are followed. Lipid biomarkers in particular have shown to be very valuable in a host of environments to detail both the sources and processing of organic materials [79], but represent a small fraction of the total organic pool.

We can use a similar scheme as above to compare the distribution of lipids during a decomposition sequence of a typical phytoplankton. Again, each compartment encompasses a suite of individual compounds, but the one might postulate that the more constrained structures would impart greater similarity in degradative rates. In this case, the total lipid pool used above (16% of POC) can be further defined by the major lipid classes. These include fatty acids as the dominant form (65% of lipid carbon) followed by phytol (25%) sterols (7%) and alkanes (2%). The remaining 1% is considered unidentified. Again, exponential first order rate constants obtained from controlled lab experiments can be employed to follow the changes in lipid distribution during the decomposition process. Although these rates may not reflect widespread field conditions, they are nevertheless reasonable approximations which more importantly allow comparative measures among different lipid fractions to be examined.

Tracing the changes in lipid composition during such a degradative sequence is shown in Fig. 4b. Similar to that for the case of broadly defined biochemical fractions, an increasing percentage of the residual organic matter is composed of compounds that elude standard methods for structural analysis. In this case over 83% of the total lipid is lost by 100 days. More importantly, by 50 days the total lipid content of POC has decreased by 60% with the fraction which is identified as lipids by traditional structural approaches constitutes only 4% of the total extractable lipid. The unidentified fraction, originally accounting for only 1% of the total, is now the majority of the extractable lipid observed.

Although such laboratory experiments have all the usual caveats concerning extrapolation to the real world, they do provide an explanation for the varied composition often seen in POC collections [80]. Under idealized conditions of largely synchronous growth and death, organic composition might be reflected by the decreasing content of particulate carbon presented by the differential losses among the various biochemical components. Yet in the environment, POC dominated by algal carbon shows varied composition, depending upon the balance between recently produced organic materials and those which have already been subject to the degradative process.

Such changes in major biochemical groups and lipid biomarker composition parallel that seen for sedimenting material, where the majority of organic matter cannot be identified at the molecular level (Wakeham and Canuel this volume, [81]). As mentioned previously for black carbon, the source in this material is often unclear. It has been suggested that a fraction of the original material may have evaded decomposition through selective preservation. Others have noted the increased presence of bacteria-specific markers in detrital material [82] and argued that it represents the replacement of carbon derived through autotrophic processes with microbial remains [83]. Depending on location, this includes a variable amount of terrestrial carbon, altering the composition and further complicating measures of its original and turnover. For the utilization of various organic biomarkers commonly used as process markers, it demonstrates that organic composition can change rapidly during decomposition, and thus assignment of source information based on organic biomarker information must be judged in the context of their temporal state – a condition which can rarely be determined with accuracy.

8
Pathways for Preservation

The changing palette of organic composition during the degradative process has often complicated the determination of organic sources, with multiple hypotheses used to explain the loss of recognizable organic structures. Based on the distributions of materials found in deep waters and often in sediments, several hypotheses and their subsequent models have attempted to explain the major diagenetic pathway that leads to organic stabilization into the macromolecular matrices that remain beyond current analytical abilities to define their molecular structure. The now classic "depolymerization – recondensation" hypothesis considers macromolecular organic matter largely as a unique material, formed after the microbial breakdown of cellular components while the remaining residues recombine into new substances only distantly related to their biological precursors [84]. This explanation requires that naturally occurring macromolecules such as polysaccharide and proteins are enzymatically depolymerized to oligo- and monomers, with the remaining fraction left to condense or polymerize through chemical or photochemical initiated cross-linking.

It is important that the classic model does not exclude the occasional biomolecules being incorporated, but the preservation of organic molecules in their native form is generally thought to be an exception rather than a common occurrence. Recent observations have suggested that most material observed in sediments and heavily degraded organic materials show similar function group arrangements for carbon and nitrogen as seen in native materials [85] suggesting that abiotic formation is not a dominant process.

In contrast, the "selective preservation" hypothesis takes essentially the opposite view, predicting that macromolecular organic matter in sediments and particles is not a new product, but rather remnant biosynthetic material which has not been degraded due to its inherent resistance to enzymatic or chemical attack [86]. Selective preservation models have gained acceptance in recent years as more sophisticated analytical techniques have made inroads into the linkages between individual components in preserved organic matter with their likely contemporary precursors. One of the better examples is the number of hydrolysis-resistant biomolecules (e.g. algaenans, suberans and cutans) which have been identified in recent years in both marine and terrestrial plants [87, 88] and in older sediments and soils. These results lend support to the idea that the winnowing of organic material during diagenesis is largely the continual loss of labile material. Recently the encapsulation of organic material within organic matrices themselves have also been suggested [89, 90] as an important mechanism as have hydrophobic interactions [91].

Hedges et al. [92] suggested that perhaps preservation does not have to be selective for the sequestration of organic matter to occur in particles (Fig. 5). Using solid-state NMR analysis of particles collected at multiple depths in sediment traps, they examined the changes in carbon linkages of particles with increasing water depth. Signal intensities of the five major carbon linkages (alkyl, amino, O-alkyl, C = C, and carboxyl) were then used to calculate the contribution of major biochemicals, allotting carbon among amino acids, lipids or carbohydrates. They then estimated the major changes occurring in biochemical composition during the most active phase of diagenesis when the majority of organic matter is recycled. Although carbohydrates showed a significant decrease, amino acids and lipids increased as a fraction of carbon in lower traps. Overall, they concluded that there were no dramatic changes in preservation potential, a point previously observed among major biochemical groups in phytoplankton in laboratory experiments [94].

An important modifier which undoubtedly impacts the preservation of organic matter as well as previously mention flux is chemisorptive attachment

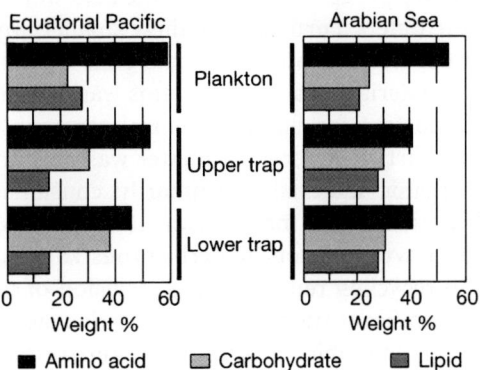

Fig. 5 Calculated weight percentages of biochemicals from sedimenting particles in the Equatorial Pacific and Arabian Sea seen by Hedges et al., (2001) [92]. Contributions of major biochemical were calculated from NMR intensities of particles collected at various depths. Although some changes were evident, overall composition showed little change with depth

to mineral surfaces. Although the emphasis on preservation has been on long term storage in sediments [94, 95], Keil et al. [96] have shown that mineral surface can be an important modifier of organic matter transport from rivers and deltas. Armstrong et al. [97] have suggest that sorption may also be an important process in the water column, with ballast minerals (including silicate and carbonate biominerals and dust) providing a critical mechanism for controlling organic matter transport. Multiple organic pools have been postulated; one tightly associated with the mineral itself and a second fraction which can be accessed and degraded in the water column. This partitioning is thought to account for the variety of degradation rates often seen in water column collections of sedimenting material as well as the variability mentioned previously on organic carbon flux estimates.

9
The Role of Microbes in Organic Matter Cycling

The important role of microbes as a key catalyst of organic matter cycling is firmly established. The concept that an unrecognized and largely unculturable group of organisms' plays a central role in the recycling of organic materials has been the subject of intense interest among microbial ecologists and biogeochemists for the last several decades [98–102]. The foundation for a central role for microbes in organic matter cycling arose from the seminal work of Pomoroy [102], who revised the paradigm of microbes as more than simple decomposers. In tandem was the observation that large phytoplankton (those typically caught in plankton nets) were not the major primary producers in the oceans, but rather smaller, autotrophic organisms less than 60 µm. This smaller size group accounted for the bulk of new organic carbon produced in euphotic waters. Furthermore, these smaller organisms than typical net plankton were also responsible for the bulk of respiration, and thus recycling of organic materials in aquatic systems was driven by microbes.

Perhaps most important for the geochemical community was that nonliving dissolved and particulate organic matter was now an important food source, and this organic material is primarily consumed by small heterotrophic microbes with diverse metabolism [101]. Although the classic idea of direct grazing on phytoplankton by herbivorous zooplankton as the major route for carbon recycling remained, the inclusion of microbes provides a mechanism for a significant fraction of both particulate and dissolved material to flow through microbial processes. Successive work solidified the concept of the "microbial loop" as an important pathway for reincorporation of dissolved organic matter into microbes as well as a path for transfer of material to higher trophic levels. The major components are shown in Fig. 6. Although this pathway is not highly efficient (with bacterial production accounting for only 10–50% of consumed carbon), it provides a mechanism for efficient recycling. In recent years it has been expanded to include the "microbial food web" with many participants and feedbacks which control organic matter recycling. Bacteria remain major consumers of dissolved materials, but are in turn consumed by bacterivorous protists, lysed by viruses or perhaps die and contribute directly to the POM pool [103].

The microbial mediation of organic matter cycling has become an important theme for geochemists interested in better defining the routes for organic mater alteration and the suite of compounds present. Although direct grazing of phytoplankton by macro zooplankton shows substantial changes in organic character, including specific biomarkers [104, 105], it appears to be a less important source (a maximum of perhaps 25%) of carbon production in

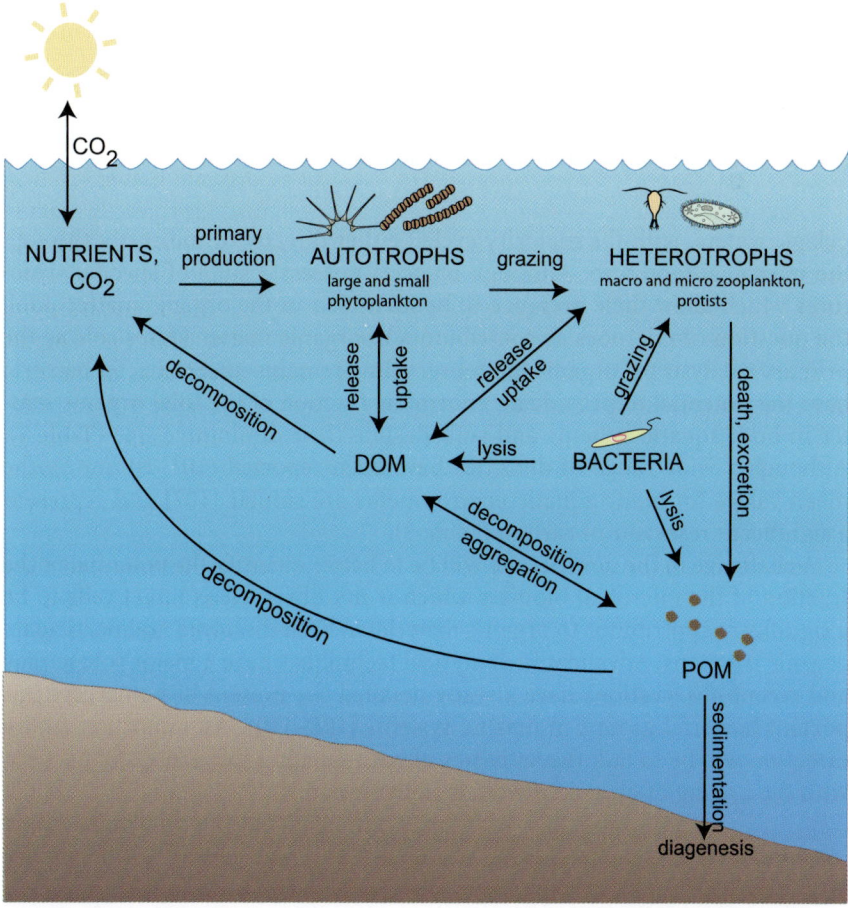

Fig. 6 Conceptual diagram of the microbial food web illustrating the major pathways for carbon recycling and transfer. The microbial food web includes both autotrophic and heterotrophic microbes, with dissolved organic matter playing a central role on the transfer of material and carbon recycling

Table 5 Number and biomass of prokaryotes in various habitats (after Whitman et al., 1998) [106]

Environment	No. of prokaryotic cells, $\times 10^{26}$	Pentagrams of carbon as prokaryotes*
Continental shelves	1.0	0.02
Ocean waters upper 200 m	360	0.72
Ocean waters below 200 m	650	1.3
Surface sediments (0–10 cm)	170	0.3
Oceanic subsurface	35 500	303
Soil	2600	26
Terrestrial subsurface	2500–25 000	22–215
Total	417–640	353–546

* calculated with the assumption of 20 fg carbon/cell for aquatic habitats and 10 fg/cell for sediments and soils. The subsurface compartments are defined as below 8 m in terrestrial systems and below 10 cm in ocean sediments.

pelagic waters, with the majority cycled either directly or indirectly through the microbial food web. Although bacteria possess a suite of specific structures which allow their presence to be identified in the organic matter pool, the question of microbes as contributors to organic matter verses role as the primary catalysts for organic matter recycling remains uncertain. Yet bacteria have the potential to provide an enormous fraction of the total organic matter in both aquatic systems and soils despite their diminutive size (Table 5). Although a wide range of densities have been reported (10^4–10^7 ml^{-1}), the mean values for many aquatic environments are similar [107] and represent a significant reservoir of carbon (Table 5).

A challenge in the next decade will be to better quantify the potential of the fraction of the microbial biomass which is not observed as intact cells to be a significant contributor to organic material in both dissolved and particulate organic fractions. Advances in analytical techniques have a major role to play and recent observations have already detailed the present of potentially important bacterial groups in marine systems [108–110]. An important future direction will be to link the activities of such unculturable groups of bacteria with the cycling of organic materials in the ocean.

10
Concluding Remarks

Although the reactivities of many organic compounds seem similar, closer examination reveals many subtlilities due to chemical structure, environmental conditions and physical matrix. The microbial food web and upper trophic levels are highly efficient at recycling the vast majority of carbon produced, yet some compounds escape to reach underlying sediments or are transported as dissolved material to deep ocean waters. Carbon age as seen in radiocarbon measurements suggest that a portion of both dissolved and particles along the size continuum are retained and recycled over long time periods, yet these fractions of organic carbon are typically those with complex or heterogenous structures which continue to elude detailed structural determination.

Much progress has been made in recent years, particularly by taking advantage of multiple approaches which can be used to constrain the age (radiocarbon), biosynthesis (isotopic) and origin (biomarker) of at least a fraction of the organic carbon pool [111, 112]. A better understanding of the contributors to the organic carbon pool together with evidence of the microbial catalysts responsible for its processing can help discern the path that organic carbon follows in the marine environment.

Acknowledgements I thank members of the MOGEL group for their input on illustrations and text and Brenda Yates for technical assistance. Generous support for much of our work has come from the Chemical Oceanography and Polar Program Divisions of the National Science Foundation. This is contribution number 3889 of the University of Maryland Center for Environmental Science.

References

1. Behrenfeld MJ, Falkowski PG (1997) Limnol Oceanogr 42:1
2. Longhurst A, Sathyendranath S, Platt T, Caverhill C (1995) J Plankton Res 17:1245
3. Berner RA (1989) Palaeogeogr Palaeoclimat Palaeoecol 73:97
4. Berner RA (2003) Nature 426:323
5. Stein R, Macdonald RW (eds) (2003) The organic carbon cycle in the Arctic ocean. Springer-Verlag, New York
6. Siefert RL, Johansen AM, Hoffmann MR (1999) J Geophys Res 104:3511
7. Fung IY, Meyn SK, Tegan I, Doney SC, John JG, Bishop JKB (2000) Global Biogeochem Cyc 14:281
8. Simoneit BRT, Cardoso JN, Robinson N (1991) Chemosphere 23:447
9. Cornell SE, Jickells TD, Cape JN, Rowland AP, Duce RA (2003) Atmospheric Env 37:2173-2191
10. Siegenthaler U, Sarmiento JL (1993) Nature 365:119
11. Fan S, Gloor M, Mahlman J, Pacala S, Sarmiento J, Takahashi T, Tans P (1998) Science 282:442
12. Schmidt MWI, Noack GA (2000) Global Biogeochem Cycles 14:777
13. Masiello CA, Druffel ERM (2001) Global Biogeochem Cycles 15:407
14. Mitra S, Bianchi TS, McKee BA, Sutula M (2002) Environ Sci Technol 36:2296
15. Middelburg JJ, Nieuwenhuize J, van Breugel P (1999) Mar Chem 65:245
16. Masiello CA, Druffel ERM (1998) Science 280:1911
17. Mannino A, Harvey HR (2004) Limnol Oceanogr 49:735

18. Jackson GA, Burd AB (2002) Deep-Sea Research II 49:193
19. Falkowski PG, Barber RT, Smetacek V (1998) Science 200:206
20. Hedges JI, Keil RG (1995) Mar Chem 49:81
21. Honjo S, Manganini SJ, Cole JJ (1982) Deep Sea Res 29:608
22. Suess E (1980) Nature 288:260
23. Martin JH, Knauer GA, Karl DM, Broenkow WW (1987) Deep-Sea Res 34:267
24. Heiskanen AS (1995) Mar Ecol Prog Ser 122:45-48
25. Yu EF, Francois R, Bacon MP, Honjo S, Fleer AP, Manganini SJ, van der Loeff MMR, Ittekot V (2001) Deep-Sea Res I 48:865
26. Lampitt RS, Antia AN (1997) Deep Sea Res 44:1377
27. Longhurst A, Sathyendranath S, Platt T, Caverhill C (1995) J Plankton Res 17:1245
28. Hebel DV, Karl DM (2001) Deep Sea Res 48:1669
29. GoZi MA, Aceves HL, Thunell RC, Tappa E, Black D, Astor Y, Varela R, Muller-Karger F (2003) Deep Sea Research I 50:781
30. Stemmann L, Jackson GA, Ianson D (2004a) Deep Sea Res 51:865
31. Stemmann L, Jackson GA, Gorsky G (2004b) Deep Sea Res 51:885
32. Armstrong RA, Lee C, Hedges JI, Honjo S, Wakeham SG (2002) Deep-Sea Res II 49:219
33. Mayer LM (1994) Chem Geol 114:347
34. Ingalls AE, Lee C, Wakeham SG, Hedges JI (2003) Deep Sea Res 50:713
35. Hedges JI, Keil RG, Benner R (1997) Org Geochem 27:195
36. Schlünz B, Schneider RR (2000) Int J Earth Sci 88:599
37. Hedges JI, Oades JM (1997) Org Geochem 27:319
38. Romankevich EA (1984) Geochemistry of Organic Matter in the Ocean. Springer, Berlin
39. Simoneit BRT, Cardoso JN, Robinson N (1991) Chemosphere 23:447
40. Zafiriou OC, Gagosina RB, Peltzer ET, Alford JB (1985) J Geophys Res 90(D1):2409
41. Duursma EK (1961) Netherlands J Sea Res 1:1
42. Suzuki Y, Sugimura Y, Itoh T (1985) Mar Chem 16:83
43. Sugimura Y, Suzuki Y (1988) Mareorol Chem 24:105
44. Farrington J (1992) Mar Chem 39:39
45. Hedges JI, Lee C (1993) Mar Chem 41:1
46. Nagata T (2000) Microbial Ecology of the Oceans. In: Kirchman DL (ed) Production mechanisms of dissolved organic matter. Wiley-Liss, New York p 121–152
47. Ogawa H, Tanoue E (2003) J Oceanogr 59:129
48. Hansell DA, Carlson CA (eds) (2002) Biogeochemistry of Marine Dissolved Organic Matter. Elsevier Science, London
49. del Giorgio PA, Duarte CM (2002) Nature 420:379
50. Carlson CA (2002) Biogeochemistry of Marine Dissolved Organic Matter. In: Hansell DA, Carlson CA (eds) Production and removal processes. Elsevier Science, London p 91
51. Baines SB, Pace ML (1991) Limnol Oceanogr 36:1078
52. Mannino A, Harvey HR (2000a) Limnol Oceanogr 45:775
53. McCarthy MD, Pratum T, Hedges JI, Benner R (1997) Nature 390:150
54. Benner R (2002) Biogeochemistry of Marine Dissolved Organic Matter. In: Hansell DA, Carlson CA (eds) Chemical composition and reactivity. Elsevier Science, London p 59
55. Amon RMW, Benner R (1994) Nature 369:549

56. Hansell DA, Carlson CA (eds) (2002) Biogeochemistry of Marine Dissolved Organic Matter. Elsevier Science, London
57. Amon RMW, Benner R (1994) Nature 369:549
58. Williams PM, Druffel ERM (1987) Nature 339:246
59. Raymond PA, Bauer JE (2001) Org Geochem 32:469
60. Raymond P, Bauer J (2001) Limnol Oceanogr 46:655
61. Harvey HR, Mannino A (2001) Org Geochem Special Issue on Estuar 32:527
62. Benner R, Kaiser K (2003) Limnol Oceanogr 48:118
63. Hernes PJ, Benner R (2003) J Geophy Res 108:3291
64. Rontani JF (2001) Phytochem 58(2):187
65. Rontani JF, Rabourdin A, Marchand D, Aubert C (2003) Lipids 38(3):241
66. Blair NE, Levin LA, DeMaster DJ, Plaia G (1996) Limnol Oceanogr 41:1208
67. Beaulieu SE (2002) In: Gibson RN, Barnes M, Atkinson RJA (eds) Accumulation and Fate of Phytodetritus on the Sea Floor. Oceanogr Mar Biol: An Annual Review 40:171
68. Skopintsev BA (1981) Decomposition of organic matter of plankton, hummification and hydrolysis. In: Duursma EK, Dawson R (eds) Marine Organic Chemistry. Elsiever, Amsterdam p 125
69. Westrich JT, Berner RA (1984) Limnol Oceanogr 29:236
70. Emerson S, Hedges JI (1988) Paleoceanogr 3:621
71. Harvey HR, Tuttle JH, Bell JT (1995) Geochem Cosmochim Acta 59:3367
72. Sun MY, Lee C, Aller RC (1993) Geochim Cosmochim Acta 57:147
73. Canuel EA, Martens CS (1996) Geochim Cosmochim Acta 60:1793
74. Hedges JI, Prahl FG (1993) Early diagenesis: consequences for applications of Molecular Biomarkers. In: Engel MH, Macko SA (eds) Organic geochemistry. Plenum Press, New York, p 237–253
75. Volkman JK (1986) Org Geochem 9:83
76. Brown MR (1991) J Exp Mar Biol Ecol 145:79
77. Harvey HR, Tuttle JH, Bell JT (1995) Geochem Cosmochim Acta 59:3367
78. Hedges JI, Eglinton G, Hatcher PG, Kichmann DL, Arnosti C, Derenne S, Evershed RP, Kögel-Knabner I, de Leeuw JW, Littke R, Michaelis W, Rullkötter J (2000) Org Geochem 31:945
79. Volkman JK, Barrett SM, Blackburn SI, Mansour MP, Sikes EL, Gelin F (1998) Org Geochem 29:1163
80. Wakeham SG, Lee C et al. (1997) Geochim Cosmochim Acta 61:5363
81. Peulvé S, de Leeuw JW, Sicre M-A, Maas M, Saliot A (1995) Geochim Cosmochim Acta 60:1239-1259
82. Dauwe B, Middleburg, JJ et al. (1999) Limnol Oceanogr 44:1809
83. McCarthy MD, Pratum T, Hedges JI, Benner R (1997) Nature 390:150
84. Tissot BP, Welte DH (1984) Petroleum formation and occurrence, 2nd edn. Springer
85. Knicker H, Hatcher PG (1997) Naturwiss 84:231
86. Hatcher PG, Spiker EC et al. (1983) Nature 305:498
87. de Leeuw JW, Largeau C (1993) A review of macromolecular organic compounds that comprise living organisms and their role in kerogen, coal and petroleum formation. In: Engel MH, Macko SA (eds) Organic geochemistry. Plenum, New York, p 23–72
88. Gelin F, Volkman JK, Largeau C, Derenne S, Sinninghe Damsté JS, de Leeuw JW (1999) Org Geochem 30:147
89. Knicker H, Hatcher PG (1997) Naturwiss 84:231
90. Nagata T, Fukuda R, Koike I, Kogure K, Kirchman DL (1998) Aquat Microb Ecol 14:29
91. Nguyen RT, Harvey HR (2001) Geochim Cosmochim Acta 65:1467

92. Hedges JI, Baldock JA, Gelinas Y, Lee C, Peterson M, Wakeham SG (2001) Nature 409:801
93. Harvey HR, Tuttle JH, Bell JT (1995) Geochem Cosmochim Acta 59:3367
94. Mayer LM (1994) Chem Geol 114:347
95. Hedges JI, Keil RG (1995) Mar Chem 49:81
96. Keil RG, Mayer LM, Quay PD, Richey JE, Hedges JI (1997) Geochim Cosmochim Acta 61:1507
97. Armstrong RA, Lee C, Hedges JI, Honjo S, Wakeham SG (2002) Deep-Sea Res II 49:219
98. Deming JW, Baross JA (1993) The early diagenesis of organic matter: bacterial activity. In: Engel M, Macko SA (eds) Organic geochemistry, principles and applications. Plenum Press, p 119–144
99. Henrichs SM (1993) Early diagenesis of organic matter: The dynamics (rates) of cycling of organic compounds. In: Engel MH, Macko SA (eds) Organic geochemistry. Plenum Press, New York, p 101–117
100. Kirchman DL (ed) (2000) Microbial Ecology of the Oceans. Wiley-Liss Inc., New York
101. Gottschalk G (1986) Metabolism, 2nd edn. Springer-Verlag New York
102. Pomeroy LR (1974) Biosci 24:499
103. Ducklow H (2000) Microbial Ecology of the Oceans. In: Kirchman DL (ed) Bacteria production and biomass in the oceans. Wiley-Liss Inc., New York p 85
104. Prahl FG, Eglinton G, Corner EDS, O'Hara SCM, Forsberg TEV (1984) J Mar Biol Assoc UK 64:317
105. Cowie GL, Hedges JI (1996) Limnol Oceanogr 41:581
106. Whitman WB, Coleman DC, Wiebe WJ (1998) Proc Natl Acad Sci 95:6578
107. Amann RI, Ludwig W, Schleifer K-H (1995) Microbiol Rev Mar 59:143
108. Jahnke LL, Summons RE, Hope JM, des Marais DJ (1999) Geochim Cosmochim Acta 63:79
109. Schouten S, Van Driel GB, Sinninghe Damsté JP, De Leeuw JW (1994) Geochim Cosmochim Acta 58:5111
110. Hinrichs KU, Hayes JM, Sylva SP, Brewer PG, Delong EF (1999) Nature 398:802
111. Boschker HTS, Middelburg JJ (2002) FEMS Microbiol Ecol 40:85
112. Pelz O, Hesse C, Tesar M, Coffin RB, Abraham WR (1997) Isotop Environ Health Stud 33:131
113. Berner RA (1989) Palaeogeogr Palaeoclimat Palaeoecol 73:97
114. Lal R (2003) Environ Int 29:437
115. Olson SJ, Garrels RM, Berner RA, Armentano TV, Dyer MI, Taalon DH (1985) The natural carbon cycle. In: JR Trabalka (ed) Atmospheric carbon dioxide and the global carbon cycle. US Dept of Energy, Washington, DC, p 175–213
116. Hansell DA, Carlson CA (1998) Net community of dissolved organic carbon. Global Biogeochem Cyc 12:443–453
117. Emerson S, Hedges JI (1988) Paleoceanogr 3:621
118. Siegenthaler U, Sarmiento JL (1993) Nature 365:119
119. Potter C, Klooster S, Myneni R, Genovese V, Tan P-N, Kumar V (2003) Global Planetary Change 39:201
120. Schimel DS, House JI, Hibbard KA, Bousquet P, Ciais P, Peylin P, Braswell BH, Apps MJ, Baker D, Bondeau A, Canadell J, Churkina G, Cramer W, Denning AS, Field CB, Friedlingstein P, Goodale C, Heimann M, Houghton RA, Melillo JM, Moore III B, Murdiyarso D, Noble I, Pacala SW, Prentikce IC, Raupach MR, Rayner PJ, Scholes RJ, Steffen WL, Wirth C (2001) Nature 414:169

121. Behrenfeld MJ, Falkowski PG (1997) Limnol Oceanogr 42:1
122. Antoine D, Andre JM, Morel A (1996) Oceanic primary production. 2. Estimation at global scale from satellite (coastal zone color scanner) chlorophyll. Global Biogeochem Cyc 10:57–69
123. Longhurst A, Sathyendranath S, Platt T, Caverhill C (1995) J Plankton Res 17:1245
124. Hedges JI, Keil RG, Benner R (1997) Org Geochem 27:195
125. Aitkenhead JA, McDowell (2000) Global Biogeochem Cycles 14:127
126. Hedges JI, Keil RG, Benner R (1997) Org Geochem 27:195
127. Romankevich EA (1984) Geochemistry of Organic Matter in the Ocean. Springer, Berlin
128. Nguyen RT, Harvey HR (1997) Org Geochem 27:115
129. Canuel EA, Martens CS (1996) Geochim Cosmochim Acta 60:1793

Lipid Markers for Marine Organic Matter

John K. Volkman

CSIRO Marine and Atmospheric Research, GPO Box 1538, 7001 Hobart, Tasmania, Australia
john.volkman@csiro.au

1	Introduction	28
2	**Lipid Extraction**	29
2.1	Extractable Lipids	29
2.2	Free and Bound Lipids	30
3	**Identification of Biomarkers in Sediments and Seawater**	31
3.1	Hydrocarbons	31
3.1.1	Straight-Chain Alkanes	31
3.1.2	Branched Acyclic Alkanes (Including Isoprenoids)	31
3.1.3	n-Alkenes	33
3.1.4	Highly Branched Isoprenoid (HBI) Alkenes	34
3.1.5	Other Branched Alkenes	35
3.1.6	Polycyclic Alkenes and Petroleum Biomarkers	36
3.2	Fatty Acids	36
3.2.1	Monocarboxylic Fatty Acids	36
3.2.2	α,ω-Dicarboxylic Fatty Acids	40
3.2.3	Monohydroxy Monocarboxylic Fatty Acids	40
3.2.4	Polyhydroxy Monocarboxylic Fatty Acids	42
3.3	Intact Esterified Lipid Classes	42
3.3.1	Triacylglycerols	43
3.3.2	Phospholipids	43
3.3.3	Glycolipids	44
3.3.4	Betaine Lipids	45
3.3.5	Wax (Alkyl) Esters	45
3.4	Alicyclic Alcohols	46
3.5	Long-Chain Alkyl Diols	46
3.6	Isoprenoid Ether Lipids	48
3.7	Aliphatic Ketones	49
3.7.1	n-Alkan-2-ones	49
3.7.2	Mid-Chain Ketones	50
3.7.3	Alkenones	50
3.8	Steroidal Compounds	51
3.8.1	Sterols (Stenols, Stanols)	51
3.8.2	Steroid Ketones	55
3.8.3	Steryl Chlorin Esters (SCEs)	56
3.8.4	Steryl Esters and Steryl Ethers	56
3.9	Triterpenoid Alcohols	57
3.9.1	Hopanoids	58
3.9.2	"Higher Plant" Triterpenoids	60

3.9.3 Other Triterpenoids 60
3.10   Chlorophyll and Carotenoid Pigments 61
3.11   Organic Sulfur Compounds 62

4   **Summary** 62

References 62

Abstract An astonishing variety of different lipids have been found in marine sediments and the water column attesting to the diversity of biosynthetic pathways employed by aquatic organisms. Many of these compounds have distinctive structures allowing them to be used as biomarkers for particular sources of organic matter in marine ecosystems. Microalgae synthesize many unusual compounds, such as long-chain alkenones, alkenoates and alkenes, long-chain alkyl diols, highly branched isoprenoid alkenes as well as distinctive sterols and unsaturated fatty acids, thus enabling inputs of microalgal organic matter to be easily recognized. The input of terrestrial organic matter to marine environments can be recognised from lipids of higher plant origin, such as long-chain alcohols, alkanes and fatty acids, and C_{29} sterols, although marine sources for some of these compounds are now recognised. Bacteria synthesize a diverse range of compounds, such as branched fatty acids, hopanoids and isoprenoids, many of which are particularly stable, for instance those that contain an ether bond. Qualitative assignments of organic matter sources are thus reasonably straightforward, although even now lipids can be found for which no source is known. However, quantitative assessments are more difficult since lipid contents vary greatly between different organisms and lipids display a wide range of reactivities. The combination of lipid biomarker data with information from stable isotopes can provide good estimates of organic matter sources, provided that the isotope signatures of the contributing sources (end-members) are known. This chapter provides a review of biomarkers commonly found in sediments together with notes on their identification and source specificity.

Keywords Biomarkers · Lipids · GC-MS · Phytoplankton · Higher plants

1
Introduction

The lipid constituents of seawater and marine sediments have been the subject of numerous investigations over several decades. Some of the first work described the use of branched alkanes to identify algal contributions to sediments [1]. The lipids in phytoplankton and zooplankton communities were studied by Jeffries [2], who showed how specific fatty acids might be useful in food-web studies. Later work examined the occurrence of lipids in the surface microlayer of the oceans [3]. From these early beginnings a large body of research has developed – variously called environmental organic geochemistry or marine organic chemistry or marine lipid chemistry – that has demonstrated the value of studying individual compounds (biomarkers) as well as the bulk organic matter in order to understand carbon sources and cycling.

Organic geochemists have discovered many new and unusual lipid components during their studies of Recent and ancient sediments. Information obtained from the distribution of components within a variety of lipid classes in such samples has led to searches for the same compounds in possible source organisms, in particular microalgae and bacteria. Indeed, the identification of many compounds in sediments occurred some years before a biological source was recognized [4]. This chapter provides an overview of the types of lipids found in marine ecosystems, and information on how to identify those lipids that are commonly found, together with some notes on their significance. Reference is made to review articles and in particular some early papers to illustrate how the field has developed.

2
Lipid Extraction

2.1
Extractable Lipids

Most studies of lipids in sediments and marine organisms use the Bligh and Dyer method of extraction based on mixtures of chloroform, methanol and water [5], or some modification of it such as the additional use of sonication to liberate more tightly bound lipids. Also used is the earlier Folch extraction method [6]. Recently, these two methods have been compared. In animal tissues containing < 2% lipids both methods give similar yields, but at higher lipid contents the Bligh and Dyer method appears to be less effective than the Folch method [7]. Some geochemical studies just use a hydrocarbon solvent such as hexane, but it should be noted that this is only suitable for extraction of hydrocarbons and even here the solubility of long-chain alkanes in hexane is quite low and this could lead to a biased distribution. Chlorinated solvents such as methylene chloride (DCM) and chloroform provide good extraction of neutral lipids, but are not suitable for the extraction of complex lipids such as phospholipids where recoveries can be as low as 60–65% [8].

The extracts obtained by solvent extraction need to be further fractionated in order to provide fractions with simpler distributions that are more amenable to analysis by gas chromatography-mass spectrometry (GC-MS) or high performance liquid chromatography. Saponification is often used to separate fatty acids from "neutral" components such as hydrocarbons, sterols, alcohols and the like. This process breaks down polar lipids such as triacylglycerols and phospholipids into their constituent fatty acids, plus glycerol, phosphate and sugar groups. Intact polar lipids can be analyzed by HPLC (see later) or by Iatroscan TLC-FID [9, 10].

Fatty acids are converted into methyl esters for GC-MS analysis: diazomethane is often used although the need to prepare the reagent each time is seen as a disadvantage. Dry methanol containing HCl gives good results, but the use of BF_3 in methanol can lead to artifacts [11]. Extracts are often further fractionated into compound classes either by column chromatography, thin-layer chromatography or in a few cases by HPLC. The book by Christie [12] provides a good overview of methodologies available for lipid analysis.

2.2
Free and Bound Lipids

A few researchers have examined both the lipids that are readily extracted into organic solvents (usually designated as free or unbound lipids) and those that are bound more tightly to the sediment matrix (bound lipids). Several approaches have been employed, but most treat the extracted sediments with base and acid (Fig. 1) to liberate esterified (OH^--labile) compounds and amide-linked (H^+-labile) compounds [13–15], while a few studies have further examined the remaining residue by pyrolysis or chemolysis with tetramethylammonium hydroxide (TMAH) to liberate ether- or polymer-bound lipids [15]. It is commonly observed that the distributions of lipids in each fraction show some significant differences, and that the bound lipids usu-

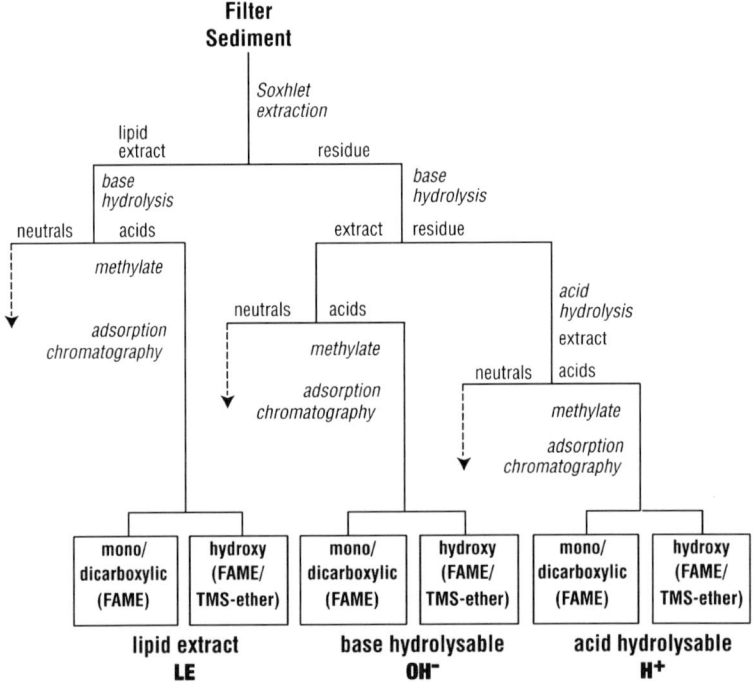

Fig. 1 Lipid analysis scheme from Wakeham [14] which provides fractions contain OH^- labile (i.e. ester-linked) and H^+-labile (i.e. amide-linked) compounds. Many variants on this scheme exist (e.g. sonication is used instead of soxhlet extraction) and so direct comparisons of data between studies can be difficult

ally contain a higher proportion of labile lipids presumably indicating that adsorption to sediments protects the lipids to some extent from microbial attack. Such studies are much more time consuming than those that involve analyses of only the extractable lipids, but they do reveal much more about the chemical forms in which lipids exist in the sample.

3
Identification of Biomarkers in Sediments and Seawater

3.1
Hydrocarbons

Hydrocarbons are those compounds that contain only carbon and hydrogen. This simple description conceals the great variety of chemical structures that can be found in marine organisms and the multitude of structures formed by degradation of functionalized lipids. For further information, the reader is referred to recent books [16–20].

3.1.1
Straight-Chain Alkanes

Alkanes isolated from marine environments typically fall into two categories. Those with odd-chains such as n-C_{15}, n-C_{17} and n-C_{19} are indicative of algal inputs [1] and are often accompanied in higher abundance by the corresponding alkenes. Long-chain (n-C_{20}–C_{35+}) alkanes that display a strong predominance of odd-chain lengths indicates a contribution from terrestrial plants. Alkanes from petroleum show little or no predominance of either odd- or even-chain lengths [21]. Erosion of ancient sediments can also be a source of similar n-alkane distributions in some aquatic sedimentary environments [22]. Hydrocarbons from eroded sediments often display distinctive sterane and hopane distributions [23].

n-Alkanes give characteristic mass spectra showing a monotonic series of C_nH_{2n+1} ions which decrease in abundance with increasing m/z value. Molecular ions are usually obvious, as is the M^+-15 ion. These distributions are usually visualized using m/z 57, 71, or 85 mass fragmentograms, but these cannot be used directly for quantification since the proportion of these ions in the total ion current decreases with chain-length.

3.1.2
Branched Acyclic Alkanes (Including Isoprenoids)

Simple branched alkanes such as 7- and 8-methylheptadecane are found in many species of cyanobacteria [24], and in algal mats and lagoonal sediments. Series of longer mid-chain monomethyl alkanes have been identified in Precambrian oils and kerogens [25], testifying to the early evolution of these compounds. Some cyanobacteria contain more complex distributions of mono-, di- and trisubstituted methyl alkanes [26–28], but these are only infrequently encountered in sediments. Long-chain *iso*- and *anteiso*-branched alkanes found in some higher plants [29], are rarely seen in sediments. An exception is the report of abundant $C_{20} – C_{30}$ *anteiso*-alkanes in some Antarctic rocks, but these are derived from cryptoendolithic microbial communities composed of microalgae, cyanobacteria, black and colorless fungi and heterotrophic bacteria [30].

The C_{19} isoprenoid alkane pristane is common in marine samples, reflecting its abundance in some zooplankton species [31]. The C_{20} isoprenoid phytane can also be found, either reflecting petroleum inputs or a contribution from Archaebacteria [32]. Two isomeric C_{25} isoprenoid alkanes are commonly encountered: 2,6,10,14,18- and 2,6,10,15,19-pentamethylicosane in sediments (Fig. 2). The former is known from halophilic bacteria [33], while the latter is usually considered to be a biomarker for methanogenic bacteria [34]. However, the finding of exceptionally light $\delta^{13}C$ values for this C_{25} isoprenoid and the C_{20} isoprenoid crocetane (2,5,10,14-tetramethylhexadecane) in reducing sediments of the Hydrate Ridge of the Cascadia continental margin implies a methanotrophic source, at least in this environment [35], while other work suggests the possibility of algal sources [36].

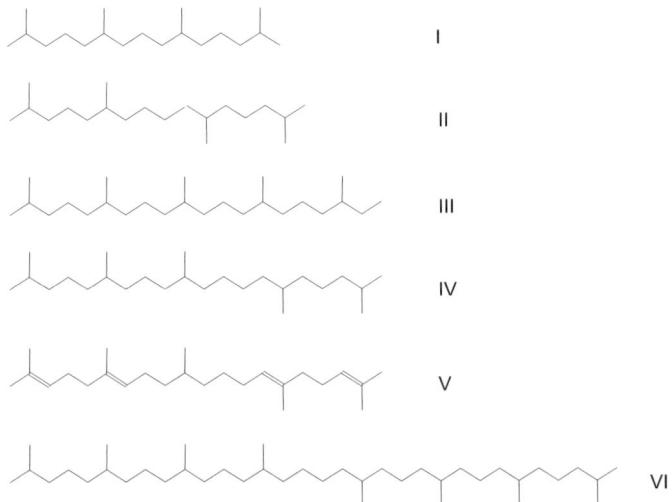

Fig. 2 Structures of some of the isoprenoid hydrocarbons found in sediments and seawater. I: pristane; II: crocetane; III: 2,6,10,14,18-pentamethylicosane; IV: 2,6,10,15,19-pentamethylicosane (PME: mainly methanogenic bacteria); V: 2,6,10,15,19-pentamethylicos-2,6,14,18-tetraene (methanogens); VI: lycopane

Polyunsaturated alkenes based on the 2,6,10,15,19-pentamethylicosane skeleton have been identified in methanogenic archaea [37]. For a review of the lipids in archaebacteria, see the comprehensive compilation of De Rosa and Gambacorta [38].

Another isoprenoid that is ubiquitous in sediments is lycopane [39]. It seems to be particularly abundant in sediments that were deposited under anoxic conditions, suggesting that the lycopane/C_{31} n-alkane ratio could be used as a proxy for oxic conditions [39]. The origin of lycopane is still unknown, but it has been suggested that a marine photoautotroph may be the source [39, 40].

Branched alkanes produce similar spectra to those of n-alkanes, but the ion due to α-cleavage is more intense so the position of branching is usually easy to identify [30]. More highly branched alkanes can also be identified from their mass spectra [26–28] (and references therein), but possible problems with co-elution must be taken into account. For example, Han et al. [24] showed that the "7,9-dimethylhexadecane" previously identified in some cyanobacteria was in fact a mixture of 7- and 8-methylheptadecane. In such cases, retention indices can be very helpful to distinguish between possible isomers [26–28, 41]. Isoprenoid alkanes are generally identified using a m/z 183 or 113 mass fragmentogram, since these ions are enhanced in abundance compared with n-alkanes. The C_{40} isoprenoid lycopane co-elutes with the n-C_{35} alkane on most non-polar capillary columns [39], but can be recognized by enhanced ions due to cleavage next to the methyl groups to give characteristic ions at m/z 477, 406/7, 336/7, 308/9 and 183.

3.1.3
n-Alkenes

A variety of unsaturated hydrocarbons (alkenes) are found in marine samples. The most common is the hexa-unsaturated alkene n-$C_{21:6}$, which is produced by many species of microalgae by decarboxylation of the 22:6(n–3) fatty acid [42]. It is particularly abundant in diatoms, which explains why these species rarely have significant amounts of the 22:6 PUFA. Conversely it is a relatively minor component in dinoflagellates which have abundant 22:6 fatty acid. It is not found in cyanobacteria or chlorophytes that lack longer-chain PUFA. In diatoms and dinoflagellates this alkene can be accompanied by the 21:5 and 21:4 alkenes [43]. The presence of such highly labile alkenes in a sediment usually points to the presence of intact (perhaps living) algal cells.

Microalgae and some prokaryotes are likely sources of long-chain alkenes in marine environments [44]. These include, *inter alia*, diunsaturated C_{31}, tri- and tetraunsaturated C_{33} alkenes (with *cis*-double bonds), in some haptophytes (e.g. [45–47]) and di- and tri-unsaturated C_{37} and C_{38} alkenes (with *trans*-double bonds) which occur in the haptophyte *Emiliania huxleyi* [45]. These alkenes are found in marine sediments [48] and fauna such as filter-feeding mussels [49]. Saturated and polyunsaturated $C_{14}-C_{31}$ hydrocarbons with a strong odd-over-even carbon predominance have been isolated from two marine *Nannochloropsis* species [50]. A North Atlantic strain of *Rhizosolenia setigera* biosynthesizes a C_{25} HBI alkene (see below) as well as C_{25} and C_{27} *n*-polyenes with six or seven double bonds [51]. These alkenes may be the precursors for C_{25} and C_{27} 2-*n*-alkylthiophenes found in sediments from palaeo upwelling regions [51]. Prokaryotes are also potential sources, although their contributions are more obvious in specialized environments such as algal mats. For example, $C_{19}-C_{29}$ alkenes occur in the cyanobacterium *Anacystis montana* [52], and green photosynthetic bacteria of the genus *Chloroflexus* contain a $C_{31:3}$ alkene all-*cis* hentriaconta-9,15,22-triene [53].

Under electron impact conditions, alkenes give mass spectra with prominent molecular ions (provided that the number of double bonds is 4 or less) and major ions at C_nH_{2n-1}. It is usually not possible to assign the position or geometry of the double bond, but derivatization techniques such as formation of DMDS adducts [53] can help identify double bond positions, although mass spectra can be difficult to interpret where the number of double bonds is greater than 3.

3.1.4
Highly Branched Isoprenoid (HBI) Alkenes

An unusual class of highly branched polyunsaturated alkenes (now termed HBI alkenes) have been recognized in many studies of marine sediments; Gearing et al. [54] seems to be the first report. HBIs have an unusual coupling of C_5 isoprene units producing a "T" shaped molecule and typically have 2–4 double bonds (for C_{25} alkenes; Fig. 3) or 4–6 double bonds (for C_{30} alkenes). A C_{35} HBI with 7 double bonds has also been observed in sediments from the Arabian Sea [55]. The parent C_{20} alkane was first noted in the Rozel Point crude oil [56], but in modern marine environments it is much more common to find C_{25} and C_{30} chain-lengths, with saturated and monounsaturated C_{20} hydrocarbons much less frequent [57]. A diatom origin was suggested by the high abundance of C_{25} HBIs in sediments and seawater from the Peru upwelling [58] and the isolation of a diunsaturated C_{25} HBI from sea-ice communities [59]. This was confirmed when high contents of C_{25} highly branched isoprenoid alkenes were identified in laboratory cultures of the diatom *Haslea ostrearia* and C_{30} HBI alkenes were found in *Rhizosolenia setigera* [43]. Since then the C_{25} and C_{30} HBI alkenes have been found in several species of *Haslea, Navicula, Pleurosigma* and *Rhizosolenia* (e.g. [51, 60–63]).

Genetic differences between strains, environmental factors and growth stage appear to be important determinants of relative abundances of the various homologues identified [64]. Recently, HBI-related monocyclic compounds containing a single cyclohexane ring (Fig. 3) have also been identified

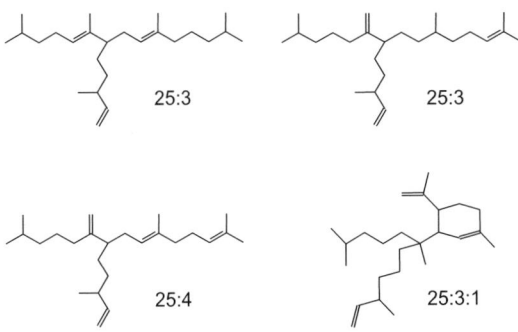

Fig. 3 Structures of a few of the many highly branched isoprenoid alkenes identified in sediments. Note that the 25:3:1 nomenclature here refers to 3 double bonds and 1 ring. The latter compound is difficult to hydrogenate and thus has been confused as a diene with 2 rings. Massé et al. [66] give full details of structures and retention times

in *Rhizosolenia setigera* [65]. Even though these are relatively minor constituents they can be readily recognized in the hydrocarbon fraction after hydrogenation. Massé et al. [66] have shown that these monocyclic alkenes also occur in a variety of sedimentary environments along with the corresponding HBI alkenes.

An extensive literature is now available to identify HBI alkenes in marine samples. Retention data are a good guide to which particular isomer is present [43, 60]. Mass spectra have been reported for the parent alkanes [60] and for most of the alkene isomers [51, 67, 68]. The paper by Massé et al. [66] provides a very good summary of RRT data and structures for HBI alkenes and their cyclohexyl analogs. NMR data for some the hydrocarbons isolated in nearly pure form are also available through the work of Rowland and co-workers [61, 62, 69, 70].

3.1.5
Other Branched Alkenes

Another class of unusual isoprenoid alkenes are the botryococcenes formed by green microalgae of the genus *Botryococcus* [71–73]. Some strains of this alga also synthesize unusual lycopadienes [74]. *Botryococcus* is found in terrestrial freshwater aquatic ecosystems, and so will not be discussed further here. However, it should be noted that compounds thought to be derived from *Botryococcus* have been detected in ancient hypersaline environments [75], so a marine source for this compound type cannot be entirely discounted.

$C_{32} - C_{36}$ polymethylene isoprenoid-like branched alkenes have been found in Black Sea sediments and attributed to an unknown photoautotroph source [76]. Their structures suggest that they were biosynthesized by methylation of an unsaturated n-C_{31} precursor at specific positions. The mass spectra of the corresponding alkanes show characteristic ion pairs due to cleavage α to the methyl group.

3.1.6
Polycyclic Alkenes and Petroleum Biomarkers

In seawater and Recent sediments one usually finds a mixture of steroidal alkenes (ster-2-enes, ster-4-enes and ster-5-enes) as degradation products of sterol and/or steroidal ketone precursors [77–79]. The presence of fully saturated steranes is indicative of thermally altered organic matter and/or petroleum. Petroleum hydrocarbons are ubiquitous in the marine environment: they originate from many sources such as shipping accidents, ships' discharge, urban and industrial effluents and natural oil seeps [21]. Sterane distributions are characterized from m/z 217 and 218 mass fragmentograms, hopanes from m/z 177, 191 and 205 mass fragmentograms and diasteranes from m/z 259 mass fragmentograms. Aromatic steroidal hydrocarbons with 1 or 3 aromatic rings are also useful compounds for fingerprinting petroleum samples.

3.2
Fatty Acids

3.2.1
Monocarboxylic Fatty Acids

Straight-chain fatty acids are often the most abundant lipids found in seawater particulate matter and in Recent marine sediments. Sources include bacteria, microalgae, higher plants and marine fauna; each of which has a distinctive fatty acid profile so that sources can usually be assigned. However, some fatty acids such as palmitic and stearic acids (16:0 and 18:0 respectively) are ubiquitous. Terrestrial plants are usually considered the main source of long-chain ($C_{20} - C_{30}$) fatty acids, showing a strong even over odd predominance, but some microalgae and/or bacteria are also suspected to be a minor source of these acids [80].

Bacteria are the major source of *iso-*, *anteiso-*, cyclopropyl and mid-chain branched fatty acids in marine ecosystems [81], but it is often overlooked that bacteria can also be a significant source of palmitoleic (16:1(n–7)) and palmitic (16:0) acids in addition to the more commonly used biomarker *cis*-vaccenic acid (18:1(n–7)) [82–84]. A typical example of a fatty acid distribution found in a modern marine sediment is shown in Fig. 4.

A fascinating recent finding is the identification in planctomycete-like bacteria capable of anaerobic ammonium oxidation of highly unusual fatty acids and related compounds containing concatenated cyclobutane rings [85]. These lipids can contain up to 5 linearly fused cyclobutane rings joined by

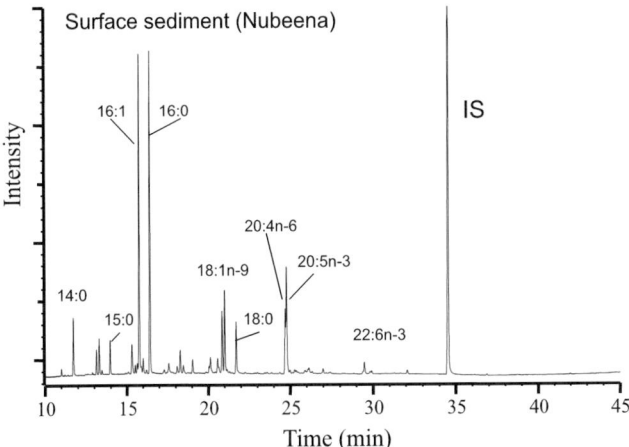

Fig. 4 Capillary gas chromatogram of total fatty acids (as methyl esters) in a surface sediment from coastal Tasmania (Nubeena), Australia. The predominance of 16:0, 16:1 and 20:5 is typical of sediments where diatoms are a major source of organic matter

cis-junctions, hence giving rise to the common name of ladderanes for these compounds [85]. These anammox bacteria contain a unique organelle called the anammoxasone in their cytoplasm [86], and it is postulated that the close packing of the ladderane lipids in the cell membrane reduces its permeability to toxic intermediates (hydrazine and hydroxylamine) produced in the anammox reaction [85]. Examples of this lipid type include fatty acid methyl esters, *sn*-2-alkyl glycerol monoethers, alcohols, *sn*-1,2-dialkyl glycerol diethers and *sn*-2-O-alkyl, *sn*-1-acyl glycerol [86]. Illustrative structures are shown in Fig. 5. Recent field work has established that coupled nitrite reduction and ammonia oxidation can be an important source of dinitrogen gas production in many aquatic systems including the Black Sea [87].

$C_{18} - C_{22}$ polyunsaturated fatty acids (PUFA) are found in almost all marine organisms, often in amounts comparable to the saturated fatty acids. Their primary source is from the phytoplankton. Marine animals obtain most of their PUFA from the food-web and further modify the distributions by chain-elongation and desaturation of the ingested fatty acids. Many studies have used fatty acid distributions to identify food items [88]; one interesting example used fatty acids in benthic foraminifera to show that these animals selectively fed on settled phytodetritus to secure their needs for PUFA [89].

The PUFA composition of microalgae has been well studied [80] and a comprehensive review of fatty acids in marine organisms has recently been published [90]. Diatoms and eustigmatophytes are rich in 20:5(*n*–3) (EPA) and produce small amounts of arachidonic acid, 20:4(*n*–6) (AA) with negligible amounts of 22:6(*n*–3) (DHA). Diatoms synthesize unusual C_{16} PUFA such as 16:4(*n*–1) and 16:3(*n*–4). In contrast, dinoflagellates have high con-

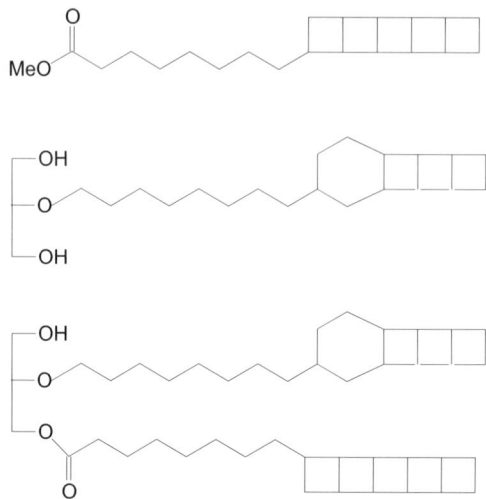

Fig. 5 Structures of ladderane lipids found in planctomycete-like anammox bacterium *Candidatus B. anammoxidans*. Modified from [86]

centrations of DHA and moderate to high proportions of EPA and precursor C_{18} PUFA [18:5(n–3) and 18:4(n–3)]. Haptophytes also contain EPA and DHA with EPA the dominant PUFA. Cryptomonads are a rich source of the C_{18} PUFA 18:3(n–3) (ALA α-linolenic acid) and 18:4(n–3) (stearidonic acid), as well as EPA and DHA. Chlorophytes (green microalgae) typically contain low concentrations of C_{20} and C_{22} PUFA, but have abundant 18:3(n–3) and 18:2(n–6), and are also able to make 16:4(n–3). Two very-long-chain highly unsaturated (C_{28}) VLC-HUFA: octacosaheptaenoic acid [28:7(n–6) (4,7,10,13,16,19,22)] and octacosaoctaenoic acid [28:8(n–3) (4,7,10,13,16,19,22,25)] were discovered in seven marine dinoflagellates [91]. More recent work has shown that the glycolipid fraction of the dinoflagellate *Karenia brevis* contains 18:5(n–3), while the phospholipid fraction contained small amounts of both 28:8(n–3) and 28:7(n–6) [92, 93].

Saturated fatty acid methyl esters (FAME) give characteristic mass spectra showing a base peak at m/z 74 and important ions at m/z 87, 143, M^+, M^+-29, M^+-31 and M^+-43. *Iso*- and *anteiso*-branched acids also show weak ions due to α-cleavage at the branching point. Thus, *anteiso*-fatty acids show an enhanced M^+-29 ion [30]. Monounsaturated FAME show a reduced abundance for the m/z 74 and 87 ions and an enhanced m/z 55 ion. The molecular ion is readily discerned as is the M^+-32 ion. The geometry of the double bond has minimal effect on the mass spectrum, but E (*trans*) and Z (*cis*) isomers can be identified from their relative retention times (E-isomers elute just after Z-isomers on non-polar columns) [94]. Much information is available from the human nutrition and clinical literature [94, 95] due to the health concerns about *trans*-fatty acids in margarines. Double bond positions can be confirmed by mass spectrometry of dimethyl disulphide (DMDS) adducts as described by [95-99]. Mass spectra of the adducts show ions due to cleavage between the two SCH$_3$ groups. E and Z isomers can be distinguished, since the erythro isomer (originally E) elutes after the threo isomer (originally Z).

Polyunsaturated fatty acids may be more difficult to identify since they can exist as different positional isomers. It is often helpful to analyse these on two capillary columns differing in polarity. Extensive sets of relative retention times (RRTs) are available [98]. A useful rule-of-thumb is that on non-polar columns, PUFA elute as pairs if they differ by one double bond and the position of the last double bond differs by three carbon atoms. Thus 20:4(n-6) elutes just before 20:5(n-3), 16:2(n-4) elutes with 16:1(n-7), 18:2(n-6) elutes just before 18:3(n-3) etc. Mass spectra are less useful for positive identification, since the intensity of the molecular ion decreases with increasing degree of unsaturation and may not observed in FAME having 4 or more double bonds (although this is somewhat instrument dependent). The base peak is usually m/z 79.

Other useful techniques are argentation TLC, column chromatography or HPLC [99, 100]. A typical TLC procedure involves double development in hexane-Et$_2$O-HOAc (94 : 4 : 2) on silica gel (7 g) plates loaded with 3% AgNO$_3$ (w/w). Bands can be visualized under 366 nm UV light after spraying with 2′,7′-dichlorofluorescein, and extracted into hexane-CHCl$_3$ (4:1, ×3). The FAME extract needs to be washed with saturated NaCl and 2 M NH$_3$ to remove Ag$^+$ and 2′,7′-dichlorofluorescein, respectively, before further analysis.

A range of derivatization methods are available for PUFA analysis, although their utilization by marine chemists and organic geochemists is still infrequent. A detailed account of the methods available is given by Christie [101]. Double bonds can be located by EI GC-MS of picolinyl esters or 4,4-dimethyloxazoline (DMOX) derivatives of the PUFA. Both techniques can be useful and should be considered complementary [101], although the picolinyl esters have long retention times and thus DMOX derivatives are often preferred. A general rule has been formulated to interpret the mass spectra of DMOX derivatives. It states that: if an interval of 12 Dalton, instead of the regular 14 Dalton, is observed between the most intense peaks of clusters of fragments containing n and $n - 1$ carbon atoms in the mass spectrum, then the double bond occurs between carbons n and $n + 1$ in the molecule (reviewed in [102]). Examples of DMOX mass spectra are given by Christie [101] and Spitzer [102].

A recent example of the application of these techniques is provided by Méjanelle et al. [103]. These authors used a combination of DMOX and DMDS derivatizations to identify the FAME in marine flagellates (smaller than 10 μm) that colonize marine particles in deep Atlantic waters. Among the compounds identified was a novel non-methylene-interrupted fatty acid 20 : 3$\Delta^{7,13,17}$, which occurred in quite high amounts.

Nichols and Davies [104] report a highly sensitive method based on analysis of 2-oxo-phenylethyl esters by high-performance liquid chromatography-mass spectrometry (LC-MS) combined with ultra violet (UV) detection. The technique was applied to identify PUFA in *Shewanella* bacteria. It is now recognized that some bacteria, including deep-sea and coastal species, can synthesize PUFA such as 20:5(n-3) and thus can be a source of PUFA in particular sedimentary environments (e.g. [105]).

3.2.2
α,ω-Dicarboxylic Fatty Acids

C_{16} and C_{18} α, ω-dicarboxylic fatty acids are major constituents of the higher plant polyester cutin and $C_{16} - C_{22}$ even-chain dicarboxylic fatty acids are often abundant in the plant biopolymer suberin [106–108]. Seagrasses also contain α, ω-dicarboxylic fatty acids, so most sediments containing higher-plant organic matter can be expected to contain these long-chain dicarboxylic acids (e.g. [109]). It is usually necessary to hydrolyze the sediments to release these esterified fatty acids [14].

Shorter-chain dicarboxylic fatty acids such as azelaic acid (C_9) occur as natural constituents of some plants and as degradation products from oxidative scission of the double bonds in unsaturated fatty acids. Azelaic and other short-chain dicarboxylic acids have been found in marine sediments and in aerosols [110]. The presence of long-chain α, ω-dicarboxylic fatty acids in Black Sea sediments in which shorter-chain α, ω-dicarboxylic fatty acids were absent suggested to Wakeham [14] that they were derived from plant matter rather than oxidation of hydroxyl fatty acids in the sediments.

3.2.3
Monohydroxy Monocarboxylic Fatty Acids

Hydroxylated fatty acids have been found in many sediments (see [44] for a review), but most reports are from lacustrine settings. Aliphatic α- and β-monohydroxy fatty acids (i.e. 2- and 3-hydroxy monocarboxylic acids) occur in a wide range of organisms [111] and are typically produced as intermediates in the α- and β-oxidation of monocarboxylic fatty acids. β-Oxidation occurs more widely than α-oxidation, although the latter is known in plants, animals and bacteria. In yeasts, α-hydroxy fatty acids are intermediates in fatty acid biosynthesis [112]. Long-chain hydroxy fatty acids are rarely reported in microalgae, although C_{22} to C_{26} saturated and monounsaturated α-hydroxy fatty acids have also been found as major lipid components of the cell wall of three marine chlorophytes [50] and a series of saturated α- and β-hydroxy acids ranging from C_{24} to C_{30} with C_{28} predominating was detected in some freshwater eustigmatophytes [113].

Shorter-chain ($C_{12} - C_{18}$) β-hydroxy fatty acids are commonly found in Gram-negative bacteria as amide-bound constituents of the lipid A component of the cell wall polysaccharide [114, 115]. The C_{14} β-hydroxy fatty acid usually predominates, but C_{15} and C_{17} *iso-* and *anteiso*-branched β-hydroxy fatty acids predominate in the common marine bacterium *Desulfovibrio desulfuricans* [116]. These biomarkers can be used to characterize microbial communities [116], and they provide a useful complement to phospholipid fatty acid analysis [117].

$C_{30} - C_{34}$ mid-chain hydroxy fatty acids were identified in several marine eustigmatophytes of the genus *Nannochloropsis* [118]. The predominant positional isomer contained an hydroxy group at the $\omega 18$ position, suggesting that the series was produced by chain-shortening or elongation of a single major precursor.

C_{16} and C_{18} ω-hydroxy fatty acids are common constituents in plant cutin while $C_{16} - C_{22}$ ω-hydroxy fatty acids are found in suberin [119]. Long-chain ω-hydroxy fatty acids and α-hydroxy fatty acids also occur in the seagrass *Zostera* [109]. ($\omega - 1$)-Hydroxy long-chain fatty acids have been identified in some marine and lacustrine environments, but their origin is still debated [116, 120]. Methane-utilizing bacteria are a possible source of C_{26}, C_{28} and C_{30} ($\omega - 1$)-hydroxy fatty acids [121], and a C_{26} ($\omega - 1$) hydroxy acid has also been found in the two cyanobacteria, *Anabaena cylindrica* [122] and *Aphanizomenon flos-aquae* [123].

Hydroxy fatty acids (as methyl esters) can be separated by silica gel TLC into different categories according to the number and position of the hydroxyl groups. These fractions are then derivatized with BSTFA (N,O-*bis*-(trimethylsilyl)-trifluroacetamide) to convert the free OH group(s) into the TMSi-ether. α- and β-hydroxy acids coelute even on the high resolution columns [106, 113], but they can be identified from their characteristic mass spectra. Mass spectra of β-hydroxy acids (as methyl esters, TMSi-ethers) show major ions at m/z 175 [$(CH_3)_3 SiO = CHCH_2CO_2CH_3$] and M^+-15 (base peak) with minor ions at m/z 103, 89, 133, 159 and M^+-31 [106]. The mass spectra of α-hydroxy fatty acids (as methyl esters, TMSi-ethers) show major ions at m/z 73 (base peak), M^+-15 and M^+-59 with minor ions at m/z 89, 103, 129, 159 and M^+-43 [106]. α- and β-hydroxy fatty acids in soils, sediments and biofilms have also been identified as the methyl ester with the OH group underivatized [117].

Quantification of α- and β-hydroxy fatty acids can be problematic since they co-elute. Volkman et al. [113] estimated the contribution of α-hydroxy acids to each peak in the hydroxy fatty acids of some freshwater eustigmatophytes from the proportion of the M^+-59 ion in the background-subtracted mass spectrum using the fact that this was about 11.5% of the total ions in the mass spectra of peaks containing only α-hydroxy acids. This value was determined from the mass spectrum of the C_{26} saturated α-hydroxy fatty acid. Note that this value may vary under different GC-MS conditions and there will be a small change in the abundance of this ion with increasing chain length.

3.2.4
Polyhydroxy Monocarboxylic Fatty Acids

$C_{16} - C_{22}$ polyhydroxy fatty acids are common constituents of higher plants where they occur inter-esterified in the cutin and suberin biopolymers. C_{16} and C_{18} polyhydroxy fatty acids are common constituents in plant cutin while $C_{16} - C_{22}$ fatty acids are found in suberin. In cutin, the most abundant ω-hydroxy fatty acid is usually 16-hydroxyhexadecanoic acid, which is formed from palmitic acid by ω-oxidation [124]. This in turn is hydroxylated at the 10- or 9-position to form 10,16- and 9,16-dihydroxy hexadecanoic acids. Oxidation of the ω-hydroxy fatty acid forms the corresponding α,ω-dicarboxylic acid. The C_{18} fatty acids are usually less abundant than the C_{16} and here the main precursor is oleic acid [18:1(n–9)] giving rise to 18-hydroxyoctadec-9-enoic acid which is further epoxidized to 18-hydroxy-9,10-epoxy octadecanoic acid which in turn is converted to 9,10,18-trihydroxyoctadecanoic acid [107, 125]. These hydroxy fatty acids, $C_{16} - C_{22}$ dicarboxylic acids and ω-hydroxy fatty acids can be excellent markers for terrestrial organic matter in sediments [106], especially when combined with compound-specific isotope data [126]. Note, however, that two dihydroxy fatty acids (identified as 15,16-dihydroxy-dotriacontanoic acid and 16,17-dihydroxytritriacontanoic acid) have been found in a marine eustigmatophyte [118], so this general type of biosynthesis is not unique to plants.

Polyhydroxy fatty acids are usually identified as their methyl ester, TMSi-ether. Strong ions occur due to α-cleavage to the hydroxy group. Thus 10,16-dihydroxyhexadecanoic acid (FAME, TMSi-ether) gives a weak molecular ion at m/z 446 and major ions at m/z M-15, M-31, M-47, m/z 373 and 275. The presence of any co-eluting 9,16-dihydroxy isomer is shown by ions at m/z 259 and 289.

3.3
Intact Esterified Lipid Classes

The fatty acids in phytoplankton occur in a variety of more complex polar lipids. Neutral lipids such as triacylglycerols (TAG; also referred to as triglycerides in the older literature), and glycolipids such as monogalactosyldiacylglycerols (MGDG), and digalactosyldiacylglycerols (DGDG) are often predominant with lesser amounts of phospholipids (PL) and so-called acetone mobile polar lipids [127]. Wax esters are very uncommon in phytoplankton, although abundant in many zooplankton [128]. Information about the polar lipids in seawater and natural phytoplankton is surprisingly sparse so we are heavily reliant on data published for microalgal species cultured in the laboratory. Often the conditions used are very different from those occurring in the field, and it is well documented that the proportions of lipid classes is very dependent on environmental conditions (e.g. [129] and references therein).

3.3.1
Triacylglycerols

Lipids of microalgae are often rich in polar lipids during logarithmic phase growth, but many species accumulate triacylglycerol during stationary phase when nitrogen is limiting [130]. The fatty acid amount and composition is dependent on both the growth conditions and the physiological state (e.g. [131]). In contrast, there are very few reports of TAG in bacteria or other prokaryotes so TAG in marine samples is almost always due to eukaryotic organisms. Triacylglycerols are also used by many marine animals as an energy reserve [132], but the proportion of PUFA is often quite low compared with either the triacylglycerols in microalgae or the phospholipids of zooplankton [133, 134]. Wax esters (WE; see later) are another lipid store, especially used by some species of zooplankton. The composition of fatty acids in these TAG or WE can provide useful clues about prey species of the animal in question (e.g. [135]).

Triacylglycerols are readily isolated by column chromatography on silica [135]. Their fatty acid composition can be determined after saponification or transesterification. If information on the composition of intact triacylglycerols is required then high temperature GC and GC-MS methods are available [134–137], although loss of the more unsaturated triacylglycerols through adsorption on the GC column can be a problem. An alternative technique is probe MS analysis of the entire TAG fraction isolated by TLC or HPLC. This was used by Boon et al. [138] to demonstrate the presence of intact TAG in a diatomaceous ooze from off-shore Namibia.

3.3.2
Phospholipids

Phospholipids are key components of biological membranes. They contain glycerol (glycerophosphatides) with a least one O-acyl, O-alkyl or O-alky-1′-enyl attached to the glycerol plus a polar head group (nitrogenous base, glycerol or inositol). Common examples containing one nitrogenous base include PC (phosphatidylcholine), PE (phosphatidylethanolamine) and PS (phosphatidyl serine), or inositol (PI: phosphatidyl inositol) or two glycerol molecules (PG: phosphatidylglycerol; and DPG: diphosphatidylglycerol or cadiolipin). Related compounds include the sphingosyl phosphatides.

The research group of Professor D.C. White pioneered the use of total phospholipid fatty acids (PLFA) as a measure of microbial biomass in sediments and other environments (e.g. [139–141]). Dobbs and Findlay [141, 142] report a set of values for converting lipid-P concentrations to carbon biomass for bacteria, cyanobacteria, yeasts and some microalgae. Viable microbes have an intact membrane which contains phospholipids and their constituent fatty acids. When these cells die, enzymes hydrolyze the phosphate group within minutes to hours leaving a diacylglycerol (diglyceride: DG) [143]. The resulting DG has the same signature fatty acids as the phospholipids (until it degrades), so a comparison of the ratio of PLFA to DG provides an indication of the proportions of viable and non-viable microbes.

A development of this approach is to analyze the intact phospholipids directly as a marker for intact viable cells in marine sediments. Rütters et al. [144, 145] have shown that intact phospholipids isolated from a sediment extract can be identified using liquid chromatography-electrospray ionization-mass spectrometry (HPLC-ESI-MS). The combined analysis of phospholipid types and their fatty acid substituents allows various groups of microorganisms living in the sediment to be differentiated. Note that phospholipids and their degradation products (free fatty acids, mono- and diacylglycerols) are rapidly recycled in sediments and so their abundances show a sharp decline with sediment depth [145, 146].

3.3.3
Glycolipids

Glycolipids are important constituents of cell membranes. Common glycolipids include monogalactosyldiacylglycerol MGDG, digalactosyldiacylglycerol DGDG, and sulfoquinovosyldiacylglycerol SQDG (e.g. [128, 147, 148]), but many other more complex glycolipids (i.e. polar molecules containing sugars) exist. New galactolipids based on a sugar galactopyranoside and 16:1 and 20:5 fatty acids have been identified in the diatom *Nitzschia* sp. [149]. Glycolipids can be purified by chromatography on DEAE-cellulose and silica-gel columns [150], since most glycolipids are less polar than most phospholipids [151]. Very few data have been published on glycolipids in marine waters or sediments, reflecting their low concentration due rapid

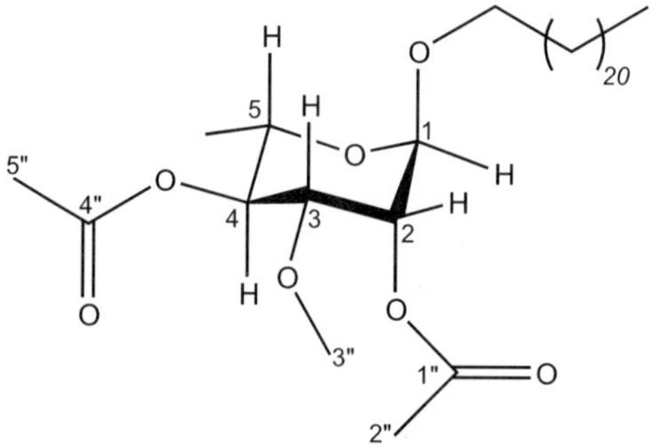

Fig. 6 Structure of the unusual glycolipid found in sediments from Ace Lake Antarctica [55]. The presence of C_{22} and C_{24} carbon chains ether-linked at position 1 of the sugar shows that some common biomarkers (in this case C_{22} and C_{24} alcohols) may be derived from quite unexpected molecules. The source organism of this compound is still unknown

hydrolysis following their liberation from the microbial cell (e.g. [151]). An interesting exception was the isolation of intact novel glycolipids from anoxic sediments in an Antarctic lake [152]. These unusual compounds were identified as docosanyl 3-O-methyl-α-rhamnopyranoside and docosanyl 3-O-methylxylopyranoside (Fig. 6). These compounds were a major and unexpected source of C_{22} and C_{24} alcohols in the sediment.

Analysis of the polar lipids in the Dead Sea biomass during a mass bloom of halophilic archaea in 1992 showed one major glycolipid to be present in the extracts, corresponding with the sulfated diglycosyl diether lipid (S-DGD-1) characteristic of the genus *Haloferax* [153]. Other glycolipids indicative of *Halobacterium sodomense* or *Haloarcula marismortui*, or other *Halobacterium* species were not found. These data indicated that the major archaea in the lake was not one of those that had been cultured.

3.3.4
Betaine Lipids

DGCC (1,2-diacylglyceryl-3-(O-carboxyhydroxymethymethylcholine)) is a betaine lipid which is a common constituent of the Haptophyceae [154]. It also occurs in dinoflagellates and in at least one diatom [154]. DGTA (1,2-diacylglyceryl-O-2'-(hydroxymethyl)-(*N,N,N*-trimethyl)-β-alanine) was detected in five out of 16 species of Haptophyceae, whereas DGTS (1,2-diacylglyceryl-O-4'-(*N,N,N*-trimethyl)homoserine) was not detected. There are no reports of betaine lipids in seawater, particulate matter or sediments, but it is doubtful that analysts would have been looking for these compounds.

3.3.5
Wax (Alkyl) Esters

Commonly used chromatographic lipid separation schemes typically provide a fraction containing long-chain wax esters, steryl ethers and steryl esters (e.g. [135, 137]), Although these quite different lipid classes have quite different origins and geochemical significance they are often analyzed together.

Wax esters (or more correctly alkyl esters) are particularly common in the marine environment as they are used as an energy store by many species of zooplankton [133, 155]. Wax esters of zooplankton origin have been identified in sediments deposited under the upwelling cells of Walvis Bay [137], the Peru margin [155], and the Arabian Sea [55]. The fatty acid composition of these wax esters generally resembles that of the phytoplankton diet and contains PUFA such as 20:5(n–3) and 22:6(n–3) [133]. The alcohols consist mainly of saturated and monounsaturated $C_{16} - C_{22}$ moieties. Indeed, elevated levels of 20:1 and 22:1 alcohols in sediments is usually indicative of hydrolysis of marine wax esters. Note that the distributions of wax esters found in lacustrine and terrestrial sediments usually show a much higher proportion of saturated fatty acids and alcohols extending to much longer chain-lengths (e.g. [156]). Some bacteria also synthesize wax esters [157], but it seems that bacteria are rarely a significant source of these lipids in aquatic environments.

Wax esters can be analyzed on low bleed non-polar capillary columns able to operate up to 350 °C. They give readily interpretable mass spectra [137, 159]. A wax ester of general formula $R_F COOR_A$ gives a strong molecular ion and cleavage ions that include the fatty acid moiety $[R_F COO + 2H]^+$ and the alcohol moiety $[R_A + H]^+$. Thus within each set of wax esters having the same overall carbon number, the various combinations of alcohols and fatty acids present can be readily identified and quantified from the abundance of the cleavage ions.

3.4
Alicyclic Alcohols

$C_{22} - C_{32}$ even-chain n-alkanols are found in most marine sediments and are usually attributed to an input from plant waxes, especially when they are accompanied by n-alkanes showing a strong predominance of odd chain-lengths. Marine sources of n-alkanols include hydrolysis of the wax esters of zooplankton which gives rise to saturated and unsaturated alcohols from C_{14} to C_{22}. These distributions often show a predominance of 16:0, 20:1, and 22:1 alcohols [128]. A specific, but unidentified, marine source of the 22:0 alcohol has also been suggested (Fig. 6; [44, 152]). An interesting finding is the presence of straight-chain monounsaturated C_{32} and C_{30} alcohols and a di-unsaturated C_{32} alcohol in acid hydrolysates of the total extract from marine eustigmatophytes [160]. The relative proportions of the $C_{30} - C_{32}$ alcohols were very similar to those of the corresponding alkyl diols, having the same carbon number but one less double bond. Free alcohols were present only in trace amounts, suggesting that the alcohols are bound to extractable polar lipids by bonds that are resistant to base hydrolysis (perhaps by amide bonds).

Alcohols are usually examined as their TMSi-ethers although some researchers prefer to use acetate derivatives because of their greater stability. The TMSi-ethers give a strong M-15 ion, weak M^+ and typically show a base peak at m/z 75 and a major ion at m/z 103.

3.5
Long-Chain Alkyl Diols

$C_{30} - C_{32}$ alkyl 1,15-diols were first discovered in Black Sea sediments [161], which was soon followed by numerous reports on their occurrence in other sediments (e.g. [162–164]). After initial suggestions that the alkyl diols were derived from cyanobacteria, it was shown that saturated and monounsaturated $C_{30} - C_{34}$ alkyl diols occur in marine eustigmatophytes from the genus *Nannochloropsis* [160] together with alcohols of the same chain-length (Fig. 7). These compounds are probably building blocks for biopolymers called algaenans found in these microalgae [50]. Algaenans such as this are quite resistant to bacterial degradation and thus they can survive into the sediment record ultimately to become an important source of hydrocarbons in some crude oils [165, 166]. The biosynthesis of these alkyl diols remain unknown although some clues are provided by the occurrence of long-chain alcohols and hydroxy fatty acids in the same species (e.g. [167]).

The distribution of alkyl diols in eustigmatophytes is not an exact match with those found in some sediments, indicating that other organisms might synthesize these compounds [160]. An additional source has now been found with the identification of C_{28}, $C_{28:1}$, C_{30}, and $C_{30:1}$ alkyl 1,14-diols, together with C_{27} and C_{29} 12-hydroxy methyl alkanoates, as major neutral lipids in rhizosolenid diatoms belonging to the widespread diatom genus *Proboscia* [168]. These components were abundant in sediment traps and sediments, especially in areas with elevated primary production such as upwelling regions of the Arabian Sea.

Alkyl diols are often accompanied by the corresponding long-chain keto-ols where a carbonyl group replaces the mid-chain hydroxy group (Fig. 7). These have no known biological source, but they occur ubiquitously in marine sediments. It seems highly likely that they are formed by oxidation of long-chain diols, both in the water column and in the sediment [169].

Alkyl diols are readily identified from their mass spectra as TMSi-ethers which show intense ions due to cleavage α to the mid-chain hydroxyl group. Illustrative mass spectra are provided by de Leeuw et al. [161] and Volkman et al. [160]. The mass spectrum of the TMSi-ether derivative of the C_{32} saturated 1,15-diol shows a molecular ion at m/z 626, and major ions at m/z 341 and 387 due to cleavage either side of the C-15 hydroxy group. The mass spectrum of the of the $C_{32:1}$ 1,15-diol (as the TMSi-ether) shows cleavage ions at m/z 339 and 387, and an M^+-90 ion at m/z 534 due to loss of the OTMSi group. Alkyl

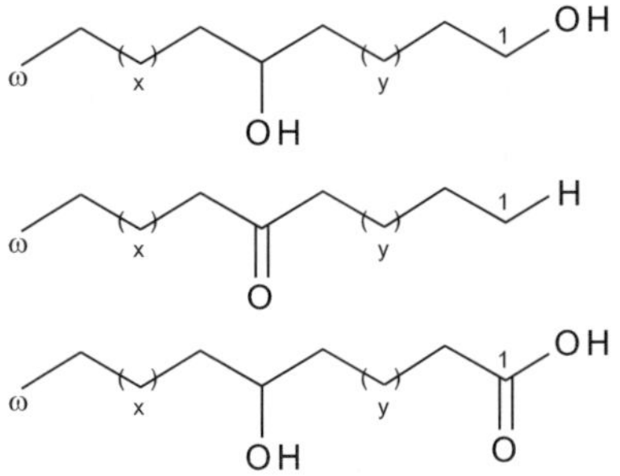

Fig. 7 Structures of alkyl keto-ols (I), alkyl diols (II) and corresponding hydroxyl fatty acids (III). In eustigmatophytes the chain-lengths are usually C_{28}, C_{30} or C_{32} and the mid-chain hydroxyl group is on C-15. In diatoms, the hydroxyl group is at C-14

keto-ols are easily identifiable from their mass spectra which show a base peak at m/z 130 and major ions at m/z 143 and M^+-15 [161].

3.6
Isoprenoid Ether Lipids

The lipids of organisms from the archaea contain ether-linked lipids [38] rather than the more common ester-linked lipids of eukaryotic organisms. A common constituent is archaeol (Fig. 8) in which C_{20} isoprenoid chains are ether-linked to the glycerol backbone. Related compounds with a hydroxyl group on the isoprenoid chain (*sn*2- or *sn*3-hydroxyarchaeols; Fig. 8) are typical constituents of the methanogenic order Methanosarcinales [170, 171]).

Glyceryl dialkyl glyceryl tetraethers (GDGTs) are specific to the archaea. A variety of chemical forms are known having different numbers of cyclopentyl rings usually designated by a numeral after GDGT. Thus GDGT-0 is the base skeleton having no cyclopentyl rings (Fig. 8). Wakeham et al. [173] showed that bicyclic and tricyclic GDGTs and their constituent ^{13}C-depleted

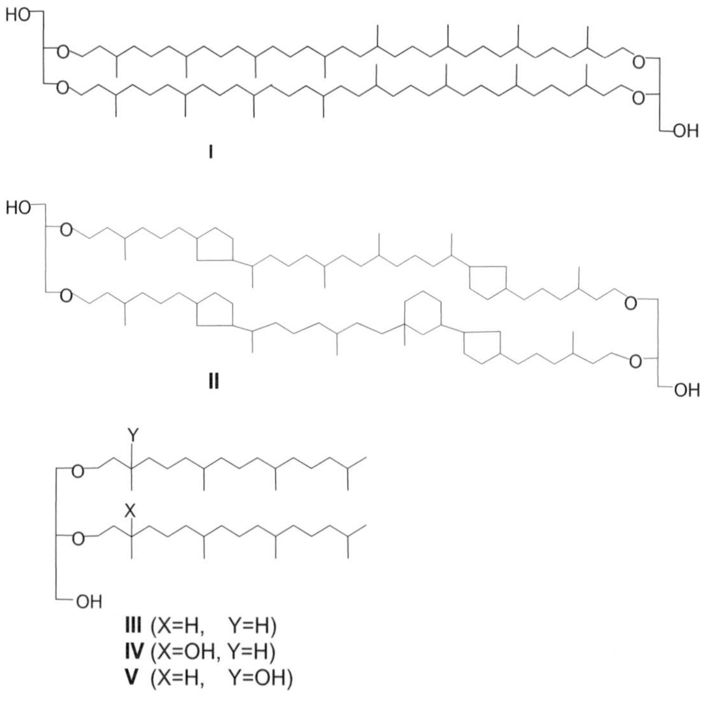

Fig. 8 Structures of some isoprenoid ether lipids of prokaryotic origin. I: GDGT-0 (from methanogenic archaea); II: Crenarchaeol; III: archaeol (X = H, Y = H) (from methanogenic archaebacteria); IV: *sn*-2-hydroxyarchaeol (X = OH, Y = H); V: *sn*-3-hydroxyarchaeol (X = H, Y = OH) (from methanogens)

monocyclic and bicyclic biphytanes (^{13}C down to −67‰) in the Black Sea were indicative of archaea involved in anaerobic oxidation of methane (AOM).

An exciting discovery was an unusual glycerol dibiphytanyl glycerol tetraether (GDGT) termed crenarchaeol in both oxic and anoxic marine waters and sediments (e.g. [174]). Its structure (Fig. 8) was identified using high field two-dimensional NMR techniques which demonstrated the presence of an unusual cyclohexane ring and four cyclopentane rings formed by internal cyclization of the biphytanyl chains [175]. Its structure is thus similar to that of GDGTs biosynthesized by (hyper)thermophilic crenarchaeota apart from the cyclohexane ring. This compound is now recognized as derived from the membrane lipids of cosmopolitan pelagic crenarchaeota. In the Black Sea, GDGT-0 and crenarchaeol dominated the distributions of ether lipids in the oxic surface and shallow anoxic waters reflecting their origin from planktonic crenarchaeota.

Compound-specific isotope analyses of the carbon skeletons suggest that planktonic archaea utilize an isotopically heavy carbon source such as algal carbohydrates and proteins or dissolved bicarbonate. Due to their high preservation potential, these lipids provide a fossil record of planktonic archaea and suggest that they have thrived in marine environments for more than 50 million years [174]. Sinninghe Damsté et al. [176] suggest that planktonic Crenarchaeota are probably facultative anaerobes and that the world's oceans contain ca. 10^{28} cells of planktonic Crenarchaeota.

Intact tetraether lipids in archaeal cell material and sediments can be analyzed by high performance liquid chromatography in combination with atmospheric pressure chemical ionization mass spectrometry [177, 178]. Non-isoprenoid dialkyl diglycerol tetraethers containing 13,16-di- or 5,13,16-trimethyloctacosanyl moieties have been identified in peats and coastal marine and lake sediments using similar techniques of HPLC-MS and high-field NMR spectroscopy [178, 179]. The ratio of these branched isoprenoids to themselves plus crenarchaeol (BIT index) has been used to estimate the relative amounts of terrestrial and marine organic inputs to marine sediments [178].

3.7
Aliphatic Ketones

A surprising diversity of aliphatic compounds containing carbonyl groups has been found in sediments and seawater. Some of these are natural products, while others are formed by diagenetic reactions.

3.7.1
n-Alkan-2-ones

Distributions of long-chain (C_{19} – C_{35}) ketones having a carbonyl at the 2-position (i.e. methyl ketones) have been found in soils [180], peats [181], lacustrine sediments [182] and some coastal marine sediments (e.g. [58]). These compounds can be derived from oxidation of n-alkanes *via* the intermediate alkan-2-ols [182], but the mismatch of chain-lengths between alkanes and ketones in coastal sediments suggests that they are not produced *in-situ*, but rather they are transported with terrestrial organic matter to the marine environment [58]. Hernandez et al. [183] reported distributions of alkan-2-ones maximizing at C_{25} in a subtropical estuary for which they proposed a seagrass source. It is not yet clear whether this could be a source of alkan-2-ones in other coastal environments. n-Alkan-2-ones are readily identified by GC-MS. Their mass spectra are similar to those of n-alkanes, but show a base peak at m/z 58 which shifts to m/z 59 with increasing chain-length [58].

3.7.2
Mid-Chain Ketones

Long-chain ketones having a carbonyl group near the centre of the chain (mid-chain ketones) are infrequently found in marine samples. Boon and de Leeuw [137] reported a series of these compounds, dominated by C_{38} and C_{40} ketones with the carbonyl at C-19, in Walvis Bay diatomaceous ooze. The origin of these compounds remains unknown, but the predominance of even carbon atoms rules out higher plants which typically contain odd-chain mid-chain ketones such as those that have been found in lacustrine settings [158].

Mid-chain ketones occur in the same chromatographic fraction as wax esters and have similar GC retention times and thus may be overlooked. A simple procedure is to re-examine the "wax ester" fraction after saponification or transesterification to see whether long-chain components are still present. Mid-chain ketones R_1COR_2 give prominent ions from cleavage α to the carbonyl group together with H and CH_3 rearrangement ions $[R_1CO]^+$, $[R_2CO]^+$, $[R_1CO + H]^+$, $[R_2CO + H]^+$, $[R_1COCH_3]^+$, $[R_2COCH_3]^+$, $[R_1COCH_3 + H]^+$, $[R_2COCH_3 + H]^+$ [137].

3.7.3
Alkenones

Very long-chain ($C_{35} - C_{40}$) unsaturated methyl and ethyl ketones termed alkenones, having a carbonyl at the 2 or 3 position respectively, are found in haptophytes including the coccolithophores *Emiliania huxleyi* and *Gephyrocapsa oceanica* [45, 48, 184]. Alkenones are straight-chain alkyl lipids with *trans* double bonds. The most abundant are C_{37} and C_{38} chain-lengths with 2 or 3 double bonds with the carbonyl group at C-2 or C-3. Recent work has demonstrated the presence of several new alkenones in these microalgae, including monounsaturated homologs, as well as the corresponding long-chain alkenols [185]. Alkenones are ubiquitous in marine sediments (e.g. [162, 186–189]), and the ratio of components has been found to vary systematically with the seawater temperature in which the microalgae grow [186, 189]. This has prompted many paleoceanographic studies that have used the ratio of concentrations of tri- to di-unsaturated C_{37} ketones in sediments (some over 100 million years old) to estimate the paleotemperature when the sediments were deposited (e.g. [187, 189, 190] and references therein). Environmental factors other than temperature such as salinity may also influence alkenone distributions (e.g. [191]), but these effects are not well understood or quantified [188].

3.8
Steroidal Compounds

Sterols and compounds derived from them by diagenetic reactions are ubiquitous in sediments. Their structures contain a number of unique features such as positions of double bonds, alkylation in the ring system and side-chain, and stereochemistry (Fig. 9) which makes them ideal for assigning sources of organic matter and for studying its short-term fate (e.g. [44, 192–194]). A diversity of steroidal compounds are found in marine waters and sediments, reflecting the variety of sterol distributions found in microalgae, plants and animals.

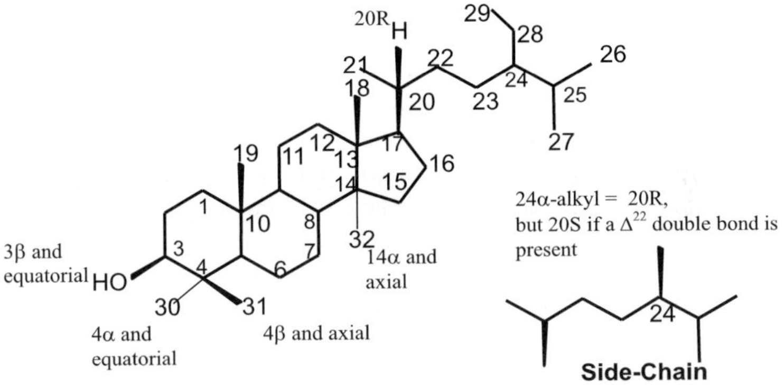

Fig. 9 Generalized structure of a 4,4-dimethyl, 24-ethyl sterol showing numbering system and stereochemistry at important centre; adapted from [194]

3.8.1
Sterols (Stenols, Stanols)

Microalgae are the primary source of sterols in the sea. Some species show a predominance of a single sterol, such as cholesterol in marine eustigmatophytes and 24-methylcholesta-5,22E-dien-3β-ol in some diatoms and haptophytes, to complex mixtures of 4-desmethyl and 4-methylsterols found in some species of dinoflagellates [192]. Some sterols are widely distributed, but others are useful chemotaxonomic markers. For example, the C_{28} sterol 24-methylcholesta-5,22E-dien-3β-ol (*epi*-brassicasterol or brassicasterol depending on the stereochemistry of the methyl group at C-24) is sometimes incorrectly thought of as a unique marker for diatoms, but it only occurs in some diatoms and it is abundant in haptophytes and cryptophytes [192, 195]. A more specific marker is 24-methylcholesta-5,24(28)-dien-3β-ol, which is abundant in some centric diatoms such as *Thalassiosira* and *Skeletonema* [196]. Some diatoms have a high content of the cholesta-5,22E-dien-3β-ol, while others have abundant 24-ethylcholesta-5,22E-dien-3β-ol, or cholesterol or cholesta-5,24-dien-3β-ol [196]. Given the diversity of sterols now known to be present in diatoms and the importance of diatoms as a source of organic matter in marine systems, it is not surprising that the sterol distributions found in many marine sediments show a variety of structures.

Other algal groups display characteristic sterol distributions. For example, 24-*n*-propylidenecholesterol seems to be a marker for some chrysophytes [197] and eustigmatophytes are a source of cholesterol [160, 167]. Green microalgae can have a wide variety of sterols, many of which have double bonds at Δ^7 [198].

Dinoflagellates are the major source of 4-methyl sterols in marine systems and the C_{30} sterol 4α,23,24-trimethyl-5α-cholest-22E-en-3β-ol (dinosterol) is often used as a biomarker for dinoflagellates (although it does occur as a minor constituent of a few diatoms [199]). However species from the genus *Pavlova* (Prymesiophyceae) also contain 4-methyl sterols and 5α(H)-stanols as well as 3,4-dihydroxy-4α-methylsterols termed pavlovols, but these rarely occur in marine sediments [200, 201]. An important consideration is that not all dinoflagellates have high contents of 4-methyl sterols [202] and some species lack dinosterol (Fig. 10). For example, *Amphidinium* spp. synthesize amphisterol (4α,24-dimethyl-5α-cholesta-8(14),24(28)-dien-3β-ol) as their major sterol [203] whereas several *Gymnodinium* species have 4α,24-dimethylcholestanol as a major sterol and contain little dinosterol [204]. An illustration of the diversity in proportions of 4-desmethyl sterols and 4-methyl sterols, even in closely related species, is shown in Fig. 10.

An uncommon example where a 4-methylsterol may not be derived from dinoflagellates comes from work by Santos et al. [205], who found high contents of 4-methylcholestanol in abyssal sediments from the Porcupine Abyssal Plain. This was also the most abundant sterol in the hindgut of holothurians (sea cucumbers), *Oneirophanta mutabilis*, suggesting that this sterol might be a marker for holothurian fecal matter in the sediments. Indeed, the significant role of benthic animals in reprocessing deposited organic matter is often overlooked in organic geochemical studies of surface sediments.

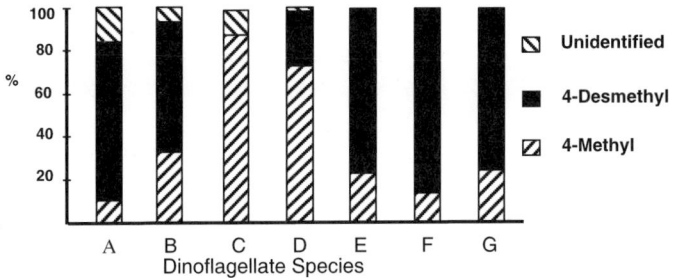

Fig. 10 Although many dinoflagellates contain a high proportion of 4-methylsterols, often dominated by dinosterol, this example shows the great range in the proportion of 4-methyl and 4-desmethylsterols in these microalgae. The dinoflagellates analysed were: A: *Gymnodinium sanguineum*; B: *Symbiodinium microadriaticum*; C: *Gymnodinium* sp.; D: *Scrippsiella* sp.; E: *Fragilidium* sp.; F: *Prorocentrum mexicanum*; G: *Prorocentrum micans*. Selected data were taken from [113] and [202]

Many microalgae synthesize C_{29} sterols and a number of examples are now known where the major sterol is 24-ethylcholesterol or 24-ethylcholesta-5,22E-dien-3β-ol [196], both of which are more commonly associated with higher plants [192]. In many near-shore sediments it is likely that much of the 24-ethylchoelsterol present is derived from plants. Figure 11 shows an example where the content of the C_{29} sterol in sediments from the Huon estuary in southern Tasmania [206] is plotted against the proportion of terrestrial organic matter determined independently from stable isotope data. The good correlation shows that little of the 24-ethylchoelsterol present in this case is derived from microalgae. However, in off-shore sediments where much of the organic matter is from phytoplankton, then algal sources can also be important contributors [58].

Sterols with a fully saturated ring system (5α(H)-stanols) occur in all marine sediments where they are thought to be formed by bacterial reduction of stenols. A typical distribution of stenols and stanols in a modern coastal sediment is shown in Fig. 12. Direct inputs from dinoflagellates are also possible [207], as well as minor inputs from diatoms (e.g. [196, 208]) and some haptophytes [200, 201]. The presence of 5β(H)-stanols in sediments is often taken as evidence for the presence of fecal-derived organic matter [209], since the 5β(H)-stanol coprostanol constitutes approximately 60% of the total sterols found in human feces. More detailed analyses of the carbon number distributions of 5β(H)-stanols indicate that human-derived fecal matter can be distinguished from other sources [210]. Note, however, that 5β(H)-stanols can also be formed in sediments under highly reducing conditions [192] and they can be derived from marine mammals [211]. Stanols with 3α(OH) stereochemistry can also be found; epicoprostanol (5β-cholestan-3α-ol) is particularly abundant in Antarctic sediments and in the feces of baleen whales [210].

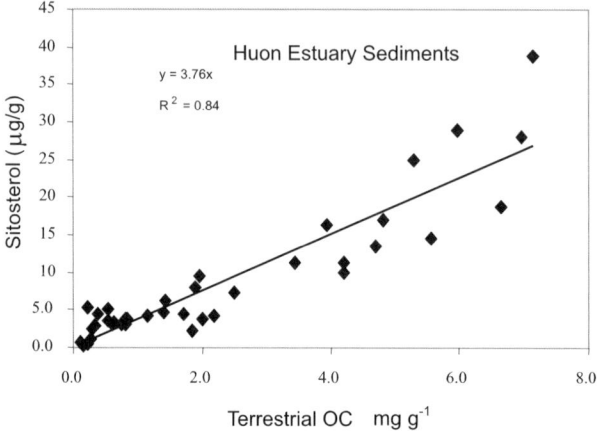

Fig. 11 Plot of sitosterol (24-ethylcholesterol) contents in sediments from the Huon estuary, Tasmania against an independent measure of terrestrial organic carbon derived from $\delta^{13}C$ values. In this instance, sitosterol provides a quantitative index of plant organic matter in the sediments, whereas in most marine sediments a contribution from microalgae may also be important. Data from Butler et al. [206]

Sterols are readily identified from their mass spectra and are usually studied as their acetate or trimethylsilyl-ether derivative, although free sterols can also produce useful mass spectra [212]. TMSi-ethers are generally preferred because they produce more diagnostic mass spectra, but are less stable than acetates [213, 214]. Acetate derivatives, however, may show a weak molecular ion, or even no molecular ion in which case the M$^+$-60 ion is abundant [214].

In analyses of TMSi-ethers, a few simple rules can help identify particular series. For example sterols with Δ^5-unsaturation show a base peak at m/z 129 and fragmentation ions at M$^+$-90 and M$^+$-129. Fully saturated sterols elute slightly later (Fig. 12) and show mass spectra with m/z 215 as the base peak (m/z 229 if a 4-methyl group is present). 5β(H)-stanols elute much earlier than 5α(H)-stanols. Sterols with Δ^7, Δ^8 or $\Delta^{9(11)}$ lack the ion at m/z 129, and show major ions at m/z 213, 229 and 255. These mass spectra are very similar [214], but retention indices can be used to identify which isomer is present [215]. The presence of a stenol/stanol pair of peaks is good indication for a Δ^5 double bond in the first eluting compound, and conversely the absence of a second peak indicates that the ring double bond must be at a position other than Δ^5. Sterols with $\Delta^{5,24(28)}$ unsaturation produce a major ion at m/z 386 (c.f. m/z 388 if the double bond at Δ^5 is absent). Sterols with $\Delta^{5,22}$ unsaturation show a base peak at m/z 255 (c.f. m/z 257 if the Δ^5 double bond is absent). Comprehensive compilations of retention time data on various GC phases are available for many of the commonly encountered structures [215, 216].

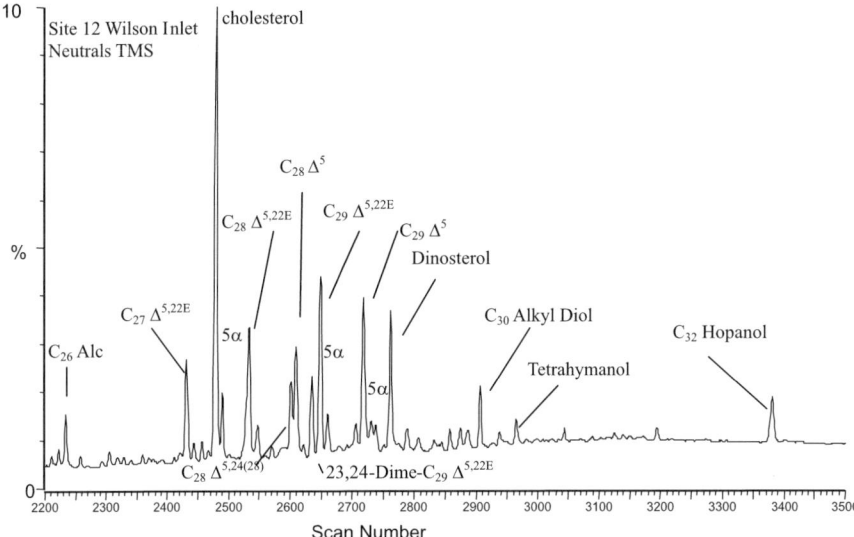

Fig. 12 Section of the chromatogram from GC-MS analysis of total neutrals in a surface sediment from Wilson Inlet, Western Australia. The sterols show a diversity of structures indicative of mixed inputs from algae, fauna, and plants. Note that each Δ^5 sterol is accompanied by smaller amounts of the corresponding 5α(H)-stanol which elutes immediately afterwards. Also shown are C_{30} alkyl diols (microalgae), tetrahymanol (protozoans) and the C_{32} triterpenoid alcohol (bacteria and/or cyanobacteria). The 4-methyl sterols are dominated by dinosterol from dinoflagellates

3.8.2
Steroid Ketones

Steroidal ketones are often found in marine environments (e.g. [217]). These distributions reflect early stage oxidation of the sterols present [217], plus direct inputs from a small subset of microalgae, such as the dinoflagellates. For example, the dinoflagellate *Scrippsiella* contains at least 21 steroidal ketones including dinosterone (4α,23,24-trimethyl-5α-cholest-22E-en-3-one), dinostanone and 4α,23,24-trimethyl-5α-cholest-8(14)-en-3-one [202, 218]. Mass spectra for these same steroid ketones isolated from a toxic dinoflagellate *Pfiesteria piscida* are provided by Leblond and Chapman [219].

Δ^5-unsaturated sterols can be converted to Δ^4-3-one stenones in sediments by microbial processes similar to those operating in the rumen [220]. These compounds are readily recognized by a base peak at m/z 124. Reduction of these unsaturated steroidal ketones can then give rise to both 5α(H)- and 5β(H)-stanols.

Steroid ketones with the carbonyl group at C-3 from oxidation of the sterol 3-hydroxy group are most common. Saturated ketones (stanones) have a base peak at m/z 231 while the 4-methyl equivalents (4-methylstanones) show a base peak at m/z 245 (see examples in [217] and [219]). The mass spectrum of Δ^{22}-unsaturated steroid ketones show strong ions due to cleavage about the side-chain. The most common example of the latter is dinosterone which has major ions at m/z 69(b), 285, 287, 314, 383 and 436 [217, 219].

Benfenati et al. [221] reported that stanones such as coprostanone (5β-cholestan-3-one) and cholestanone can be readily converted to the silylated enol ether with the silylating reagent MSTFA (N-methyl-N-trimethylsilyltrifluoroacetamide). The more commonly used reagent BSTFA is less effective and often leads to a mixture of silylated and non-silylated forms. These enol derivatives give a strong molecular ion, M-15 and M-29 ions and a major ion at m/z 143. This side-reaction needs to be considered when analyzing a total neutral extract containing sterols and steroidal ketones after silylation. Coprostanone is found in animal feces. On non-polar GC columns it elutes just before the n-C_{28} alcohol and is readily distinguished from the 5α(H)-isomer by the presence of an M-70 ion in its mass spectrum [210].

3.8.3
Steryl Chlorin Esters (SCEs)

Steryl esters of the chlorophyll a transformation product pyropheophorbide a (SCEs; also known as phorbin steryl esters – PSEs) occur in most Recent sediments and to a much lesser degree in the particulate matter in seawater [222, 223]. It is has demonstrated that they are formed from zooplankton grazing on microalgae. For example, the copepod *Calanus helgolandicus* fed the marine diatom *Thalassiosira weissflogii* produced fecal pellets containing SCEs in which all of the animal and algal sterol components were found, although in different proportions [223]. When similar experiments were carried out with the dinoflagellates *Prorocentrum micans* and *Alexandrium tamarense*, there was a clear discrimination against the uptake of 4-methyl sterols into the SCEs [224]. SCEs degrade more slowly than free sterols thus providing a means for preserving phytoplankton-derived sterols in sediments [224].

Intact SCEs are analyzed by HPLC and identified by HPLC-MS using APCI. Chlorins can be identified by electronic (UV-visible) spectroscopy and the sterols by GC-MS [223, 224].

3.8.4
Steryl Esters and Steryl Ethers

There are relatively few reports of steryl esters in seawater and marine sediments, and in some cases their presence has been inferred by isolation of sterols from the "wax ester" fraction after saponification (e.g. [161]). Intact steryl esters have been found in sediment trap samples from the North Atlantic [134]. The most abundant constituents were C_{16} and C_{18} fatty acids esterified to cholesterol, which suggests a primary source from zooplankton. Many microalgae also synthesize steryl esters (although amounts vary greatly relative to free sterols depending on the species analyzed), so this should also be considered as a possible source in surface waters and some sediments (e.g. [137]).

Steryl ethers found in marine sediments tend to be dominated by cholesteryl ethers with a $C_8 - C_{10}$ alkyl chain ether-linked at C-3 of the cholesterol molecule. Reports of steryl ethers are very limited [55, 137] and, indeed, there are few reports of their occurrence in the biota. Trace amounts have been found in bovine cardiac muscle [225], although quite a bit is known of their chemical properties from compounds synthesized to study sterol absorption in animals (e.g. [172]). The noticeable occurrence of these compounds in sediments underlying waters with high abundances of diatoms and the disparity between the sterol moiety of the ethers compared to the free sterols studied by Schouten et al. [55] led these authors to suggest that diatoms may be a direct source of sterol ethers, but this still remains unproven despite the numerous lipid studies of diatoms.

Steryl ethers isolated from marine sediments elute in the same general region of the chromatogram as wax esters [137]. The major ions are the molecular ion and two ions at m/z 368 and 370 due to loss of the cholesteryl group. A prominent ion at m/z 329 occurs due to cleavage of the A-ring. Boon and de Leeuw [137] provide a spectrum of 3-nonoxycholest-5-ene, while Schouten et al. [55] show a spectrum for the C_{10} analog.

Steryl esters elute much later than wax esters on non-polar GC columns and thus high temperature columns able to operate to 370 °C or higher are required for direct analysis [134, 136]. GC-MS analysis is not straightforward due to this temperature limitation, but has nonetheless been used to identify steryl esters in sediment trap material [134]. Steryl esters rarely give molecular ions in EI or CI mode, but useful fragment ions can be obtained using CI with methane as the reagent gas [134].

3.9
Triterpenoid Alcohols

This is a very broad collection of compounds and so the discussion here will be brief. The term "triterpenoid" alcohols is usually restricted to those compounds with a 6-membered E-ring (e.g. compounds based on the oleanane or ursane skeletons), but the term can be taken to include those with a 5-membered ring such as hopanoids (Figs. 13 and 14 [194]). Triterpenoids have proven to be particularly useful to assign sources of organic matter in sediments and petroleum.

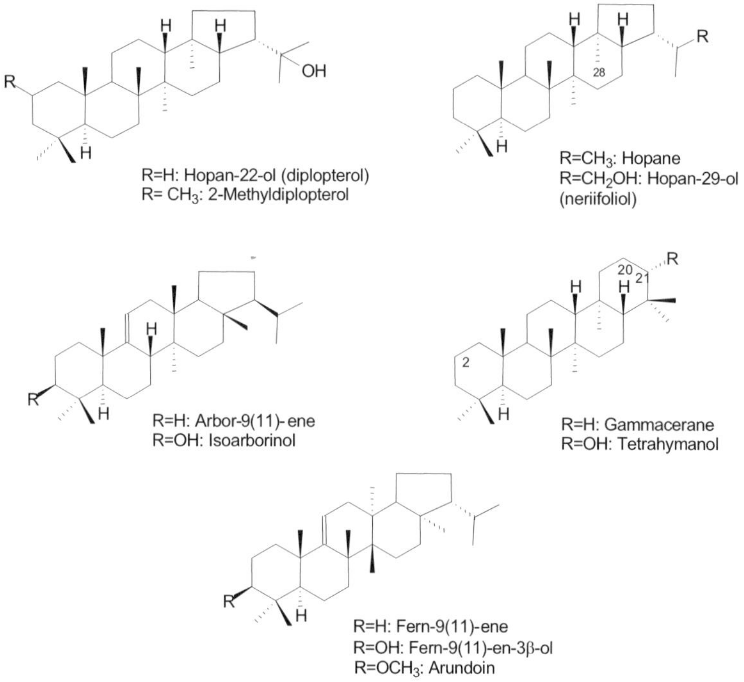

Fig. 13 Structures of some triterpenoids derived primarily from prokaryotes found in sediments

3.9.1
Hopanoids

The most common triterpenoid class in sediments is the hopanoids. Indeed, they may even be the most abundant natural product on earth [226]. Hopanoids are synthesized in bacteria from squalene which is cyclized by squalene-hopene cyclase (e.g. [227]). This reaction shares considerable similarities with that of the conversion of oxidosqualene to sterols (e.g. [194]). Hopanoids can occur in a variety of functionalized forms including alkenes, ketones, acids and alcohols (Fig. 13). Most are derived either directly from bacterial sources and have an extended alkyl chain at C-22 (e.g. [228]), or are diagenetic products. These diagenetic reactions occur very early in the diagenetic sequence in sediments (e.g. [229, 230]). For example, sediments of the highly productive lake (Priest Pot, UK) contain not only the parent bacteriohopanetetrol (BHT), but a series of hopanediols and triols such as trishomohopane-32,33-diol and bishomohopane-30,31,32-triol which represent intermediate stages in the conversion of biologically produced hopanoids to the diagenetic products, hopanes [231]. This process can be mimicked in the laboratory by oxidizing the biohopanoid in the presence of cupric chloride in pyridine [232].

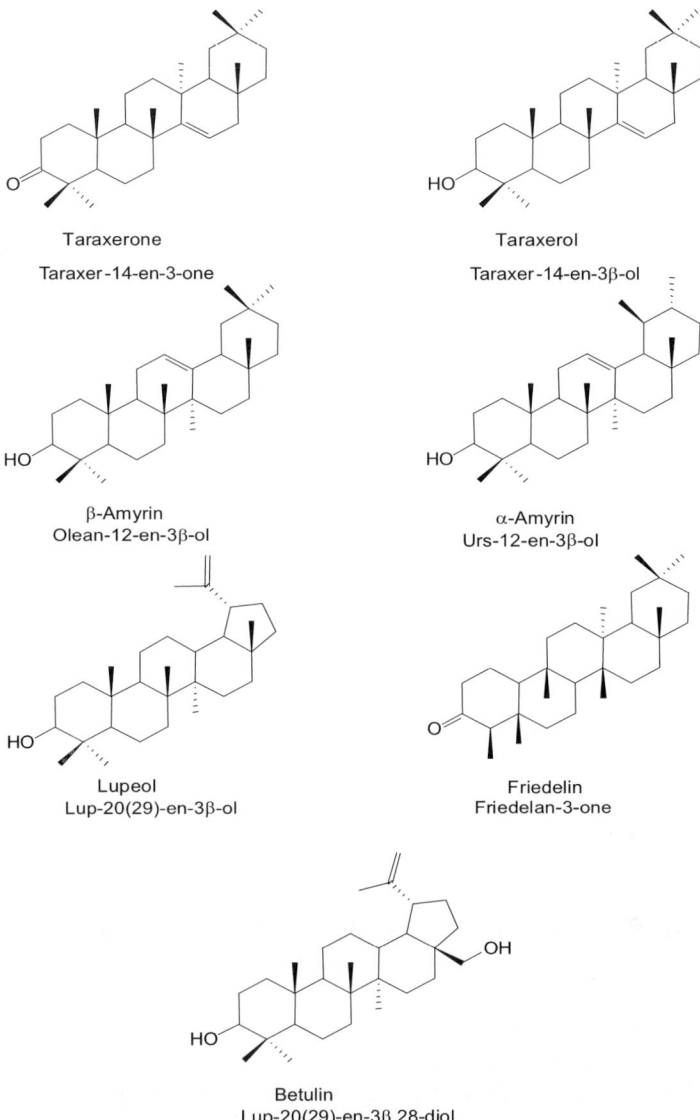

Fig. 14 Structures of some triterpenoids commonly found in higher plants

Unsaturated hydrocarbons (hopenes) such as hop-22(29)-ene, hop-17(21)-ene and 17β(H)-hop-21-ene [233] and C_{31} and C_{32} ββ-isomers of diagenetically formed hopanes derived from polyhydroxyhopanoids of bacterial origin [234] are often present in immature sediments. Some unusual hopenes with the "moretane" stereochemistry such as 17β(H)-moret-22(29)-ene have also been reported in sediments [235]. Fully saturated hopanes are rare in

Recent sediments and seawater and in most cases these are attributable to ancient organic matter or petroleum contamination [21], rather than biological sources. This is most clearly shown when the carbon number range extends from C_{27} to C_{35} and several 17,21 α- and β-isomers are present. A full discussion of these compounds would take a chapter in their own right, so the reader is referred to leading reviews [18–20].

Most hopanoids can be identified directly by straightforward GC-MS techniques (e.g. [216]), and are usually characterized with a m/z 191 mass fragmentogram (or m/z 205 for hopanoids with a methyl group in the A ring or m/z 177 for demethylated hopanes). The direct detection of most intact biohopanoids is not possible due to their highly functionalized and amphiphilic nature. Talbot et al. [236] have developed a new reversed-phase high-performance liquid chromatography method for the direct analysis of acetylated, intact bacteriohopanepolyols and applied it to solvent extracts of methanotrophic bacteria. This method was suitable for identifying the four biohopanoids: bacteriohopanetetrol and aminobacteriohopanetriol, -tetrol and -pentol.

3.9.2
"Higher Plant" Triterpenoids

A major source of triterpenoid alcohols in modern sediments is from the angiosperms, which are the most abundant plant group on the earth today. Angiosperms contain triterpenoids of the β-amyrin (oleananoid) type, which on diagenesis ultimately produces the C_{30} triterpenoid hydrocarbon oleanane. The occurrence of oleanane in the geological record broadly correlates with the emergence of angiosperms in the Early Cretaceous and radiation in the Late Cretaceous and Tertiary (e.g. [237]).

Triterpenoids with a single hydroxyl group such as α- and β-amyrin, germanicol and taraxerol (Fig. 14), are widely used as terrestrial plant markers. For example, Killops and Frewin [238] used β-amyrin and taraxerol and their degradation products to trace organic matter from mangroves (*Rhizophora mangle*) in sediments from Florida Bay. Koch et al. [239] have had similar success in using such compounds as tracers. Volkman et al. [240] used a dihydroxylated triterpenoid betulin (Fig. 14) and other triterpenoid alcohols to show that sediments from the Wadden Sea contained organic matter from eroded peats. Note, however, that these compounds are not source-specific and can be found in a variety of plants [241], so care must be taken when using them to trace organic matter from specific plants.

3.9.3
Other Triterpenoids

Unfortunately, higher plants are not the only source of triterpenoid skeletons. For example, Smetanina et al. [242] identified the 3β-methoxyolean-18-ene (miliacin – the methoxy ether derivative of germanicol) in a marine fungus *Chaetomium olivaceum* (Fig. 13). α-Amyrin has been positively identified in the spores from 16 species of arbuscular mycorrihizal fungi (AMF) belonging to the order Glomales [243]. Triterpene methyl ethers have been reported in a few sediments (e.g. [244]), but their origins remain obscure. Methyl ethers of other common triterpene series (e.g. cylindrin – isoarborinol methyl ether; sawamilletin - taraxerol methyl ether; isosawamilletin – β-amyrin methoxy ether) are also known from a variety of plants, but most commonly from grasses [245, 246].

Fernene isomers of bacterial origin have also been observed in some Recent sediments (e.g. [233, 247]). The distribution of fernenes appears to follow progressive isomerization from the Δ^7 to the thermodynamically more stable Δ^8 and $\Delta^{9(11)}$ isomers (structures in Fig. 13) with depth, although the latter can also have natural sources [248]. These C_{30} triterpenes have a prominent m/z 243 ion in their mass spectra. Fernenes and fernenols are abundant in contemporary ferns (e.g. [249, 250]), but their most likely source in most sediments is from bacteria [233, 251]. Fern-7-ene and fern-9(11)-ene have been isolated from a purple, non-sulfur bacterium *Rhodomicrobium vanniellii* [248].

Another interesting class of triterpenoids is based on the isoarborinol skeleton (Fig. 13). Isoarborinol has been found in quite high concentration in sediments from the Permian and Triassic [252], but reports in contemporary environments are rare. It has been suggested that it is derived from a microbial source [253–255].

Gammacerane has been found in sediments from the Late Proterozoic which also contain some of the earliest examples of fossil protozoans [256]. It is presumed to be derived from tetrahymanol (gammaceran-21α-ol) which is found in ciliates, ferns, fungi and bacteria [257, 258]. Tetrahymanol is both common and abundant in recent marine sediments (e.g. [259]).

3.10
Chlorophyll and Carotenoid Pigments

Microalgae contain a diverse array of carotenoid and chlorophyll pigments, including a number of newly discovered components. Some pigments are specific to one or a limited number of algal classes and thus can be used as biomarkers in the same way as lipids. Indeed, a valuable development has been the software program CHEMTAX [260], which uses specific pigment ratios to ascertain the abundance of microalgal groups in seawater samples. A comprehensive review of pigments is provided in the SCOR-UNESCO publication on phytoplankton pigments in oceanography [261]. This information is updated and summarized in the chapter by Wright and Jeffrey. The combination of pigment markers with lipid biomarkers such as fatty acids has proven to be very useful for elucidating sources of organic matter in the sea and for studying their rates of degradation [262].

3.11
Organic Sulfur Compounds

A great variety of organic sulfur compounds (OSCs) have been identified in marine sediments and in anoxic seawater. Most of these compounds result from incorporation of sulfur into functionalized biolipids, especially when the content of reactive iron in the sediments is low. Both intra- and intermolecular incorporation of sulfur has been documented. These OSCs thus contain biomarker information about the originally deposited organic matter. Further details can be found in recent papers [263–265].

4
Summary

A wide range of chemically distinct organic compounds can be found in marine sediments and seawater. The structures of many of these have been elucidated, although it seems that every year brings the discovery of some new class of compound. Some chemical structures can be related to specific groups of organisms, or even to a particular Class, Order or genus and thus they can be used as biomarkers for elucidating the sources of the organic matter. Many biomarkers degrade over time as a result of chemical and biological processes. The distributions of these degradation products can thus also provide a history of the breakdown of organic matter in the sediments or water column. Advances in analytical and chemical identification techniques now make it possible to identify very small amounts (ng to mg) in aquatic environment even when present in complex mixtures. Moreover, it is now possible to determine the ^{13}C and ^{14}C content in many compounds, thus providing additional information about their source and age.

Acknowledgements I wish to pay tribute to my colleagues for their assistance, advice and fruitful conversations over many years which has led to this overview of organic geochemical biomarkers. Funding from CSIRO Marine Research, the Fisheries Research and Development Corporation, Aquafin CRC, Antarctic Science Advisory Committee and the Australian Research Council is gratefully acknowledged.

References

1. Han J, McCarthy ED, Van Hoeven W, Calvin M, Bradley WH (1968) Proc Natl Acad Sci USA 59:29
2. Jeffries HP (1970) Limnol Oceanogr 15:419
3. Larsson K, Odham G, Södergren A (1974) Mar Chem 2:49
4. Albrecht P, Ourisson G (1971) Angew Chem Int Ed Engl 10:209
5. Bligh EG, Dyer WJ (1959) Can J Biochem Physiol 37:912

6. Folch J, Lees M, Sloane Stanley GH (1957) J Biol Chem 226A:497
7. Iverson SJ, Lang SLC, Cooper MH (2001) Lipids 36:283
8. Delmas RP, Parrish CC, Ackman RG (1984) Anal Chem 56:1272
9. Parrish CC, Ackman RG (1983) J Chromatogr 262:103
10. Volkman JK, Everitt DA, Allen DI (1986) J Chromatogr 356:147
11. Lough AK (1964) Nature 202:795
12. Christie WW (2003) Lipid analysis: isolation, separation, identification and structural analysis of lipids, 3rd edn. Oily Press, England
13. de Leeuw JW, Rijpstra WIC, Schenck PA, Volkman JK (1983) Geochim Cosmochim Acta 47:455
14. Wakeham SG (1999) Org Geochem 30:1059
15. Garcette-Lepecq A, Largeau C, Bouloubassi I, Derenne S, Saliot A, Lorre A, Point V (2004) Org Geochem 35:959
16. Johns RB (ed) (1986) Biological markers in the sediment record. Methods in geochemistry and geophysics, vol 24. Elsevier, Amsterdam
17. Engel MH, Macko SA (eds) (1993) Organic geochemistry. Principles and applications. Plenum, NY
18. Philp RP (1985) Fossil fuel biomarkers. Applications and spectra. Methods in geochemistry and geophysics, vol 23. Elsevier, Amsterdam
19. Moldowan JM, Albrecht P, Philp RP (1992) Biological markers in sediments and petroleum. Prentice Hall, Englewood Cliffs, NJ
20. Killops SD, Killops VJ (1993) An introduction to organic geochemistry. Longman Scientific and Technical, NY
21. Volkman JK, Revill AT, Murray AP (1997) Applications of biomarkers for identifying sources of natural and pollutant hydrocarbons in aquatic environments. In: Eganhouse RP (ed) Molecular markers in environmental geochemistry. ACS Symposium Series 671. American Chemical Society, NY p 110
22. Giger W, Schaffner C, Wakeham SG (1980) Geochim Cosmochim Acta 44:119
23. Rowland SJ, Maxwell JR (1984) Geochim Cosmochim Acta 48:617
24. Han J, McCarthy ED, Calvin M, Benn MH (1968) J Chem Soc C 2785
25. Höld IM, Schouten S, Jellema J, Sinninghe Damsté JS (1999) Org Geochem 30:1411
26. Köster J, Volkman JK, Rullkötter J, Scholz-Böttcher BM, Rethmeier J, Fischer U (1999) Org Geochem 30:1367
27. Kenig F, Sinninghe Damsté JS, Dalen ACK, Rijpstra WIC, Huc AY, de Leeuw JW (1995) Geochim Cosmochim Acta 59:2999
28. Kenig F, Simons DJH, Crich D, Cowen JP, Ventura GT, Rehbein-Khalily T, Brown TC, Anderson KB (2003) Proc Natl Acad Sci USA 100:12554
29. Nishimoto S (1974) J Sci Hiroshima Univ A 38:165
30. Matsumoto GI, Friedmann EI, Watanuki K, Ocampo-Friedmann R (1992) J Chromatogr 598:267
31. Blumer M, Mullin MM, Thomas DW (1963) Science 140:974
32. Volkman JK, Maxwell JR (1986) Acyclic isoprenoids as biological markers. In: Johns RB (ed) Biological markers in the sedimentary record. Elsevier, Amsterdam, p 1
33. Waples DW, Haug P, Welte DH (1974) Geochim Cosmochim Acta 38:381
34. Brassell SC, Wardroper AMK, Thomson ID, Maxwell JR, Eglinton G (1981) Nature 290:693
35. Elvert M, Suess E, Whiticar MJ (1999) Naturwiss 86:295
36. Freeman KH, Wakeham SG, Hayes JM (1994) Org Geochem 21:629
37. Sinninghe Damsté JS, Schouten S, van Vliet NH, Huber R, Geenevasen JAJ (1997) Tetrahedron Lett 38:6881

38. De Rosa M, Gambacorta A (1994) Archaeal lipids. In: Goodfellow M, O'Donnell AG. Chemical methods in prokaryotic systematics. John Wiley and Sons, Chichester p 197
39. Sinninghe Damsté JS, Kuypers MMM, Schouten S, Schulte S, Rullkötter R (2003) Earth Planet Sci Letts 209:215
40. Wakeham SG, Freeman KH, Pease TK, Hayes JM (1993) Geochim Cosmochim Acta 57:159
41. Kissin YV, Feulmer GP, Payne WB (1986) J Chrom Sci 24:164
42. Lee RF, Loeblich III AR (1971) Phytochem 10:593
43. Volkman JK, Barrett SM, Dunstan GA (1994) Org Geochem 21:407
44. Volkman JK, Barrett SM, Blackburn SI, Mansour MP, Sikes EL, Gelin F (1998) Org Geochem 29:1163
45. Volkman JK, Eglinton G, Corner EDS, Forsberg TEV (1980) Phytochem 19:2619
46. Volkman JK, Smith DJ, Eglinton G, Forsberg TEV, Corner EDS (1981) J Mar Biol Assoc UK 61:509
47. Rieley G, Teece MA, Peakman TM, Raven AM, Greene KJ, Clarke TP, Murray M, Leftley JW, Campbell CN, Harris RP, Parkes RJ, Maxwell JR (1998) Lipids 33:617
48. Volkman JK, Eglinton G, Corner EDS, Sargent JR (1980) Novel unsaturated straight-chain $C_{37}-C_{39}$ methyl and ethyl ketones in marine sediments and a coccolithophorid *Emiliania huxleyi*. In: Douglas AG, Maxwell JR (eds) (1979) Advances in organic geochemistry. Pergamon, Oxford, p 219
49. Rowland SJ, Volkman JK (1982) Mar Environ Res 7:117
50. Gelin F, Boogers I, Noordeloos AAM, Sinninghe Damsté JS, Riegman R, de Leeuw JW (1997) Org Geochem 26:659
51. Sinninghe Damsté JS, Rijpstra WIC, Schouten S, Peletier H, van der Maarel MJEC, Gieskes WWC (1999) Org Geochem 30:95
52. Gelpi E, Oró J, Schneider HJ, Bennett EO (1968) Science 161:700
53. van der Meer MTJ, Schouten S, Ward DM, Geenevasen JAJ, Sinninghe Damsté JS (1999) Org Geochem 30:1585
54. Gearing P, Gearing JN, Lytle TF, Lytle JS (1976) Geochim Cosmochim Acta 40:1005
55. Schouten S, Hoefs MJL, Sinninghe Damsté JS (2000) Org Geochem 31:509
56. Yon DA, Maxwell JR, Ryback G (1982) Tetrahedron Lett 23:2143
57. Hird SJ, Evens R, Rowland SJ (1992) Mar Chem 37:117
58. Volkman JK, Farrington JW, Gagosian RB, Wakeham SG (1983) Lipid composition of coastal marine sediments from the Peru upwelling region. In: Bjorøy M et al. (eds) Advances in organic geochemistry 1981. John Wiley, Chichester p 228
59. Nichols PD, Volkman JK, Palmisano AC, Smith GA, White DC (1988) J Phycol 24:90
60. Rowland SJ, Robson JN (1990) Mar Environ Res 30:191
61. Belt ST, Massé G, Allard WG, Robert J-M, Rowland SJ (2001) Org Geochem 32:1271
62. Belt ST, Allard WG, Massé G, Robert J-M, Rowland SJ (2001) Tetrahedron Lett 42:5583
63. Schouten S, Klein Breteler WCM, Blokker P, Rijpstra WIC, Grice K, Baas M, Sinninghe Damsté JS (1998) Geochim Cosmochim Acta 62:1397
64. Rowland SJ, Allard WG, Belt ST, Massé G, Robert J-M, Blackburn S, Frampton D, Revill AT, Volkman JK (2001) Phytochem 58:717
65. Belt ST, Massé G, Allard WG, Robert JM, Rowland SJ (2003) Tetrahedron Lett 44:9103
66. Massé G, Belt ST, Allard WG, Lewis CA, Wakeham SG, Rowland SJ (2004) Org Geochem 35:813.
67. Wraige EJ, Belt ST, Lewis CA, Cooke DA, Robert J-M, Massé G, Rowland SJ (1997) Org Geochem 27:497

68. Wraige EJ, Johns L, Belt ST, Massé G, Robert J-M, Rowland S (1999) Phytochem 51:69
69. Belt ST, Cooke DA, Hird SJ, Rowland SJ (1994) J Chem Soc Chem Comm 2077
70. Belt ST, Allard WG, Massé G, Robert J-M, Rowland SJ (2000) Geochim Cosmochim Acta 64:3839
71. Cox RE, Burlingame AL, Wilson DM, Eglinton G, Maxwell JR (1973) J Chem Soc Chem Comm 284
72. Metzger P, Casadevall E, Coute A (1988) Phytochem 27:1383
73. Okada S, Murakami M, Yamaguchi K (1997) Phytochem Anal 8:198
74. Metzger P, Allard B, Casadevall E, Berkaloff C, Cout A (1990) J Phycol 26:258
75. Grice K, Schouten S, Nissenbaum A, Charrach J, Sinninghe Damsté JS (1998) Org Geochem 28:195
76. Sinninghe Damsté JS, Erkes AMWEP, Rijpstra WIC, de Leeuw JW, Wakeham SG (1995) Geochim Cosmochim Acta 59:347
77. Mackenzie AS, Brassell SC, Eglinton G, Maxwell JR (1982) Science 217:491
78. Peakman TM, ten Haven HL, Rechka JR, de Leeuw JW, Maxwell JR (1989) Geochim Cosmochim Acta 53:2001
79. Brassell SC, McEvoy J, Hoffmann CF, Lamb NA, Peakman TM, Maxwell JR (1984) Isomerisation, rearrangement and aromatisation of steroids in distinguishing early stages of diagenesis. In: Schenck PA, de Leeuw JW, Lijmbach GWM (eds) (1983) Advances in organic geochemistry. Pergamon, Oxford, p 11
80. Volkman JK, Jeffrey SW, Nichols PD, Rogers GI, Garland CD (1989) J Exp Mar Biol Ecol 128:219
81. Perry GJ, Volkman JK, Johns RB, Bavor HJ (1979) Geochim Cosmochim Acta 43:1715
82. Fulco AJ (1983) Prog Lipid Res 22:133
83. Volkman JK, Johns RB (1977) Nature 267:693
84. Ivanova EP, Zhukova NV, Svetashev VI, Gorshkova NM, Kurilenko VV, Frolova GM, Mikhailov VV (2000) Curr Microbiol 41:341
85. Sinninghe Damsté JS, Strous M, Rijpstra WIC, Hopmans EC, Geenevasen JAJ, van Duin ACT, van Niftrik LA, Jetten MSM (2002) Nature 419:708
86. Jetten MSM, Sliekers O, Kuypers M, Dalsgaard T, van Niftrik L, Cirpus I, van de Pas-Schoonen K, Lavik G, Thamdrup B, Le Paslier D, Op den Camp HJM, Hulth S, Nielsen LP, Abma W, Third K, Engström P, Kuenen JG, Jørgensen BB, Canfield DE, Sinninghe Damsté JS, Revsbech NP, Fuerst J, Weissenbach J, Wagner M, Schmidt I, Schmid M, Strous M (2003) Appl Microbiol Biotechnol 63:107
87. Kuypers MMM, Sliekers AO, Lavik G, Schmid M, Jørgensen BB, Kuenen JG, Sinninghe Damsté JS, Strous M, Jetten MSM (2003) Nature 422:608
88. Napolitano GE, Pollero RJ, Gayoso AM, MacDonald BA, Thompson RJ (1997) Biochem Syst Ecol 25:739
89. Suhr SB, Pond DW, Gooday AJ, Smith CR (2003) Mar Ecol Prog Ser 262:153
90. Dalsgaard J, St John M, Kattner G, Muller-Navarra D, Hagen W (2003) Adv Mar Biol 46:225
91. Mansour MP, Volkman JK, Holdsworth DG, Jackson AE, Blackburn SI (1999) Phytochem 50:541
92. Leblond JD, Chapman PJ (2000) J Phycol 36:1103
93. Leblond JD, Evans TJ, Chapman PJ (2003) Phycologia 42:324
94. Shantha NC, Napolitano GE (1992) J Chromatogr 624:37
95. Moss CW, Daneshvar MI (1992) J Clin Microbiol 30:2511
96. Dunkelblum E, Tan SH, Silk PJ (1985) J Chem Ecol 11:265
97. Nichols PD, Guckert JB, White DC (1986) J Microbiol Met 5:49
98. Ackman RG, Burgher RD (1964) J Fish Res Bd Canada 21:319

99. Ghosh A, Hogue M, Dutta J (1972) J Chromatogr 69:207
100. Dobson G, Christie WW, Nikolovada-Myanova B (1995) J Chromatogr B Biomed Appl 671:197
101. Christie WW (1998) Lipids 33:343
102. Spitzer V (1996) Prog Lipid Res 35:387
103. Méjanelle L, Laureillard J, Saliot A (2002) J Microbiol Met 48:221
104. Nichols DS, Davies NW (2002) J Microbiol Met 50:103
105. Skerratt JH, Bowman JP, Nichols PD (2002) Int J Syst Evol Microbiol 52:2101
106. Eglinton G, Hunneman DH, Douraghi-Zadeh K (1968) Tetrahedron 24:5929
107. Kolattukudy PE (1980) Cutin, suberin, and waxes. In: Stumpf PK (ed) The biochemistry of plants. A comprehensive treatise, vol 4, Lipids: Structure and Function. Academic, New York, p 571
108. Kolattukudy PE (1984) Can J Bot 62:2918
109. Volkman JK, Johns RB, Gillan FT, Perry GJ, Bavor HJ (1980) Geochim Cosmochim Acta 44:1133
110. Kawamura K, Gagosian RB (1990) Naturwiss 77:25
111. Downing DT (1961) Rev Pure Appl Chem 11:196
112. Fulco AJ (1967) J Biol Chem 242:3608
113. Volkman JK, Barrett SM, Blackburn SI (1999) J Phycol 35:1005
114. Edlund A, Nichols PD, Roffey R, White DC (1985) J Lipid Res 26:982
115. Kaltashov IA, Doroshenko V, Cotter RJ, Takayama K, Qureshi N (1997) Anal Chem 69:2317
116. Boon JJ, de Leeuw JW, van der Hoek GJ, Vosjan JH (1977) J Bacteriol 129:1183
117. Keinanen MM, Korhonen LK, Martikainen PJ, Vartiainen T, Miettinen IT, Lehtola MJ, Nenonen K, Pajunen H, Kontro MH (2003) J Chromatogr B 783:443
118. Gelin F, Volkman JK, de Leeuw JW, Sinninghe Damsté JS (1997) Phytochem 45:641
119. Kolattukudy PE, Croteau R, Walton TJ (1975) Plant Physiol 55:875
120. Fukushima K, Kondo H, Sakata S (1992) Org Geochem 18:913
121. Skerratt JH, Nichols PD, Bowman JP, Sly LI (1992) Org Geochem 18:189
122. Abreu-Grobois FA, Billyard TC, Walton TJ (1977) Phytochem 16:351
123. de Leeuw JW, Rijpstra WIC, Mur LR (1992) Org Geochem 18:575
124. Kolattukudy PE, Walton TJ (1972) Biochemistry 11:1897
125. Walton TJ, Kolattukudy PE (1972) Biochem Biophys Res Comm 46:16
126. Huang YS, Lockheart MJ, Logan GA, Eglinton G (1996) Org Geochem 24:289
127. López Alonso D, Belarbi E-H, Rodríguez-Ruiz J, Segura CI, Giménez A (1998) Phytochem 47:1473
128. Sargent JR, Gatten RR, Henderson RJ (1981) Pure Appl Chem 53:867
129. Pernet F, Tremblay R, Demers E, Roussy M (2003) Aquaculture 221:393
130. Dunstan GA, Volkman JK, Barrett SM, Garland CD (1993) J Appl Phycol 5:71
131. Siron R, Giusti G, Berland B (1989) Mar Ecol Prog Ser 55:95
132. Mayzaud P, Errhif A, Bedo A (1998) J Mar Syst 17:391
133. Lee RF, Nevenzel JC, Paffenhofer G-A (1971) Mar Biol 9:99
134. Wakeham SG (1982) Geochim Cosmochim Acta 46:2239
135. Wakeham SG, Farrington JW, Volkman JK (1983) Fatty acids, wax esters, triacylglycerols, and alkyldiacylglycerols associated with particles collected in sediment traps in the Peru upwelling. In: Bjorøy M et al (eds) (1981) Advances in organic geochemistry. John Wiley, Chichester, p 185
136. Wakeham SG, Frew NM (1982) Lipids 17:831
137. Boon JJ, de Leeuw JW (1979) Mar Chem 7:117

138. Boon JJ, Rijpstra WIC, de Leeuw JW, Burlingame AL (1980) Geochim Cosmochim Acta 44:131
139. Bobbie RJ, White DC (1980) Appl Environ Microbiol 39:1212
140. White DC (1993) Phil Trans Roy Soc Lond A 344:59
141. Dobbs FC, Findlay RH (1993) Quantitative description of microbial communities using lipid analysis. In: Kemp PF, Sherr BF, Sherr EB, Cole JJ (eds) Handbook of methods in aquatic microbial ecology. Lewis Publishers, Boca Raton, p 271
142. Findlay RH, Dobbs FC (1993) Analysis of microbial lipids to determine biomass and detect the response of sedimentary microorganisms to disturbance. In: Kemp PF, Sherr BF, Sherr EB, Cole JJ (eds) Handbook of methods in aquatic microbial ecology. Lewis Publishers, Boca Raton, p 347
143. White DC, Davis WM, Nickels JS, King JD, Bobbie RJ (1979) Oecologia 40:51
144. Rütters H, Sass H, Cypionka H, Rullkötter J (2002) J Micro Met 48:149
145. Rütters H, Sass H, Cypionka H, Rullkötter J (2002) Org Geochem 33:803
146. Laureillard J, Pinturier L, Fillaux J, Saliot A (1997) Deep Sea Res II 44:1085
147. Okuyama H, Morita N, Kogame K (1992) J Phycol 28:465
148. Budge SM, Parrish CC (1999) Phytochem 52:561
149. Son BW, Cho YJ, Choi JS, Lee WK, Kim DS, Choi HD, Choi JS, Jung JH, Im KS, Choi WC (2001) Nat Prod Letts 15:299
150. Batrakov SG, Nikitin DI, Pitryuk IA (1996) Biochim Biophys Acta-Lipids Lipid Metab 1303:39
151. Gillan FT, Sandstrom MW (1985) Org Geochem 8:321
152. Sinninghe Damsté JS, van Dongen BE, Rijpstra WIC, Schouten S, Volkman JK, Geenevasen JAJ (2001) Org Geochem 32:321
153. Oren A, Gurevich P (1993) FEMS Microbiol Ecol 12:249
154. Kato M, Adachi K, Hajiro-Nakanishi K, Ishigaki E, Sano H, Miyachi S (1994) Phytochem 37:279
155. ten Haven HL, Littke R, Rullkötter J, Stein R, Welte DH (1990) Accumulation rates and composition of organic matter in Late Cenozoic sediments underlying the active upwelling area off Peru. In: Suess E, von Huene R et al. (eds) Proceedings of the Ocean Drilling Program, scientific results, vol 112. Ocean Drilling Program, College Station, p 591
156. Cranwell PA, Volkman JK (1981) Chem Geol 32:29
157. Russell NJ, Volkman JK (1980) J Gen Microbiol 118:131
158. Cranwell PA (1984) Org Geochem 6:115
159. Aasen AJ, Hofstetter HH, Iyengar BTR, Holman RT (1971) Lipids 6:502
160. Volkman JK, Barrett SM, Dunstan GA, Jeffrey SW (1992) Org Geochem 18:131
161. de Leeuw JW, Rijpstra WIC, Schenck PA (1981) Geochim Cosmochim Acta 45:2281
162. Nichols PD, Johns RB (1986) Org Geochem 9:25
163. Versteegh GJM, Bosch HJ, de Leeuw JW (1997) Org Geochem 27:1
164. Versteegh GJM, Jansen JHF, de Leeuw JW, Schneider RR (2000) Geochim Cosmochim Acta 64:1879
165. Tegelaar EW, Matthezing RM, Jansen JBH, Horsfield B, de Leeuw JW (1989) Nature 342:529
166. Largeau C, de Leeuw JW (1995) Insoluble, non-hydrolysable aliphatic macromolecule constituents of microbial cell walls. In: Jones JG (ed) Advances in microbial ecology, vol 14. Plenum, New York, p 77
167. Volkman JK, Barrett SM, Blackburn SI (1999) Org Geochem 30:307
168. Sinninghe Damsté JS, Rampen S, Rijpstra WIC, Abbas B, Muyzer G, Schouten S (2003) Geochim Cosmochim Acta 67:1339

169. Ferreira AM, Miranda A, Caetano M, Baas M, Vale C, Sinninghe Damsté JS (2001) Org Geochem 32:271
170. Schouten S, Rijpstra WIC, Kok M, Hopmans EC, Summons RE, Volkman JK, Sinninghe Damsté JS (2001) Geochim Cosmochim Acta 65:629
171. Sprott GD, Dicaire CJ, Choquet CG, Patel GB, Ekiel I (1993) Appl Environ Microbiol 59:912
172. Paltauf F (1983) Ether lipids containing sterol moieties. In: Mangold HK and Paltauf F, Ether lipids: biochemical and biomedical aspects. Academic, NY, p 177
173. Wakeham SG, Lewis CM, Hopmans EC, Schouten S, Sinninghe Damsté JS (2003) Geochim Cosmochim Acta 67:1359
174. Hoefs MJL, Schouten S, de Leeuw JW, King LL, Wakeham SG, Sinninghe Damsté JS (1997) Appl Environ Microbiol 63:3090
175. Sinninghe Damsté JS, Schouten S, Hopmans EC, van Duin ACT, Geenevasen JAJ (2002) J Lipid Res 43:1641
176. Sinninghe Damsté JS, Rijpstra WIC, Hopmans EC, Prahl FG, Wakeham SG, Schouten S (2002) Appl Environ Microbiol 68:2997
177. Hopmans EC, Schouten S, Pancost RD, van der Meer MTJ, Sinninghe Damsté JS (2000) Rapid Comm Mass Spectrom 14:585
178. Hopmans EC, Weijers JWH, Schefuß E, Herfort L, Sinninghe Damsté JS, Schouten S (2004) Earth Planet Sci Lett 224:107
179. Sinninghe Damsté JS, Hopmans EC, Pancost RD, Schouten S, Geenevasen JAJ (2000) Chem Comm 1683
180. Morrison RI, Bick W (1966) Chem Ind April 2, p 596
181. Lehtonen K, Ketola M (1990) Org Geochem 15:275
182. Cranwell PA, Eglinton G, Robinson N (1987) Org Geochem 11:513
183. Hernandez ME, Mead R, Peralba MC, Jaffé R (2001) Org Geochem 32:21
184. Volkman JK, Barrett SM, Blackburn SI, Sikes EL (1995) Geochim Cosmochim Acta 59:513
185. Rontani J-F, Marchand D, Volkman JK (2001) Org Geochem 32:1329
186. Marlowe IT, Brassell SC, Eglinton G, Green JC (1984) Org Geochem 6:135
187. Sikes EL, Volkman JK, Robertson LG, Pichon JJ (1997) Geochim Cosmochim Acta 61:1495
188. Sikes EL, Sicre MA (2002) Geochem Geophys Geosyst 3: art no. 1063
189. Prahl FG, Wakeham SG (1987) Nature (Lond) 330:367
190. Brassell SC, Eglinton G, Marlowe IT, Pflaumann U, Sarnthein M (1986) Nature (Lond) 320:129
191. Rosell-Melé A (1998) Paleoceanogr 13:694
192. Volkman JK (1986) Org Geochem 9:83
193. Volkman JK (2003) Appl Micro Biotech 60:495
194. Volkman JK (2004) Org Geochem (in press)
195. Goad LJ, Holz GGJ, Beach DH (1983) Phytochem 22:475
196. Barrett SM, Volkman JK, Dunstan GA, LeRoi JM (1995) J Phycol 31:360
197. Moldowan MJ, Fago FJ, Lee CY, Jacobson SR, Watt DS, Slougui N-E, Jeganathan A, Young DC (1990) Science 247:309
198. Patterson GW (1971) Lipids 6:120
199. Volkman JK, Barrett SM, Dunstan GA, Jeffrey SW (1993) Org Geochem 20:7
200. Volkman JK, Kearney P, Jeffrey SW (1990) Org Geochem 15:489
201. Volkman JK, Farmer CL, Barrett SM, Sikes EL (1997) J Phycol 33:1016
202. Mansour MP, Volkman JK, Jackson AE, Blackburn SI (1999) J Phycol 35:710
203. Withers NW, Goad LJ, Goodwin TW (1979) Phytochem 18:899

204. Withers N (1987) Dinoflagellate sterols. In: Taylor FJR (ed) The biology of dinoflagellates. Biological Monographs, vol 21. Blackwell Scientific, Oxford, p 316
205. Santos VLCS, Billett DSM, Wolff GA (2000) Aquat Eco Health Management 3:397
206. Butler E, Parslow J, Volkman J et al. (2000) Huon Estuary Study – Environmental research for integrated catchment management and aquaculture. FRDC Final Report Project No. 96/284. Fisheries Research and Development Corporation, Canberra
207. Robinson N, Eglinton G, Brassell SC, Cranwell PA (1984) Nature (Lond) 308:439
208. Nishimura M, Koyama T (1977) Geochim Cosmochim Acta 41:379
209. Nishimura M (1982) Geochim Cosmochim Acta 46:423
210. Leeming R, Latham V, Rayner M, Nichols P (1997) Detecting and distinguishing sources of sewage pollution in Australian inland and coastal waters and sediments. In: Eganhouse RP (ed) ACS Symposium Series No. 617. Molecular markers in environmental geochemistry. American Chemical Society, NY p 306
211. Venkatesan MI, Santiago CA (1989) Mar Biol 102:431
212. Knights BA (1967) J Gas Chromatog 5:273
213. Brooks CJW, Henderson W, Steel G (1973) Biochim Biophys Acta 296:431
214. Rahier A, Benveniste P (1989) Mass spectral identification of phytosterols. In: Nes WD, Parish EJ (eds) Analysis of sterols and other biologically significant steroids. Academic, San Diego, p 223
215. Itoh T, Tani H, Fukishima K, Tamura T, Matsumoto T (1982) J Chromatogr 234:65
216. Jones GJ, Nichols PD, Shaw PM (1994) Analysis of microbial sterols and hopanoids. In: Goodfellow M, O'Donnell AG (eds) Chemical methods in prokaryotic systematics. John Wiley and Sons, Chichester, p 163
217. Gagosian RB, Smith SO (1979) Nature 277:287
218. Harvey RH, Bradshaw SA, O'Hara SCM, Eglinton G, Corner EDS (1988) Phytochem 27:1723
219. Leblond JD, Chapman PJ (2004) J Phycol 40:104
220. Parmentier G, Eyssen H (1974) Biochim Biophys Acta 348:279
221. Benfenati E, Cools E, Fattore E, Fanelli R (1994) Chemosphere 29:1393
222. King LL, Repeta DJ (1994) Geochim Cosmochim Acta 58:4389
223. Talbot HM, Head RN, Harris RP, Maxwell JR (1999) Org Geochem 30:1163
224. Talbot HM, Head RN, Harris RP, Maxwell JR (2000) Org Geochem 31:871
225. Fumasaki J, Gilbertson JR (1968) J Lipid Res 9:766
226. Ourisson G, Albrecht P (1992) Acc Chem Res 25:398
227. Kannenberg EL, Poralla K (1999) Naturwiss 86:168
228. Rohmer M, Bouvier-Navé P, Ourisson G (1984) J Gen Microbiol 130:1137
229. Rodier C, Llopiz P, Neunlist S (1999) Org Geochem 30:713
230. Tritz JP, Herrmann D, Bisseret P, Connan J, Rohmer M (1999) Org Geochem 30:499
231. Watson DF, Farrimond P (2000) Org Geochem 31:1247
232. Bisseret P, Rohmer M (1995) Tetrahedron Lett 36:7077
233. Volkman JK, Allen DI, Stevenson PL, Burton HR (1986) Org Geochem 10:671
234. Ourisson G, Albrecht P, Rohmer M (1984) Sci Am 251:44
235. Uemura H, Ishiwatari R (1995) Org Geochem 23:675
236. Talbot HM, Watson DF, Murrell JC, Carter JF, Farrimond P (2001) J Chromatogr A 921:175
237. Moldowan JM, Dahl J, Huizinga BJ, Fago FJ, Hickey LJ, Peakman TM, Taylor DW (1994) Science 265:768
238. Killops SD, Frewin NL (1994) Org Geochem 21:1193
239. Koch BP, Rullkötter J, Lara RJ (2003) Wetlands Ecol Management 11:257

240. Volkman JK, Rohjans D, Rullkötter J, Scholz-Böttcher BM, Liebezeit G (2000) Cont Shelf Res 20:1139
241. Abe I, Rohmer M, Prestwich GD (1993) Chem Rev 93:2189
242. Smetanina OF, Kuznetzova TA, Denisenko VA, Pivkin MV, Khudyakova YV, Gerasimenko AV, Popov DY, Il'in SG, Elyakov GB (2001) Russ Chem Bull 50:2463
243. Grandmougin-Ferjani A, Dalpé Y, Hartmann M-A, Laruelle F, Sancholle M (1999) Phytochem 50:1027
244. Grimalt JO, Yruela I, Saiz-Jimenez C, Toja J, de Leeuw JW, Albaigés J (1991) Geochim Cosmochim Acta 55:2555
245. Jacob J, Disnar J-R, Boussafir M, Albuquerque ALS, Sifeddine A, Turcq B (2005) Org Geochem 36:449
246. Ohmoto T, Nikaido T, Nakadai K, Tohyama E (1970) Yakugaku Zasshi 90:390
247. ten Haven HL, Rullkötter J, Stein R (1989) Preliminary analysis of extractable lipids in sediments from the Eastern North Atlantic (Leg 108): comparison of a coastal upwelling area (Site 658) with a nonupwelling area (Site 659). In: Ruddiman W, Sarnthein M et al. (eds) Proceedings of the Ocean Drilling Program, scientific results, vol 108. Ocean Drilling Program, College Station, p 351
248. Howard DL, Simoneit BRT, Chapman DJ (1984) Arch Microbiol 137:200
249. Ageta H, Shiojima K, Arai Y, Masuda K (1987) J Pharm Sci 76:S213
250. Shiojima K, Sasaki Y, Ageta H (1993) Chem Pharm Bull 41:268
251. Brassell SC, Eglinton G (1983) Steroids and triterpenoids in deep sea sediments as environmental and diagenetic indicators. In: Bjorøy M et al. (eds) Advances in organic geochemistry 1981. John Wiley, Chichester, p 684
252. Hauke V, Adam P, Trendel JM, Albrecht P, Schwark L, Vliex M, Hagemann H, Püttmann W (1995) Org Geochem 23:91
253. Hauke V, Graff R, Wehrung P, Trendel JM, Albrecht P, Schwark L, Keely BJ, Peakman TM (1992) Tetrahedron 48:3915
254. Hauke V, Graff R, Wehrung P, Trendel JM, Albrecht P, Riva A, Hopfgartner G, Gülaçar FO, Buchs A, Eakin PA (1992) Geochim Cosmochim Acta 56:3595
255. Jaffé RI, Hausmann KB (1995) Org Geochem 22:231
256. Summons RE, Brassell SC, Eglinton G, Evans E, Horodyski RJ, Robinson N, Ward DM (1988) Geochim Cosmochim Acta 52:2625
257. Kleemann G, Poralla K, Englert G, Kjøsen H, Liaaen-Jensen S, Neunlist S, Rohmer M (1990) J Gen Microbiol 136:2551
258. Bravo J-M, Perzl M, Härtner T, Kannenberg EL, Rohmer M (2001) Eur J Biochem 268:1323
259. Harvey HR, McManus GB (1991) Geochim Cosmochim Acta 55:3387
260. Mackey MD, Mackey DJ, Higgins HW, Wright SW (1996) Mar Ecol Prog Ser 144:265
261. Jeffrey SW, Mantoura RFC, Wright SW (1997) Phytoplankton pigments in oceanography: guidelines to modern methods. UNESCO Paris
262. Tolosa I, Vescovali I, Leblond N, Marty JC, De Mora S, Prieur L (2004) Mar Chem 88:103
263. Werne JP, Lyons TW, Hollander DJ, Formolo MJ, Sinninghe Damsté JS (2003) Chem Geol 195:159
264. Aycard M, Derenne S, Largeau C, Mongenot T, Tribovillard N, Baudin F (2003) Org Geochem 34:701
265. Sinninghe Damsté JS, Schouten S, de Leeuw JW, van Duin ACT, Geenevasen JAJ (1999) Geochim Cosmochim Acta 63:31

Pigment Markers for Phytoplankton Production

Simon W. Wright[1] (✉) · S. W. Jeffrey[2]

[1] Australian Antarctic Division and Antarctic Climate and Ecosystems CRC,
Channel Hwy, Kingston, Tasmania 7050, Australia
Simon.Wright@aad.gov.au

[2] CSIRO Marine Research, GPO Box 1538, Hobart, Tasmania 7001, Australia
Shirley.Jeffrey@csiro.au

1	Introduction	72
2	**Pigment Markers**	73
2.1	Chlorophylls	74
2.2	Carotenoids	75
2.3	Phycobiliproteins	77
2.4	Pigment Diversity and Chemotaxonomy	78
3	**Methods of Analysis**	80
3.1	Collection and Storage of Field Samples	82
3.2	Extraction of Pigments	82
3.3	Choice of HPLC Methods	83
3.4	Peak Detection and Integration	88
3.5	Peak Identification and Quantitation	88
3.6	Data Quality	89
4	**Taxonomic Interpretation and Quantitative Analysis of Pigment Data**	90
4.1	Hierarchical Guide to Interpreting Pigment Data	90
4.2	Mathematical Tools for Interpretation of Pigment Data Sets	90
4.2.1	Multiple Regression	90
4.2.2	Inverse Simultaneous Equations	93
4.2.3	Matrix Factorization (CHEMTAX Software)	93
4.2.4	Characteristics of Computational Methods	93
4.3	Comparison of Pigment and Microscopic Analysis	95
4.4	Selected Bibliography of Recent Field Studies	96
5	**Estimating Biomass from Pigment Content**	96
6	**Conclusions**	99
	References	99

Abstract Chlorophylls and carotenoids are commonly used as quantitative biomarkers for the composition and biomass of marine phytoplankton. This chapter provides an overview of the molecular diversity of pigment markers, their distribution across algal taxa (based on theories of plastid diversity through endosymbiosis), and their environmental variability. Three new methods for analysis of pigments by HPLC are compared with the original SCOR method. Guidelines for interpreting HPLC pigment chromatograms from field samples are given to determine the likely algal types present, thus enabling optimal computational analysis. Mathematical techniques for analysis of complex pigment data sets (multiple linear regression, inverse simultaneous equations and matrix factorization methods, using CHEMTAX software) are discussed. Methods for converting pigment data to carbon biomass are considered, with suggested strategies for improving biomass estimates.

Keywords Carotenoid · Chemotaxonomy · Chlorophyll · Computational methods · Phytoplankton

Abbreviations

But-fuco	19′-butanoyloxyfucoxanthin
Chl	Chlorophyll
Chl c_2-MGDG	chlorophyll c_2-monogalactosyl diacylglyceride ester. A suffix, e.g. [18 : 4/ 14 : 0], denotes the chain lengths (18, 14) and the number of double bonds (4, 0) of the two esterified fatty acids, respectively
Chlide	chlorophyllide
DMF	dimethyl formamide
DV	divinyl
Hex-fuco	19′-hexanoyloxyfucoxanthin
HPLC	high performance liquid chromatography
IS	internal standard
4-k Hex-fuco	4-keto-19′-hexanoyloxyfucoxanthin
MgDVP	magnesium divinyl pheoporphyrin a_5 monomethyl ester
MV	monovinyl
Neox	neoxanthin
Np	non-polar
SCOR	Scientific Council for Oceanic Research
TBAA	Tetrabutylammonium acetate
tr	trace
Unk	unknown

1
Introduction

Chromatographic analysis of algal pigments is a powerful tool for characterization of phytoplankton in field populations. Chlorophylls, carotenoids and phycobiliproteins have many favourable characteristics as chemotaxonomic markers. They are present in all photosynthetic algae, but not in most bacteria, protozoa or detritus, allowing phytoplankton to be distinguished from other components of the microbial community. Many pigments are limited to particular classes or even genera, allowing the taxonomic composition of the phytoplankton to be determined to class level or better. They are strongly coloured, and in the case of chlorophylls and phycobiliproteins, fluorescent at visible wavelengths, allowing them to be sensitively detected. Finally, they are labile and are rapidly degraded after the death of the cell, thus distinguishing living from senescent cells.

Pigments also suffer several disadvantages as markers. Their lability means that special conditions must be employed to preserve them, as they are sensitive to light, heat, oxygen, acids and alkalis, as well as spontaneously forming families of isomers in solution. Their distribution is complex, with few unambiguous markers. Their expression is variable, even within a particular class, and their content per cell varies with environmental factors such as irradiance and nutrients. Due to the fact that some pigments span many algal classes [1], interpretation of pigment data is difficult. It is important that other techniques (e.g. microscopy or genetic analysis) are used on representative samples to identify the algal types present.

Pigment analysis, however, offers the best technique for mapping phytoplankton populations and monitoring their abundance and composition. Modern automated HPLC analysis of pigments makes it feasible to analyse several hundred phytoplankton samples from a single oceanographic cruise, a task that would be completely impractical by microscopy. Pigment analysis is a powerful means of recognizing nano- and pico-planktonic organisms, which are normally unrecognizable by light microscopy (unless they possess special features such as fluorescent phycobilins), and are often difficult to preserve. Flow cytometry can be used to rapidly count cells but it is poor at identification. DNA analysis is powerful for identifying the composition of phytoplankton, but is still slow and complex for analyzing mixed populations.

A detailed account of the theory and practice of pigment chromatography is given in the 1997 UNESCO monograph *Phytoplankton Pigments in Oceanography: Guidelines to Modern Methods* [2], a widely used publication that gives a comprehensive account of the field to 1996 (history, applications, specific recommendations for techniques, and data for identifying and quantifying pigments). Subsequent reviews [3, 4] highlight new developments in pigment chromatography. This chapter updates the current understanding of pigment analytical methods and data interpretation.

2
Pigment Markers

Phytoplankton contain three types of pigments involved in light harvesting and photoprotection: chlorophylls, carotenoids and biliproteins. Their chemical structures and properties have been extensively reviewed [5–8], as have their metabolism [9] and applications in oceanography [10]. Comprehensive data and graphics sheets were compiled for 47 of the most important chlorophylls and carotenoids found in marine algae [11]. Marker pigments discovered since 1996 are described in a recent review [1].

2.1
Chlorophylls

All photosynthetic phytoplankton contain one or more types of chlorophylls as part of the light-harvesting complexes in their chloroplasts. Chlorophylls are magnesium coordination complexes of conjugated cyclic tetrapyrroles with a fifth isocyclic ring and often an esterified long-chain alcohol [7, 12].

Figure 1 shows the structure and numbering system for chlorophyll *a* (Chl *a*), with four rings (A–D) of the tetrapyrrole macrocycle, a cyclopentanone ring (E) conjoint with ring C, and a propionic acid side-chain at C-17, esterified to the C_{20} alcohol, phytol. The central magnesium atom is bound to the nitrogen atoms of the pyrrole rings, but can also bind to electron donors on either side of the plane of the macrocycle: water, proteins, or the 13-keto group of another chlorophyll molecule.

Other chlorophylls differ according to the oxidation state of the macrocycle, the type of side-chains, and the type of esterifying alcohol, if present. The macrocycle may be a porphyrin, with rings A–D all fully unsaturated (Chl *c* family); a chlorin (17,18-dihydroporphyrin), with ring D reduced (Chls *a*, *b*, and *d*); or a bacteriochlorin (7,8,17,18-tetrahydroporphyrin), with rings B and D reduced (bacteriochlorophylls *a*, *b*, *g* [7]). Chl *b* differs from Chl *a*

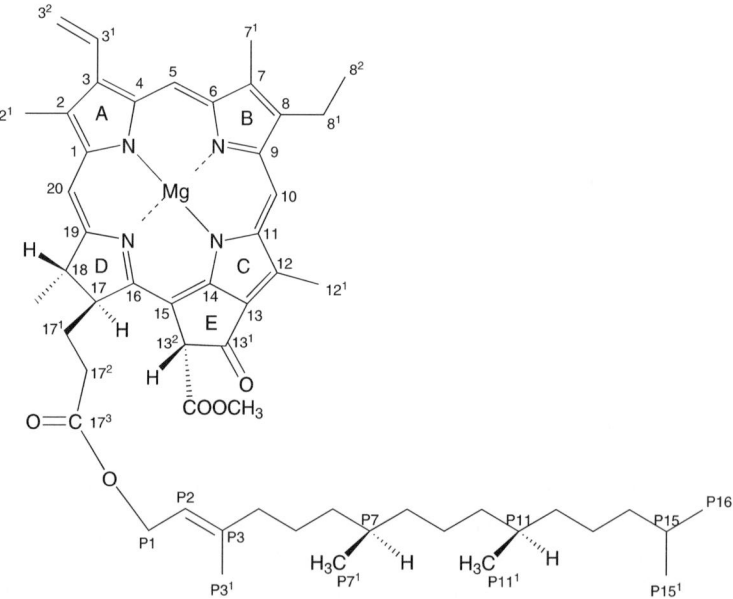

Fig. 1 Structure of Chl *a*, with numbering scheme, showing four rings (A–D) of the tetrapyrrole macrocycle, a cyclopentanone ring (E), a propionic acid side chain at C-17, esterified to phytol, from Data Sheets in [11]

by having an aldehyde rather than a methyl group at position C-7 of ring B, which alters its spectral properties and increases both its polarity and stability to photooxidation. Divinyl (DV) forms of both Chl a and Chl b are found in prochlorophytes, in which the C-8-ethyl group is replaced by a second vinyl group (in addition to that at C-3).

Nine members of the Chl c family have been identified in phytoplankton, mostly from the algal division Haptophyta [13]. In addition to being porphyrins, they differ from Chl a by having an acrylic, rather than a propionic acid side chain at C-17, in ring D (except MgDVP and Chl c_{cs-170}, which retain the propionic acid). In Chls a and b, this very acidic carboxyl group is esterified to phytol, but in the Chl c family, it is usually unesterified and can significantly affect the chromatographic properties of the molecule depending on pH and the presence of counter-ions. Chls c_1 and c_2 differ in having an ethyl group (c_1) or a vinyl group (c_2) at C-8 of ring B of the macrocycle [14]. Chl c_3 has a carbomethoxy group at position C-7 on ring B, as well as a vinyl group at C-8 [15, 16]. Monovinyl (MV) Chl c_3 has recently been identified in haptophytes [17]. Chl c_{cs-170} [18] is thought to be the propionate derivative of Chl c_3 [11]. In several non-polar Chl c pigments the C-17 acrylic acid is esterified to a massive galactolipid side chain (Chl c_2-MGDG [19]).

Many chlorophyll derivatives are found both naturally and as artefacts of extraction. These may have lost the Mg atom (pheophytins), the phytol chain (chlorophyllides), both Mg and phytol (pheophorbides), and/or the C-13^2 carbomethoxy group (pyro-derivatives), and they may also spontaneously rearrange (epimers) or oxidize (allomers) [9]. Analytical techniques should be optimized to prevent the formation of these artefacts.

2.2
Carotenoids

Carotenoids are a diverse family of yellow, orange or red isoprenoid, polyene pigments—the carotenes (hydrocarbons) and xanthophylls (oxygenated carotenoid derivatives). Many are involved in light-harvesting, with the ability to absorb light of blue and green wavelengths (420–550 nm), bridging the gap between Chl a and b absorption bands. Certain carotenoids are involved in photoprotection, notably diadinoxanthin and diatoxanthin in chromophytes and violaxanthin, antheraxanthin, and zeaxanthin in green algae. These pigments are taxonomically useful but quantitatively variable since their abundance can change dramatically in response to irradiance. Carotenoids may also help to stabilize the photosynthetic apparatus [8]. Most have a C_{40} skeleton with alternating single and double bonds that form the chromophore, responsible for the spectral characteristics of the molecule [20]. The IUPAC nomenclature of carotenoids [21] (see http://www.chem.qmw.ac.uk/iupac/carot/) differs from general IUPAC rules [22]. The base name for a carotene depends on the end groups, four types of which are found in phytoplankton (Fig. 2). IUPAC names specify both end groups, in order of their appearance in the Greek alphabet. For instance, Fig. 3a shows β,ε-carotene, along with the numbering system for carotenes and their derivatives. Trivial names were used for carotenes in older literature and are still sometimes encountered.

Common carotenoid modifications include the degree of unsaturation of the isoprenoid skeleton (with rearrangements including acetylenic and allenic units), oxygen functional groups (e.g. hydroxyl, ketones, and epoxides), and esterification of hydroxyl derivatives with acyl or large (705 Daltons) glycosidic groups. Many of these structural groups are found in 19′-hexanoyloxyfucoxanthin (Hex-fuco, Fig. 3b), formally (3S,5R,6S,3′S,5′R,6′S)-5,6-epoxy-3,3′,5′,19′-tetrahydroxy-6′,7′-didehydro-5,6,7,8,5′,6′-hexahydro-β,β-caroten-8-one 3′-acetate19′-hexanoate. Loss of in-chain carbons may result in shortened skeletons, e.g. the C_{37} skeleton of the abundant light-harvesting dinoflagellate carotenoid, peridinin.

Over 600 carotenoids are known in nature [23–25]. Many of the enzymatic pathways required for their synthesis are taxonomically restricted (particu-

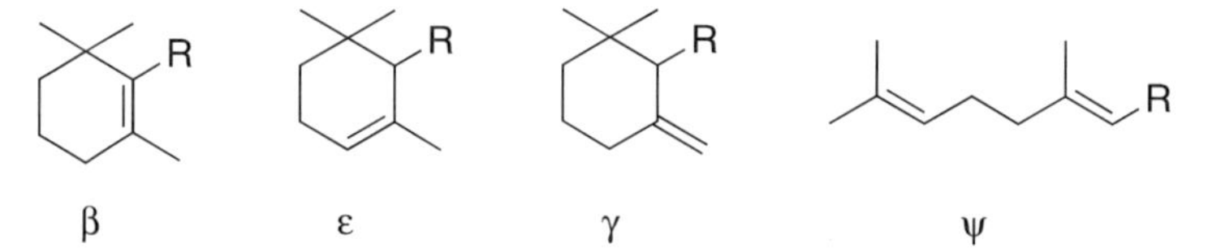

Fig. 2 Structure of the four types of end groups commonly found in carotenoids from phytoplankton, modified from Bjørnland [22]

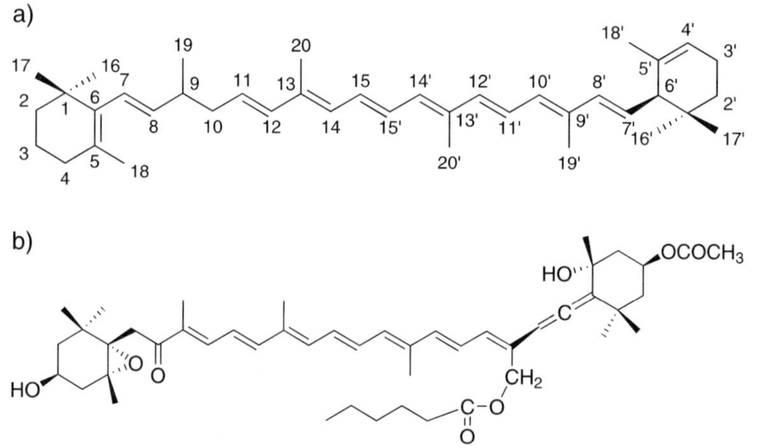

Fig. 3 Structures of carotenoids: **a** β,ε-carotene, with the numbering system for carotenes and their derivatives, **b** 19′-hexanoyloxyfucoxanthin, modified from Data Sheets in [11]

larly ε-cyclization, 5,6 epoxidation, allenic and acetylenic bond formation, 19-hydroxylation, 8-keto formation, acetylation, allelic 3-hydroxylation [26]).

Carotenoids are particularly labile molecules. Pigments with 5,6-epoxides, like diadinoxanthin and violaxanthin, readily form 5,8 furanoxides, especially under acidic conditions. Esters may be hydrolyzed. Carotenoids also spontaneously rearrange in solution. Nearly all carotenoids found in phytoplankton (with the notable exception of 9'-cis-neoxanthin), have the carbon atoms arranged around the skeletal double bonds in the *trans* form, producing linear molecules such as those shown in Fig. 3a. Once extracted into an organic solvent, carotenoids will form a stereoisomeric equilibrium mixture of all-*trans* and *cis* carotenoids. The *cis* forms are bent, exposing their (generally) hydrophobic mid-chains and usually retarding their elution in reverse-phase HPLC systems, producing additional peaks that complicate analyses.

Carotenoid absorption spectra generally have three peaks (depending on the solvent used), labelled I, II and III from short to long wavelengths. Often peaks I and III are reduced to shoulders or hidden altogether by overlapping of the peaks due to interference of functional groups with the carotenoid chromophore. The shape and position of the peaks vary in different solvents, and are useful in identification [11, 27]. The %III/II ratio is also useful, being the relative heights of the III and II peaks over the valley between them (not the baseline, see Data Sheets [11]). Identification of carotenoids is facilitated by descriptions of UV/visible spectra [20, 27], mass spectra [28], other data compilations [11], and analytical approaches [29].

2.3
Phycobiliproteins

Phycobiliproteins are the third type of light harvesting pigment found in cyanobacteria, rhodophytes and cryptophytes. Three main subtypes are found—the phycoerythrobilins, phycocyanobilins and the phycourobilins. The chromophore consists of an open-chain tetrapyrrole, which does not contain a central metal ligand like the chlorophylls, but is bound covalently to an *apo*protein [9].

Although phycobiliproteins are taxonomically restricted, they are generally not used as chemical markers, probably because the algal classes that contain them are so easily recognized and counted by techniques that detect biliprotein fluorescence directly in situ. These include epifluorescence microscopy [30, 31], flow cytometry [32–34] and delayed fluorescence excitation spectroscopy [35]. Since biliproteins are water soluble, they are not extracted by organic solvents used in analysis of chlorophylls and carotenoids. Their chemical analysis will not be considered further in this chapter.

2.4
Pigment Diversity and Chemotaxonomy

A recent review of pigment characteristics of microalgal classes [1] considered patterns of 56 pigments across 32 algal groups. The complexity of these pigment patterns is better understood through recent developments in endosymbiotic theories of the origins of plastid diversity [36-39].

These theories suggest that plastids originally arose by ingestion of a previously free-living cyanobacterium by a non-photosynthetic protist of unknown origin. Permanent symbioses that developed during subsequent evolution produced three major primary lineages, each monophyletic: the Glaucocystophyta, the Chlorophyta and the Rhodophyta radiations [40]. Modern Cyanophyta evolved from the ancestral cyanobacterium without further endosymbioses. Cyanophyta (e.g. *Richelia intracellularis*) may be found residing in modern-day algal taxa (e.g. the diatom *Rhizosolenia clevei*) in an association, in which the cyanophyte is not reduced to an organelle [41].

Further diversity was introduced by secondary and tertiary endosymbioses derived from engulfment of members of primary lineages by other protists. Thus plastids from the diatoms, brown algae, chrysophytes, haptophytes, cryptophytes and most dinoflagellates were derived by ingestion of rhodophytes by non-photosynthetic protists [1, 37]. Similarly euglenophytes, chlorarachniophytes and green dinoflagellates acquired their plastids from the chlorophytes. In some groups, multiple plastid losses and replacements have occurred—particularly in the dinoflagellates, which are now known to possess five different pigment suites, reflecting the various origins of their plastids.

In spite of the great diversity of pigments across algal taxa [1], relatively few are unambiguous markers. Many pigments are shared across taxa, making it necessary to consider suites of pigments when interpreting field data.

The most complete list of pigment distributions currently available (Table 20 in Jeffrey and Wright [1]) distinguished multiple major pigment suites (defined as Types 1 to n) within various algal divisions: Cyanophyta (two Types), Prochlorophyta (three), Prasinophyta (three), Chrysophyta (three), Dinophyta (five), and Haptophyta (eight). The remaining divisions (Chlorophyta, Euglenophyta, Rhodophyta, Cryptophyta, Bacillariophyta, Bolidophyta, Raphidophyta, and Eustigmatophyta) contained only one major pigment type. A key problem is that many taxa share the same pigment patterns, and thus they often cannot be distinguished on the basis of pigments alone. Microscopy or other methods should be used to assist identification of phytoplankton types in field samples.

Table 1 summarizes the pigment distribution of commonly encountered phytoplankton groups, using the type definitions of Jeffrey and Wright described above.

Environmental factors strongly influence pigment composition of microalgae. These include irradiance [42-45], spectral distribution of light [46-48],

Pigment Markers for Phytoplankton Production 79

Table 1 Definitive pigments in selected algal taxa are listed using the Type definitions from Table 20 in Jeffrey and Wright [1]. Thirteen chemotaxonomically useful pigment suites are presented in *bold*, while eight single pigments unique to specific algal Types are marked with an *asterisk*

Algal type	Definitive pigments
Cyanophyta (e.g. *Synechococcus* sp.)	**Chl *a*, zeaxanthin**
Prochlorophyta (e.g. *Prochlorococcus marinus*)	**DV Chl *a**, DV Chl *b****, β,ε-carotene, zeaxanthin
Chlorophyta (e.g. *Chlorella* sp.) Prasinophyta Type 1 (e.g. *Tetraselmis* sp.)	**Chl *a*, Chl *b*, lutein**, neoxanthin, violaxanthin, zeaxanthin (tr)
Prasinophyta Type 2 (e.g. *Pyramimonas amylifera*)	**Chl *a*, Chl *b*, MgDVP, siphonaxanthin ester***, neoxanthin, violaxanthin (minor)
Prasinophyta Type 3 (e.g. *Micromonas pusilla*) Dinophyta Type 5 (e.g. *Gymnodinium chlorophorum*)	**Chl *a*, Chl *b*, MgDVP, prasinoxanthin*, uriolide, micromonal**
Cryptophyta (e.g. *Rhodomonas*) Dinophyta Type 4 (e.g. *Dinophysis norvegica*)	Chl *a*, Chl c_2, **alloxanthin***, crocoxanthin, monadoxanthin
Bacillariophyta Type 1 (e.g. *Phaeodactylum tricornutum*) Haptophyta Type 1 (e.g. *Pavlova lutheri*) Dinophyta Type 3 (e.g. *Kryptoperidinium foliaceum*)	**Chl *a*, Chl c_1, Chl c_2, Fucoxanthin**, diadinoxanthin
Haptophyta Type 6 (e.g. *Emiliania huxleyi*)	**Chl *a*, Chl c_2, Chl c_3, MV-Chl c_3 (tr)*, Hex-fuco**, fucoxanthin, diadinoxanthin, 4-keto-Hex-fuco
Haptophyta Type 7 (e.g. *Chrysochromulina polylepis*)	**Chl *a*, Chl c_2, Chl c_3, Chl c_2-MGDG[14:0/14:0]*, Hex-fuco**, fucoxanthin, diadinoxanthin, 4-keto-Hex-fuco
Haptophyta Type 8 (e.g. *Phaeocystis antarctica*)	**Chl *a*, Chl c_2, Chl c_3,** Chl c_2-MGDG[18:4/14:0], **Hex-fuco, But-fuco**, fucoxanthin, diadinoxanthin

Table 1 (continued)

Algal type	Definitive pigments
Chrysophyta Type 3 (Pelagophytes) (e.g. *Pelagococcus subviridis*)	Chl *a*, Chl c_2, **Chl c_3**, **But-fuco**, fucoxanthin, diadinoxanthin
Dinophyta Type 1 (e.g. *Amphidinium carterae*)	Chl *a*, Chl c_2, **peridinin***, diadinoxanthin
Dinophyta Type 2 (e.g. *Gymnodinium galatheanum*)	Chl *a*, Chl c_2, **Chl c_3**, **Hex-fuco**, **gyroxanthin diester***, fucoxanthin, diadinoxanthin

daylength [49], diurnal cycle [50], nutrient status [43, 45], iron concentration [51, 52], growth phase [44, 53], and strain differences [13, 54]. This variability is usually limited to changes in the total pigment quantity per cell rather than the type of pigments present, although in senescent or nutrient-limited populations secondary pigments may be produced.

Pigment concentrations within taxa may also vary regionally due to strain variations. For instance, most strains of *Phaeocystis* spp. contain significant quantities of Hex-fuco [13, 44, 55], but many northern European strains lack this pigment [56–58]. It is therefore advantageous to have cultures of local strains for reference.

3
Methods of Analysis

Many methods of varying accuracy are available for chlorophyll analysis, with HPLC the most powerful. If an estimate of total phytoplankton biomass is all that is required, then spectrophotometric or fluorometric analysis of Chl *a* may be appropriate. The accuracy of spectrophotometry, fluorometry and HPLC were compared in the SCOR/UNESCO volume [59]. Briefly, it was found that the "Jeffrey and Humphrey (1975)" spectrophotometric method [60] and the "Holm-Hansen et al. (1965) extracted" fluorometric method [61] were reasonably accurate if chlorophyll degradation products were absent. The acid-spectrophotometric and fluorometric methods reduced interference from pheophytins and pheophorbides, but chromatography was required to accurately assess chlorophylls in the presence of degradation products. If analysis of marker pigments is required, then HPLC is the best method.

In situ and in vivo fluorometry are useful aids to HPLC pigment sampling, giving continuous fine scale resolution of phytoplankton populations

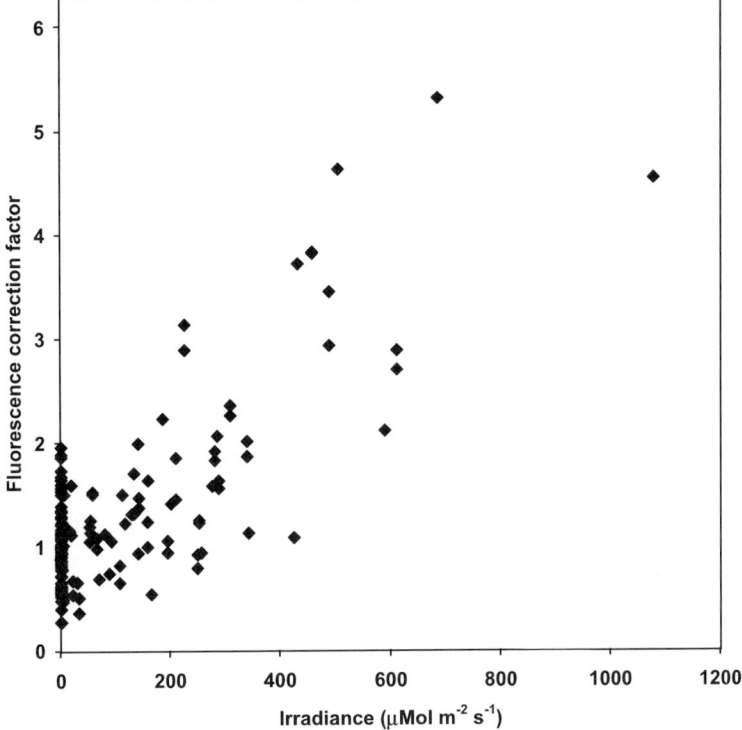

Fig. 4 Change in HPLC Chl *a*: in vivo fluorescence correction factor correlated with ambient irradiance. The correction factor equals unity for night time samples, and increases at high irradiances due to fluorescence quenching

that cannot be matched by discrete sampling regimes. In situ profiles by fluorometers attached to CTDs allow directed sampling of stratified populations. Surface fluorometry also provides immediate (but approximate) underway data.

Fluorometry, being non-discriminate, suffers from interferences from compounds other than Chl *a*. In estuarine samples, fluorometry sometimes overestimates Chl *a* by nearly two orders of magnitude compared to HPLC, probably due to fluorescent compounds abundant in dissolved organic matter [62]. The in vivo fluorescence response per unit Chl *a* is altered by irradiance, and the response curve changes markedly during the day. A fivefold variation was recently found in chlorophyll fluorescence (in vivo) in Southern Ocean phytoplankton due to diurnal changes in irradiance (Wright, unpublished; Fig. 4). It is, therefore, desirable to collect at least five HPLC samples per day, throughout the 24-hr cycle, to calibrate underway Chl *a* fluorescence.

3.1
Collection and Storage of Field Samples

Water samples may be collected from surface samplers, Niskin bottles or a clean seawater line. From the moment of collection, the pigment ratios will begin to change due to the altered light environment. In multidisciplinary cruises, water samples may not be available from Niskin bottles for 40 minutes or more, while gas samples are taken, during which time pigments of the violaxanthin or diadinoxanthin cycles can change markedly.

Suitable filtration equipment and recommended procedures are described elsewhere [63]. If the volume of seawater sample is limited, overall method sensitivity can be increased by using small diameter glass fibre filters that can subsequently be extracted in a small volume of solvent. The filters should be folded, blotted and stored immediately in liquid nitrogen. Alternatively, they may be stored for several days at – 20 °C or several weeks at – 90 °C [64]. Samples should not be freeze-dried since this causes degradation and reduces extractability of the pigments [64]. It is important to blot remaining seawater reproducibly and as completely as possible from the folded filter, since any remaining water will dilute the extraction solvent, altering its effectiveness and potentially altering HPLC retention times. Representative samples should also be preserved for light and electron microscopy.

3.2
Extraction of Pigments

Accurate analysis of phytoplankton pigments depends on the effectiveness of the extraction technique yet, despite the importance of this step, several extraction techniques of varying effectiveness remain in current use. There is disagreement over the best physical techniques of sample disruption (grinding, bath sonication, high-powered probe sonication, or soaking) and the most suitable solvents [acetone, methanol, dimethyl formamide (DMF)]. Three criteria are important—the ability to completely extract all pigments from field samples irrespective of the phytoplankton species composition, compatibility with the chromatographic technique (the ability to produce sharp peaks), and stability of pigments in the extraction solvent (since samples must often wait many hours in an autosampler before injection).

Common methods and solvents for extraction are reviewed in the UNESCO monograph [65]. The relative performance of different solvents varies according to the extraction methods used and the type of alga being extracted. DMF was the best solvent for pigment extraction when used with a high-powered probe sonicator, but the toxicity and ease of skin absorption of DMF for the operators make it too hazardous for safe use at sea. Sonication in 100% methanol was recommended as the second best method. The JGOFS protocol instead recommended extraction in acetone using bath sonication and soaking overnight in a freezer (– 18 °C), and this technique is used by many groups.

Methanol has some advantages for extraction. It has lower volatility than acetone, and produces sharper HPLC peaks than acetone extracts, particularly for polar pigments [66]. A disadvantage is that methanol promotes allomerization of Chl *a* [67], and recent data [68] suggests that holding methanol-extracted samples in an autosampler at 4 °C produces significant pigment degradation over 24 hours, more than occurs in samples which are extracted in acetone.

Experiments comparing extraction protocols in our laboratory (Wright and van den Enden, unpublished) using local estuarine seawater, confirmed the superiority of methanol for extracting Chl *b* [65]. Pigment degradation was not regularly observed in either methanol or acetone. Occasional losses of pigments were seen in both solvents, suggesting that variations in sample composition, perhaps lipid content, may have affected the losses of the pigments, through coprecipitation and/or adherence to surfaces.

Clearly more work is required to better characterize different protocols and to determine factors that affect pigment stability in the various solvents. The type of microalgae being extracted is also important.

3.3
Choice of HPLC Methods

HPLC analysis of algal pigments presents a major challenge due to the diversity of large molecules spanning a wide range of polarities, but also including many that have closely similar chemical structures, some differing only by the position of a double bond.

Routine pigment analysis of field samples became feasible with the development of HPLC methods in the 1980s [69–74] with automated analysis and quantitation of pigments, and the possibility of on-line identification using diode-array detection. All current methods use reverse-phase chemistry, in which compounds are resolved primarily on their polarity, with C_8–C_{30} stationary phases and gradient elution. Resolution of acidic chlorophylls, for which buffering and ion-pairing or ion-suppression reagents are required, has until recently been achieved using ammonium acetate, sometimes coupled with tetrabutylammonium acetate (TBAA) [75, 76].

The Wright et al. (1991) technique recommended in the UNESCO monograph [77] used a reverse-phase C_{18} monomeric column and ammonium acetate modifier that provided good resolution of 40 algal carotenoids and 12 chlorophylls and their derivatives, but lacked resolution of monovinyl and divinyl chlorophyll pairs, in particular Chl c_1/c_2, and the important MV/DV forms of Chls *a*, *b*, and c_3. This method remains in common use but it has been superseded by methods that exploit the subtle polarity differences between MV/DV Chl pairs on monomeric C_8 columns or use molecular shape selectivity of pigments on polymeric stationary phases. Such approaches have been recently evaluated and reviewed [3, 4].

Two new methods using monomeric C_8 columns have improved resolution by replacing ammonium acetate with either pyridine or TBAA modifier:

- The Zapata et al. (2000) method [78] uses a gradient from aqueous methanol/acetonitrile to methanol/acetonitrile/acetone, with pyridine modifier (as the acetate, pH 5.0), achieving resolution of seven polar Chl c derivatives, the Chl a/DV Chl a pair and partial resolution of Chl b/DV Chl b.
- The Van Heukelem and Thomas (2001) method [76] uses an aqueous methanol to methanol gradient at 60 °C with TBAA modifier (pH 6.5).

Both of these C_8 methods have greater resolution of the Chl c family than the C_{18} Wright et al. (1991) method [77], and are recommended for routine analysis, but neither is perfect. Relative differences in their resolution of significant pigment pairs are summarized in Table 2, and two representative chromatograms are presented in Fig. 5a,b.

Two other methods offer particular advantages:

- The Garrido and Zapata (1997) method [75] employs molecular shape selectivity on a polymeric C_{18} column with an aqueous methanol/acetonitrile to acetone gradient with pyridine modifier at 15 °C. It achieves excellent resolution of MV/DV pairs, but poorer resolution of carotenoids than the other methods due to the slower mass transfer characteristics of polymeric coatings. The retention order of several pigments differs from other methods, offering a useful means of confirming pigment identities [79].
- The Airs et al. (2001) method [80] achieves excellent resolution of bacteriochlorophylls and should be used whenever photosynthetic bacteria are present.

Pigment retention times of these three methods are compared with those of the Wright et al. (1991) method in Table 3. Whichever HPLC method is em-

Table 2 Relative differences in the resolution of significant pigment pairs in the Zapata et al. (2000) and Van Heukelem and Thomas (2001) methods. Resolution factors (R) are the difference in retention times divided by the average of peak widths [134]

Pigment pair	Resolution factors (R)	
	Zapata et al. (2000)	Van Heukelem and Thomas (2001)
MgDVP/Chl c_2	Good ($R > 1$)	Poor ($R < 0.5$)
Chl b/DV-Chl b	Poor ($R < 0.5$)	Partial ($R = 0.8$)
4-k Hex/9-*cis* Neox	Complete ($R > 1.25$)	Not resolved
Lutein / Zeaxanthin	Partial ($R = 0.8$)	Good ($R = 1$)

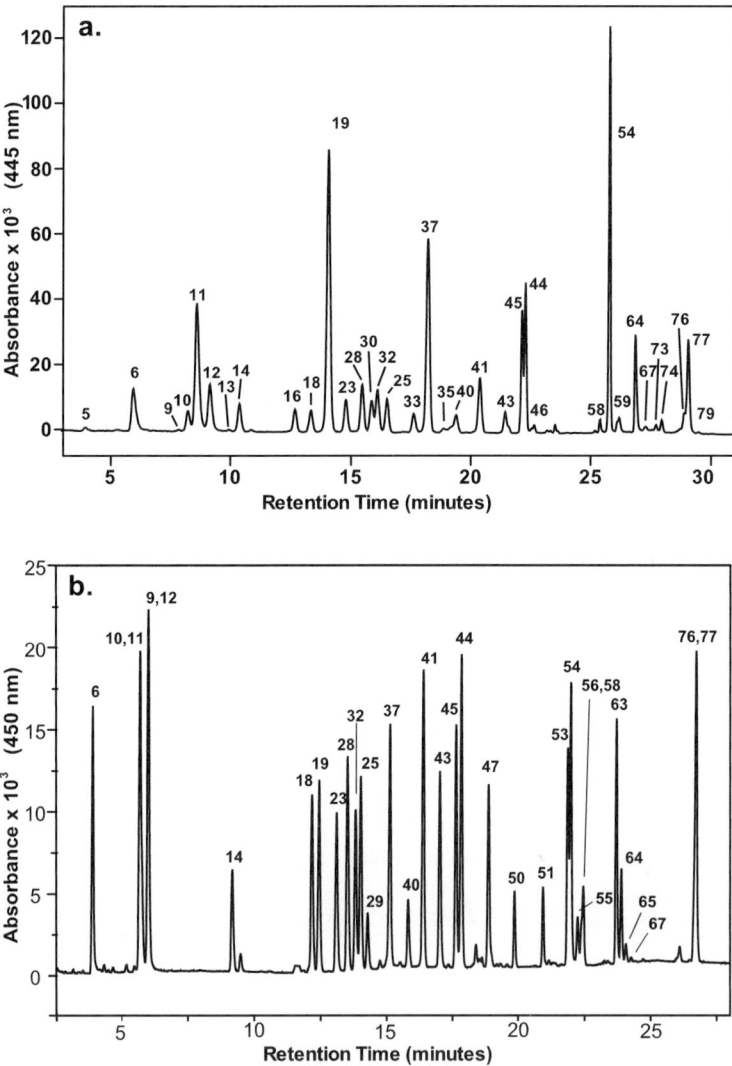

Fig. 5 Example chromatograms from HPLC methods: **a** Zapata et al. (2000) method [78]: from mixed cultures (see Sect. 3.3, S. Wright, unpublished); **b** Van Heukelem and Thomas (2001) method [76]: mixed pigments. For peak identities, see Table 3

ployed, it is important that the HPLC system is correctly optimized to produce the best results (see guidelines [81]).

Table 3 Comparisons of pigment retention times (RT) between four chromatographic methods. Pigment peak numbers correspond to those in Fig. 5a,b. Species names as suffixes refer to the original sources of the pigments: *Pavlova gyrans*, *Prymnesium parvum* and *Micromonas pusilla*

Peak No.	Pigment	RT (min) in various Methods			
		Wright et al. (1991) [77]	Zapata et al. (2000) [78]	Van Heukelem & Thomas (2001) [76]	Garrido & Zapata (1997) [75]
0	(solvent front)	2.56	1.93		
1	Chlide b	4.48	5.43		
2	Carotenoid P468	4.50			
3	Peridininol		6.06		
4	Methyl Chlide b		7.19		
5	Carotenoid P457	5.11			
6	Chl c_3	5.38	7.94	3.88	13.85
7	Chl c *P. gyrans*		8.27		8.42
8	MV Chl c_3	5.38	8.66	4.14	13.31
9	Chlide a	5.01	10.46	6.06	4.89
10	MgDVP	6.40	11.01	5.81	10.59
11	Chl c_2	6.40	11.44	5.70	14.94
12	Chl c_1	6.40	12.14	6.05	13.04
13	Methyl Chlide a	6.40	13.13		
14	Peridinin	7.42	14.20	9.32	7.88
15	Siphonaxanthin	8.11	14.76		
16	Uriolide		17.03		9.51
17	4-Keto-fucoxanthin		17.60		
18	But-fuco	8.11	17.94	12.31	8.15
19	Fucoxanthin	8.70	18.87	12.63	9.23
20	*trans*-Neoxanthin	9.11			
21	*cis*-But-fuco	9.12			
22	Neochrome	9.21			
23	9′ *cis*-Neoxanthin	9.31	19.62	13.29	11.14
24	4-Keto-Hex-fuco	9.31	20.92	13.31	8.96
25	Hex-fuco	9.31	21.75	14.16	10.05
26	*cis*-Fucoxanthin	9.68			
27	*cis*-Hex-fuco	9.97			
28	Prasinoxanthin	10.20	20.46	13.74	12.77
29	Micromonol		20.99		13.58
30	Phaeophorbide a	10.39			
31	Violaxanthin	10.59	21.32	13.99	14.12
32	Micromonal		23.24		15.75
33	Phaeophorbide a-like	10.62			
34	Dinoxanthin	10.76	25.22	15.49	14.67
35	*cis*-Prasinoxanthin	11.11			
36	Diadinoxanthin	11.61	24.11	15.23	
37	Diadinochrome I	11.79	23.27	15.02	

Table 3 (continued)

Peak No.	Pigment	RT (min) in various Methods			
		Wright et al. (1991) [77]	Zapata et al. (2000) [78]	Van Heukelem & Thomas (2001) [76]	Garrido & Zapata (1997) [75]
38	Diadinochrome II	11.96			
39	Antheraxanthin	12.24	25.38	15.99	17.93
40	Alloxanthin	12.51	26.25	16.53	19.83
41	Monadoxanthin	12.78	27.07	17.22	25.53
42	Diatoxanthin	13.08	26.90	17.12	
43	Lutein	13.36	27.65	17.98	20.37
44	Zeaxanthin	13.59	27.49	17.79	21.46
45	Dihydrolutein	13.83	28.00		
46	Canthaxanthin	14.00		19.07	
47	trans-β-apo-8'-Carotenal	14.00			
48	Siphonein	14.36	29.37		
49	Gyroxanthin diester			19.94	
50	Gyroxanthin diester			21.00	
51	Chl b allomer	14.85	31.28		
52	DV Chl b	15.15	31.58	21.92	
53	Chl b	15.15	31.62	22.03	23.9
54	DV Chl b epimer			22.29	
55	Chl b epimer		31.87	22.50	
56	Ethyl 8'-β-apocarotenoate	15.43			
57	Crocoxanthin	15.87	31.11	22.42	
58	Chl c_2-MGDG[18:4/14:0]		32.18		31.51
59	Np-Chl (c_1 like) P. parvum		32.44		28.79
60	Chl a allomer	15.87	32.63	23.30	
61	Np-Chl c_2			23.53	
62	DV Chl a	16.15	32.83	23.76	
63	Chl a	16.15	33.15	23.96	27.16
64	DV Chl a epimer	16.53		24.13	
65	Chl c_2-MGDG[14:0/14:0]		33.50		34.49
66	Chl a epimer	16.53	33.48	24.33	28.47
67	Echinenone	16.74			
68	Lycopene	17.59			
69	Phaeophytin b	17.68			
70	cis-Lycopene	17.84			
71	Phaeophytin a	18.56	35.41		
72	β,γ-Carotene	18.26	34.25		
73	Unk carotenoid M. pusilla		34.95		30.42
74	ε,ε-Carotene	18.40	35.52		
75	β,ε-Carotene	18.64	35.74	26.65	32.59
76	β,β-Carotene	18.76	35.95	26.71	32.86
77	cis-β,ε-Carotene	18.83			
78	cis-β,β-Carotene	18.94	36.26		

3.4
Peak Detection and Integration

For samples containing a variety of pigments, a diode-array detector is essential to allow identification of peaks from their spectra collected during elution.

The following wavelengths are useful for routine detection and integration

- 435 nm—detects all common pigments except phaeophytin *a*, phaeophorbide *a* and their derivatives;
- 470 nm—detects carotenoids, Chl *b*, and Chls *c*, without interference from Chl *a* derivatives;
- 665 nm—detects Chl *a*, phaeophytin *a*, phaeophorbide *a* and their derivatives.

Sensitivity can be increased in most systems by including a channel that sums a range of wavelengths (e.g. 427–464 nm). Such "wavelength bunching" improves the signal to noise ratio, but reduces the selectivity of the detection.

Fluorescence detection is a valuable addition to diode-array detection, due to its sensitivity and selectivity. Broad excitation and emission bandwidths are better than narrow bandwidths since sensitivity and detection of all chlorophyll derivatives are increased [81].

3.5
Peak Identification and Quantitation

Peaks are identified by comparison of their retention times and spectra with those of standard pigments. Many are commercially available, but expensive, and not always pure. It is far more efficient (and cheaper) to employ pigments from well-characterized reference phytoplankton cultures [82].

A standard mixture of pigments is injected (after a blank column conditioning cycle) before each batch of samples. This is prepared by mixing extracts of (typically) *Pavlova lutheri*, *Pelagococcus subviridis*, *Micromonas pusilla*, *Dunaliella tertiolecta*, *Amphidinium carterae*, and *Chroomonas salina*. Individual algae are analyzed first, then mixed so that the peak heights of major pigments are approximately equal. Aliquots (0.5 ml) of the mixture are dispensed into cryotubes that are immediately frozen in liquid nitrogen. A freshly thawed sample is injected each day, providing an invaluable monitor of system performance as well as the basis of a retention time table.

Peaks can be quantified using either the internal standard (IS) or external standard methods [83]. Using an IS gives increased accuracy and precision, since it accounts for any volume changes due to evaporation or dilution, and it also provides a check on the injection status. It is thus recommended for routine oceanographic samples, but for unfamiliar samples it is prudent to run a sample without an IS in case there are pigments that co-chromatograph with it. The most commonly used commercially available internal standards are ethyl 8'-apo-β-carotenoate, 8'-apo-β-carotenal and canthaxanthin, the first of which is most suitable due to its stability and non-occurrence in natural systems.

3.6
Data Quality

Diagnoses of general HPLC instrumental problems are beyond the scope of this chapter, but several problems may be encountered during pigment analysis that can reduce data quality.

Significant peaks of chlorophyllide a (Chlide a) are often seen in chromatograms of some diatoms because chlorophyllase enzymes are activated when the cell is damaged, e.g. during filtration, storage or extraction [63, 84, 85]. Significant degradation of chlorophyll may occur if the cells are left too long on the filter, frozen too slowly or not cold enough, or extracted in a solvent that does not inactivate the chlorophyllase. Chlorophyllases are inactivated more rapidly by 100% acetone than 90% acetone [84, 86]. Sonication in methanol can produce some Chlide a (although the resultant methyl-Chlide a is clearly recognizable as an extraction artefact [87]). Chlide a concentration is generally included in the total Chl a fraction for biomass estimation [88].

Chromatography problems may be caused by animal lipids in a sample, e.g. from a copepod on the filter. These are extracted along with the pigments and may cause smeared chromatographic peaks (eluting later, with lower peak heights and severe tailing). Any animals visible on the filter should be removed before freezing the sample [63]. In extracting symbiotic microalgal pigments from animal tissue, the microalgae should first be isolated before pigment extraction to prevent such artefacts [89].

Peak integration errors may be introduced by reproducible baseline changes that occur during gradient elution due to refractive index changes in the solvents. These are often most pronounced in the region of Chl a elution [at least in the Wright et al. (1991) and Zapata et al. (2000) methods] and may interfere with the integration of Chl a in samples with low concentrations. It is thus preferable to use a fluorescence detection channel for such samples as this is unaffected by refractive index.

For large peaks, an optimized HPLC system can achieve a precision of about 1% [90], but the uncertainty increases as the peak size is reduced and detector noise becomes more significant. A recent intercalibration exercise [91] found average percentage differences of 7.0% for total Chl a between laboratories, which was reduced to 5.5% after excluding very low values, standardizing quantitation and accurately accounting for DV-Chl a. The authors suggested that all pigment data submitted to databases should include information on the limits of detection for each pigment.

4
Taxonomic Interpretation and Quantitative Analysis of Pigment Data

This section describes a systematic approach for determining the likely taxonomic components of field samples from HPLC pigment chromatograms, and the various mathematical techniques available for estimating the relative abundance of algal types. The merits of these mathematical techniques are then considered in a summary of some recently published quantitative studies.

4.1
Hierarchical Guide to Interpreting Pigment Data

Due to the large number of pigments involved, a formal approach should be used to interpret chromatograms, based on knowledge of the most recent pigment distributions [1]. After considering Chl a, an index of total phytoplankton biomass (excluding prochlorophytes), one works through a pigment hierarchy, starting with unambiguous markers for algal types, then deducing the algal composition from the content of other pigments. Table 4 suggests a possible approach. Having derived the likely taxonomic composition of field samples from the HPLC chromatograms, quantitative analysis of pigment data can be undertaken. Due to the environmental variability of pigment composition, one cannot simply use pigment ratios from cultures and apply them to field populations. Pigment : Chl a ratios must be determined from analysis of the field data [3], using the analytical tools below.

4.2
Mathematical Tools for Interpretation of Pigment Data Sets

Until recently, most applications of pigment methods to oceanography [10] were semi-quantitative and based on the use of single marker pigments for particular taxa. Qualitative analysis was introduced through the application of multiple regression, inverse simultaneous equations, and matrix factorization through CHEMTAX software.

4.2.1
Multiple Regression

Multiple regression allows a statistically sound analysis of the relationship between various marker pigments and total Chl a [70]. It does not allow for shared pigments between algal taxa, so that while the contribution of "fucoxanthin-containing" algae, for example, may be accurately determined, it is not known whether the fucoxanthin came from diatoms, haptophytes, chrysophytes or mixed populations. The contribution of minor groups may be ignored altogether if they are swamped by noise in the data [92]. As an ex-

Table 4 Hierarchical guide for interpreting pigment field data

Pigment	Significance
Chl a	An index of total algal biomass, excluding prochlorophytes
Unambiguous markers for algal types	
DV-Chl a	An index of prochlorophyte biomass
DV-Chl b	Unambiguous marker for prochlorophytes
Siphonaxanthin esters	Unambiguous marker for Type 2 prasinophytes [135]
Prasinoxanthin	Unambiguous marker for Type 3 prasinophytes
Peridinin	Type 1 dinoflagellates
Alloxanthin	Cryptophytes
Gyroxanthin diester	Dinoflagellates Type 2
Chl c_2 MGDG [14:0/14:0]	*Chrysochromulina* spp. (Haptophyte Type 7) [13]
Chl b	Distinguishes "green algae" (chlorophytes, prasinophytes, euglenophytes and green dinoflagellates) from all other algal types. The relative proportion of these groups can be deduced [111] from the proportions of the following major carotenoids [44, 45, 90, 105]. Types 2 and 3 prasinophytes are distinguished by the presence of **siphonaxanthin esters** and **prasinoxanthin**, respectively (see above). Chlorophytes and Type 1 prasinophytes can be identified by their relative ratios of **Lutein** to **Chl b** (Lut : Chl b = 0.30–1.77, 0–0.18, respectively). Euglenophytes are difficult to distinguish since their major carotenoid, **diadinoxanthin**, is a major component of the chromophytes. Green dinoflagellates have no known distinguishing pigments and must be identified microscopically before fixation.
Chl c series	Distinguishes chromophyta from all other algal types
Chl c_1	Widely distributed. A useful marker for diatoms in populations dominated by Type 6 and Type 8 haptophytes [108]
Chl c_2	The major Chl c component in chromophyte algae
Chl c_3	A significant component of Haptophytes Types 4–8 [13] including coccolithophorids (Type 6), *Chrysochromulina* sp. (Type 7), and *Phaeocystis* sp. (Type 8). Chl c_3 is also present in Chrysophytes Type 3 (pelagophytes), bolidophytes, and some diatoms [136], notably the harmful bloom-forming genus, *Pseudonitzschia* [137].
Chl c_2 MGDG	Haptophytes Type 3–8 contain Chl c_2 MGDG [18:4/14:0] [13]. See also Chl c_2 MGDG [14:0/14:0], above.
MgDVP	A marker for prasinophyte Types 2 and 3, but it occurs in trace amounts in most algae
Chl c_2 *P. gyrans* type	(Minor pigment) chrysophytes Type 1 and haptophytes Type 2

Table 4 (continued)

Pigment	Significance
Np-Chl c_1 like	(Minor pigment) haptophytes Type 4 [13]
MV-Chl c_3	(Minor pigment) haptophytes Type 6 (coccolithophorids)
Fucoxanthin and derivatives	
Fucoxanthin	Erroneously regarded as a unique marker for diatoms. Also present in haptophytes, chrysophytes, raphidophytes, bolidophytes and some dinoflagellates [1].
Hex-fuco	Restricted to haptophytes Types 6–8 (but see Sect. 2.4) and dinoflagellates Type 2 (which also have **gyroxanthin diester**, see above).
But-fuco	Restricted to haptophytes Type 8 (with traces in haptophytes Types 6 and 7), where it always co-occurs with Hex-fuco. Pelagophytes contain But-fuco, but no Hex-fuco.
Other pigments	
Zeaxanthin	A useful marker for cyanobacteria when they are a major component of the population; widespread in low concentrations in prochlorophytes, chlorophytes, prasinophytes, euglenophytes, chrysophytes, raphidophytes and eustigmatophytes.
Vaucheriaxanthin esters	Markers for eustigmatophytes [1] and chrysophytes Type 1 (Jeffrey et al., unpublished data).
Loroxanthin	An occasional component of chlorophytes [138]; a useful marker for chlorophytes in Antarctic waters [108].
Bacteriochlorophyll a	A marker for photosynthetic proteobacteria [139, 140].
Diadinoxanthin and Diatoxanthin	Major pigments of the chromophyte algae are found in most oceanic HPLC chromatograms. While they are not definitive taxonomically, their role in the light-regulated epoxide cycle [9] allows their ratio to Chl a to be used as an index of the light history [141] of the algae in the water column. This can be used to compute vertical mixing velocities [142]. The two pigments are interconverted on a timescale of minutes [143–145], too fast for conventional oceanographic sampling methods.

ploratory tool, multiple regression is excellent for pigment data since it does not require assumptions about the composition of the phytoplankton population, and it remains in common use [92–94].

Even if more advanced mathematical techniques are to be employed, it is helpful to first plot or run multiple regressions of the concentrations of major pigments against Chl a to determine the likely main components. Similarly, running multiple regressions of green algal pigments (lutein, prasinoxanthin, siphonaxanthin esters) against Chl b, and fucoxanthin and its derivatives against Chl c_2 may be useful. An important check for interpreting such relationships is to test the correlation between two pigments that are not found in the same organism, e.g. Hex-fuco and Chl b. This will show to what extent there is a correlation between unrelated taxa, such as might occur following a nutrient incursion. Where such correlation exists, other correlations must be treated with caution.

4.2.2
Inverse Simultaneous Equations

This approach [95–98] uses a series of simultaneous equations in which the contribution to total Chl a by each algal group is calculated from the concentration of each marker pigment, choosing the appropriate marker pigment : Chl a ratios (see Sect. 4.2.4) and including proportionate subtraction of markers shared between groups. Each equation includes the chosen values for each marker pigment : Chl a ratio and the proportion of shared markers. These ratios are modified by inverse methods to find the least squares best solution to the total concentration of Chl a. This approach is less flexible than matrix factorization (see the following section).

4.2.3
Matrix Factorization (CHEMTAX Software)

CHEMTAX analysis [90, 99] is fundamentally similar to the simultaneous equations method, except that instead of building a series of simultaneous equations, the operator constructs a matrix of the algal types and their pigment content (from microscopy and examination of chromatograms in the data set, see Sect. 4.1). A second matrix constrains how far each pigment ratio can change. Each element of the pigment ratio matrix is iteratively modified to optimize the agreement between observed and computed pigment abundance, to estimate the marker pigment : Chl a ratios in the field samples.

Entering pigment data as a matrix is more flexible for adjusting the taxonomic makeup and more suited to handling shared markers than building sets of equations. It readily accommodates hypothesis testing whereby several different models of a field population can be compared. The study by Havskum et al. [100] provides an excellent approach to using CHEMTAX on field samples.

4.2.4
Characteristics of Computational Methods

Simultaneous equations and CHEMTAX are both able to distinguish broad algal groups within a phytoplankton population, making it possible to map the distribution geographically or in relation to oceanographic features. Both approaches require the pigment ratios to be stable within the data set. Thus any known factors affecting pigment ratios must be eliminated as far as possible, e.g. by breaking the data set into subsets by water masses (controlling nutrient status) and sample depth (controlling irradiance [101, 102]). The range of pigment : Chl *a* ratios calculated by CHEMTAX in field studies is smaller than observed in culture (Higgins and Descy, in preparation), probably because phytoplankton in the mixed layer never get the chance to adapt their pigmentation to a constant irradiance, and develop a composition closer to the median.

Unlike multiple regression, the computational methods do not ignore algal groups with low abundance, but in general pigment ratios for such groups are not optimized and the groups are quantified on the basis of the chosen initial ratio, upon which the accuracy of the final determination depends.

CHEMTAX was originally tested using synthetic data sets based on culture analyses [90], with which it performed well, and then with field data from a surface transect across the Southern Ocean [99]. In the latter case, CHEMTAX could distinguish two populations of haptophytes with identical pigment compositions; the northern population, considered to be mainly coccolithophorids, showed a distribution consistent with a previous microscopic study. The southern populations were considered to be *Phaeocystis sp*. In a recent study of the Rio de La Plata [103], CHEMTAX was able to distinguish four categories of haptophytes as well as pelagophytes and diatoms.

Choosing starting ratios for CHEMTAX (or inverse simultaneous equations) remains the biggest problem due to insufficient knowledge of algal pigment ratios in the field. If local data are not available, then ratios recommended in the CHEMTAX manual [90] supplemented by recent surveys [13, 104, 105] will serve as a guide. However, these ratios are known to vary according to region, and it is far better to use pigment data from cultures of local isolates, grown under an appropriate range of conditions.

It is worthwhile performing multiple runs of CHEMTAX using a range of initial marker pigment : Chl *a* ratio matrices. CHEMTAX works by iteratively minimizing the pigment residual (i.e. the difference between observed and calculated concentrations of the pigment to be optimized, usually Chl *a*). It is apt to find a local minimum rather than the overall minimum for the data set, particularly if there is a poor signal-to-noise ratio in the data. During testing of CHEMTAX Version 2, Wright (unpublished) performed multiple CHEMTAX runs on pigment data from Antarctic picoplankton using 28 pigment ratio tables that had been multiplied by a scaled random number to adjust each ratio up to ±50%. Each calculation produced a slightly different result. When the results were sorted in order of decreasing pigment residual, the taxonomic estimates were found to converge towards stable values. So although CHEMTAX was encountering local minima, it was tending towards the overall minimum in the data. Performing such multiple estimates from randomized starting points gives the analyst an indication of the stability of the results and its confidence limits.

4.3
Comparison of Pigment and Microscopic Analysis

Several recent studies have compared pigment chemotaxonomy and microscopic analysis of phytoplankton groups, using individual pigments [106], linear regression [58, 92], simultaneous equations [107], or CHEMTAX [44, 45, 100, 103, 108–112]. The microscopic studies attempted cell identifications and in many of them carbon cell biomass was calculated from cell volume measurements of each species group in the sample. Most studies compared patterns and proportions of algal groups by the two methods—cell volume/carbon equivalent (microscopy) and µg Chl *a* (HPLC analysis)—rather than comparing biomass in the same units. Major threads from these studies include:

- Microscopic and pigment analyses are difficult to compare quantitatively due to the poor precision of cell counts. Counting precision was not considered explicitly in most studies, but (where given) cell numbers normally ranged from 100 total to 100 per category, implying counting precision ranging from ±50% to ±20%, respectively for five categories [113]. Pigment analysis can achieve 1% precision of concentration estimates, but estimating taxonomic proportions from such data is less precise.
- Microscopic identification of cells and estimation of cell biomass was time-consuming and difficult [44, 92, 100, 107, 112], particularly when hampered by poor cell preservation, because the best fixatives were not used in all cases.
- Correlation of pigments with cell biomass was improved if very large diatoms were removed from the analysis [100, 111].
- Correlation was improved if the pigment data set was split into similar regions [92] or depths [107], in order to have stable pigment ratios within the data set (see Sect. 4.2).
- Pigments gave good estimates of groups with well-defined markers [92, 100, 103], but multiple regression worked well only for groups that had a well-defined pigment marker that was present in reasonable concentration [92]. Pigments sometimes detected taxa having unambiguous markers that were missed by microscopy [70, 100, 112].
- Dinoflagellates presented particular problems when non-pigmented species [112] or those with major pigments other than peridinin [108] were present.
- Minor pigments, in particular Chls *c*, were very useful in discriminating algal types [103, 108].
- Groups with similar pigment patterns were often confused, e.g. *Phaeocystis pouchetti* was erroneously attributed to diatom Chl *a*, suggesting that it contained unexpectedly high fucoxanthin concentrations [112]; confusion between *Synechococcus* sp. and *Trichodesmium* sp. was also noted [114].

- In a mesocosm study [100], cyanobacteria biomass was negatively correlated with estimates from pigment data in treatments with added nutrients (but not the control, which was positively correlated). This was ascribed to variations in the zeaxanthin : Chl a ratio, perhaps due to different light environments between treatments.
- CHEMTAX was sometimes found to be sensitive to erroneous initial pigment ratios [45] and insensitive in other cases [44]. A thorough knowledge of the phytoplankton community is necessary for trustworthy results [45], as shown in the excellent study of the Urdaibai Estuary [110].
- No clear relationship was found between CHEMTAX community analysis and estimates of carbon derived from microscopic cell counts from two years of data from the English Channel [112]. Much of the variability appeared to be due to problems with microscopic analysis as well as the changing pigment ratios due to changes in irradiance, nutrient concentrations and community composition within the data set. Nevertheless, computed pigment : Chl a ratios were similar to those of other studies [44, 45, 115], giving some confidence in the results. Other authors [92, 103, 107] found that computed field pigment ratios were within the range of literature values.

4.4
Selected Bibliography of Recent Field Studies

Most studies of community composition still employ pigments as independent chemotaxonomic markers, using them as indicators of particular algal taxa. This can be very useful in distinguishing patterns of distribution without determining contributions to Total Chl a [116, 117]. Where formal computational methods are applied for quantitative analysis, CHEMTAX is a popular choice. A summary of recent publications employing multiple linear regression, simultaneous equations or CHEMTAX analysis of quantitative pigment HPLC data of phytoplankton populations is given in Table 5, along with the focus of the study and the geographical region in which it was conducted.

The objective in most cases was to determine the community composition of phytoplankton. Excellent examples include a cross-Atlantic survey [118], and a South Pacific transect [119] that provided impressive large scale synopses of phytoplankton distributions.

5
Estimating Biomass from Pigment Content

Chl a concentration is a useful index of phytoplankton biomass when modeling primary production, but carbon content is preferable for studies of

Table 5 Examples of recent studies employing HPLC pigment analysis for quantitative phytoplankton biomass estimation, with the method of data analysis—multiple linear regression (MLR), simultaneous equations (SE), or CHEMTAX—as well as the focus of each study and geographic region in which it was performed

Authors	Data analysis method	Focus	Region	Ref.
Abrahamsson et al. 2004	CHEMTAX	Halocarbons	Southern Ocean	[146]
Andersson et al. 2003	MLR	Grazing	Baltic	[147]
Ansotegui et al. 2003	CHEMTAX	Size distribution	Urdaibai estuary, France	[110]
Bergmann et al. 2002	CHEMTAX	Photosynthetic efficiency	Neuse Estuary, USA	[148]
Breton et al. 2000	MLR	Community composition	English Channel	[58]
Carreto et al. 2003	CHEMTAX	Springtime communities	Argentina	[103]
Di Tullio et al. 2003	CHEMTAX	Community composition	South Pacific	[119]
Ediger et al. 2001	MLR	Community composition	Galway Bay, Ireland	[149]
Furuya et al. 2003	CHEMTAX	Community dynamics	East China Sea	[150]
Garibotti et al. 2003	MLR	Community composition	Southern Ocean	[92]
Gibb et al. 2001	CHEMTAX+MLR	Community composition	NE Atlantic	[151]
Gin et al. 2003	SE	Community composition	Singapore	[152]
Havskum et al. 2004	SE	Method comparison	Mesocosms, Denmark	[100]
Henriksen et al. 2002	CHEMTAX	Light and nutrients	Danish waters	[45]
Li et al. 2002	MLR	Community composition	East China Sea	[153]
Llewellyn et al. 2005	CHEMTAX	Community composition	English Channel	[112]
Mackey et al. 2002	CHEMTAX	Community composition	Equatorial Pacific	[154]
Riegman & Kraay 2001	CHEMTAX	Community composition	Faroe-Shetland Channel	[155]
Rodriguez et al. 2002	CHEMTAX	Community composition	Southern Ocean	[108]
Rodriguez et al. 2003	CHEMTAX	Community composition	Bay of Biscay	[109]
Schlüter et al. 2000	CHEMTAX	Light and nutrients	Danish waters	[44]
Steinberg et al. 2001	SE	Time series	Bermuda	[156]
Suzuki et al. 2005	CHEMTAX	Iron fertilization	Subarctic Pacific	[157]
Veldhuis & Kraay 2004	CHEMTAX	Community composition	Subtropical Atlantic	[114]
Vidussi et al. 2000	SE	Community composition	Mediterranean	[98]

ecosystem dynamics or carbon flux. Conversion is not straightforward. The carbon-to-Chl a ratio (C : Chl a) of phytoplankton varies widely, from < 10–200 g C g^{-1} Chl a in culture [120–123]. The range observed in the field is similar [92, 112, 124] but varies seasonally [45]. A wider range has been observed for prochlorophytes, from 450 g C g^{-1} DV-Chl a at the surface to 15 g C g^{-1} DV-Chl a at 150 metres depth [114].

A model of C : Chl a versus irradiance, daylength, temperature, and nutrients [125] predicted values of C : Chl a that matched field studies. Application of this model to a one-dimensional model of phytoplankton production dynamics [123] predicted a range of 20–160 g C g^{-1} Chl a with lowest values predicted at the top of the nutricline within the seasonal thermocline and highest values in the nutrient-depleted surface mixed layer in mid-summer. The model produced excellent correlations with observed data for a variety of cultures ($r^2 = 0.75–0.89$), but is not immediately applicable to field studies since it includes a term K_I, the saturation parameter for the growth irradiance curve, that depends on α, the initial slope of the photosynthesis irradiance curve. This is species-dependent and varies with environmental conditions. Nevertheless, the model predicts a linear relationship between C : Chl a and irradiance, suggesting that the vertical change in C : Chl a can be interpolated between representative samples in field studies.

The carbon content per cell for species observed in field samples is generally determined by microscopic measurements of individual cells [126] and application of a cellular carbon-to-volume ratio (C_c : V_c) that may range from 0.04–0.4 pg C μm^{-3} for typical phytoplankton [127]. Recent examples of this approach [100, 112] summarize previous literature values. A new C_c : V_c relationship has recently been developed for dinoflagellates, diatoms, and other protists [128]. C : Chl a can also be measured by incorporation of ^{14}C [114, 120, 129–131]. A promising flow cytometric method [114] uses fluorescent labeling of DNA, which has a strong correlation with cellular carbon.

Estimates of biovolume may have standard deviations of 15 to 50% [53], and C_c : V_c may vary 5-fold depending on the choice of regression model [132]. C_c : V_c also varies markedly according to size [58, 128].

Possible avenues for reducing uncertainties in estimates of chlorophyll or carbon biomass include:

- Size-fractionating representative field samples to get more accurate estimates of Pigment : Chl a and C : Chl a for different size classes.
- Improved knowledge of the pigment and carbon content of major taxa in field samples through further isolation and culture studies under different environmental conditions.
- Exploring the use of particular species to indicate the history of the light environment, using the following rationale. Certain taxa, notably cyanobacteria, cryptophytes and prochlorophytes, have distinctive pigment markers (Table 1) and can also be reliably distinguished and accu-

rately counted in field samples using flow cytometry. For these species it should be possible to determine the content of each pigment per cell in field samples, which varies according to light environment, as well as other factors (see Sect. 2.4). The light history of these cells could then be deduced by comparison with culture data and used to constrain possible values of marker pigments : Chl a and C : Chl a, in CHEMTAX and in biomass calculations, respectively.

Finally, it may be preferable to calculate the biovolume of the algal class directly from the concentrations of suites of marker pigments rather than indirectly through computed Chl a concentrations [133].

6
Conclusions

HPLC pigment analysis is currently the best means of mapping phytoplankton populations in the oceans and gaining an in-depth view of the dynamics of such populations. Much more work should be done to get basic parameters of key species by culturing them and determining their pigment content, particularly under simulated field conditions, such as rapidly varying irradiance and nutrient stress.

Acknowledgements Figure 5b was kindly provided by Laurie Van Heukelem. This work was supported by the Australian Government's Cooperative Research Centres Programme through the Antarctic Climate and Ecosystems Cooperative Research Centre (S.W. Wright), and CSIRO Marine Research (S.W. Jeffrey).

References

1. Jeffrey SW, Wright SW (2005) Photosynthetic Pigments in Marine Microalgae. In: Subba Rao DV (ed) Algal Cultures, Analogues of Blooms and Applications. Science Publishers, New Hampshire, USA (in press)
2. Jeffrey SW, Mantoura RFC, Wright SW (1997) (eds) Phytoplankton pigments in oceanography: Guidelines to modern methods. UNESCO, Paris
3. Jeffrey SW, Wright SW, Zapata M (1999) Mar Freshwater Res 50:879
4. Garrido JL, Zapata M (2005) Chlorophyll analysis by new high performance liquid chromatography methods. In: Grimm B, Porra RJ, Rüdinger W, Scheer H (eds) Chlorophylls and bacteriochlorophylls: biochemistry, biophysics and biological function. Kluwer Academic, Dordrecht (in press)
5. Rowan KS (1989) Photosynthetic Pigments of Algae. Cambridge University Press, Cambridge
6. Young A, Britton G (1993) Carotenoids in Photosynthesis. Chapman & Hall, London
7. Scheer H (ed) (1991) Chlorophylls. CRC Press, Boca Raton, FL
8. Scheer H (2003) The pigments. In: Green BR, Parsons WE (eds) Light-harvesting antennas in photosynthesis. Advances in Photosynthesis and Respiration, vol 13. Kluwer Academic Publishers, Dordrecht p 29

9. Porra RJ, Pfündel EE, Engel N (1997) Metabolism and function of photosynthetic pigments. In: Jeffrey SW, Mantoura RFC, Wright SW (eds) Phytoplankton pigments in oceanography: Guidelines to modern methods. UNESCO, Paris, p 85
10. Jeffrey SW (1997a) Application of pigment methods to oceanography. In: Jeffrey SW, Mantoura RFC, Wright SW (eds) Phytoplankton pigments in oceanography: Guidelines to modern methods. UNESCO, Paris, p 127
11. Jeffrey SW, Mantoura RFC, Bjørnland T (1997) Data for the identification of 47 key phytoplankton pigments. In: Jeffrey SW, Mantoura RFC, Wright SW (eds) Phytoplankton pigments in oceanography: Guidelines to modern methods. UNESCO, Paris, p 447
12. Vernon LP, Seely GR (1966) The Chlorophylls. Academic Press, New York
13. Zapata M, Jeffrey SW, Wright SW, Rodríguez F, Garrido JL, Clementson L (2004) Mar Ecol Prog Ser 270:83
14. Strain HH, Cope BT, McDonald GN, Svec WA, Katz JJ (1971) Phytochem 10:1109
15. Jeffrey SW, Wright SW (1987) Biochim Biophys Acta 894:180–188
16. Fookes CJR, Jeffrey SW (1989) J Chem Soc Chem Comm 23:1827
17. Garrido JL, Zapata M (1998) J Phycol 34:70
18. Jeffrey SW (1989) Chlorophyll c pigments and their distribution in the chromophyte algae. In: Green JC, Leadbeater BSC, Diver WL (eds) The Chromophyte Algae: problems and perspectives. Clarendon Press, Oxford, p 13
19. Garrido JL, Otero J, Maestro MA, Zapata M (2000) J Phycol 36:497
20. Bjørnland T (1997) UV-vis spectroscopy of carotenoids. In: Jeffrey SW, Mantoura RFC, Wright SW (eds) Phytoplankton pigments in oceanography: guidelines to modern methods, UNESCO monographs on oceanographic methodology, vol 10. UNESCO, Paris p 578
21. IUPAC Commission on Nomenclature of Organic Chemistry (1975) Pure Appl Chem 41:407
22. Bjørnland T (1997) Structural relationships between algal carotenoids. In: Jeffrey SW, Mantoura RFC, Wright SW (eds) Phytoplankton pigments in oceanography: guidelines to modern methods, UNESCO monographs on oceanographic methodology, vol 10. UNESCO, Paris, p 572
23. Straub O, Pfander H, Gerspacher M, Rychener M, Schwabe R (1987) Key to Carotenoids. Birkhäuser, Basel
24. Liaaen-Jensen S (1985) Pure Appl Chem 57:649
25. Mercandente AZ (1999) Pure Appl Chem 71:2263
26. Bjørnland T, Liaaen-Jensen S (1989) Distribution patterns of carotenoids in relation to chromophyte phylogeny and systematics. In: Green JC, Leadbeater BSC, Diver WL (eds) The Chromophyte algae: Problems and perspectives. Clarendon Press, Oxford, p 37
27. Britton G (1995) UV/visible spectroscopy. In: Britton G, Liaaen-Jensen S, Pfander H (eds) Carotenoids, vol 1B Spectroscopy. Birkhäuser, Basel, p 13
28. Enzell CR, Back S (1995) Mass spectrometry. In: Britton G, Liaaen-Jensen S, Pfander H (eds) Carotenoids, vol 1B Spectroscopy. Birkhäuser, Basel, p 261
29. Liaaen-Jensen S (1995) Combined approach: identification and structure elucidation of carotenoids. In: Britton G, Liaaen-Jensen S, Pfander H (eds) Carotenoids, vol 1B Spectroscopy. Birkhäuser, Basel, p 343
30. Johnson PW, Sieburth JM (1979) Limnol Oceanogr 24:928
31. Waterbury JB, Watson SW, Guillard RRL, Brand LE (1979) Nature 277:293
32. Li WKW, Wood M (1988) Deep-Sea Res 35:1615

33. Chisholm SW, Olson RJ, Zettler ER, Goericke R, Waterbury JB, Welshmeyer NA (1988) Nature 334:340
34. Olson RJ, Chisholm SW, Zettler ER, Armbrust EV (1990) Limnol Oceanogr 35:45
35. Bodemer U (2004) J Plankton Res 26:1147
36. Bhattacharya D (1997) Origins of algae and their plastids. Springer, Berlin Heidelberg New York
37. Delwiche CF (1999) Am Nat 154:S164
38. McFadden GI (2001) J Phycol 37:951
39. Palmer JD (2003) J Phycol 39:4
40. Moreira D, LeGuyader H, Phillipè H (2000) Nature 405:69
41. Hallegraeff GM, Jeffrey SW (1984) Mar Ecol Prog Ser 20:59
42. Johnsen G, Samset O, Granskog L, Sakshaug E (1994) Mar Ecol Prog Ser 105:149
43. Goericke R, Montoya JP (1998) Mar Ecol Prog Ser 169:97
44. Schlüter L, Møhlenberg F, Havskum H, Larsen S (2000) Mar Ecol Prog Ser 192:49
45. Henriksen P, Riemann B, Kaas H, Sørensen HM, Sørensen HL (2002) J Plankton Res 24:835
46. Wood MA (1985) Nature 316:253
47. Bidigare RR, Schofield O, Prezelin BB (1989) Mar Ecol Prog Ser 56:177
48. Partensky F, Hoepffner N, Li WKW, Ulloa O, Vaulot D (1993) Plant Physiol 101:285
49. Sakshaug E, Andresen K (1986) J Plankton Res 8:619
50. Tukaj Z, Matusiak-Mikulin K, Lewandowska J, Szurkowski J (2003) Plant Physiol Biochem 41:337
51. Wilhelm SW, Maxwell DP, Trick CG (1996) Limnol Oceanogr 41:89
52. van Leeuwe MA, Timmermans KR, Witte HJ, Kraay GW, Veldhuis MJW, de Baar HHW (1998) Mar Ecol Prog Ser 166:43
53. Wilhelm C, Manns L (1991) J Appl Phycol 3:305
54. Stolte W, Kraay GW, Noordeloos AAM, Riegman R (2000) J Phycol 36:529
55. Jeffrey SW, Wright SW (1994) Photosynthetic pigments in the Haptophyta. In: Green JC, Leadbeater BSC (eds) The Haptophyte Algae. Clarendon Press, Oxford, p 111
56. Claustre H, Poulet SA, Williams R, Marty JC, Coombs S, Ben Mlih F, Hapette AM, Martin-Jezéquel V (1990) J Mar Biol Assoc UK 70:197
57. Vaulot D, Birrien J-L, Marie D, Casotti R, Veldhuis MJW, Kraay GW, Chrétiennot-Dinet M-J (1994) J Phycol 30:1022
58. Breton E, Brunet C, Sautour B, Brylinski JM (2000) J Plankton Res 22:1423
59. Mantoura RFC, Jeffrey SW, Llewellyn CA, Claustre H, Morales CE (1997) Comparison between spectrophotometric, fluorometric and HPLC methods for chlorophyll analysis. In: Jeffrey SW, Mantoura RFC, Wright SW (eds) Phytoplankton pigments in oceanography: Guidelines to modern methods. UNESCO, Paris, p 361
60. Jeffrey SW, Humphrey GF (1975) Biochem Physiol Pflanzen 167:191
61. Holm-Hansen O, Lorenzen CJ, Holmes RW, Strickland JDH (1965) J Cons Perm Int Explor Mer 30:3
62. Jacobsen TR (1978) Arch Hydrobiol Beih Ergebn Limnol 16:35
63. Wright SW, Mantoura RFC (1997) Guidelines for collection and pigment analysis of field samples. In: Jeffrey SW, Mantoura RFC, Wright SW (eds) Phytoplankton pigments in oceanography: Guidelines to modern methods. UNESCO, Paris, p 429
64. Mantoura RFC, Wright SW, Jeffrey SW, Barlow RG, Cummings DE (1997) Filtration and storage of pigments from microalgae. In: Jeffrey SW, Mantoura RFC, Wright SW (eds) Phytoplankton pigments in oceanography: Guidelines to modern methods. UNESCO, Paris, p 283

65. Wright SW, Jeffrey SW, Mantoura RFC (1997) Evaluation of methods and solvents for pigment extraction. In: Jeffrey SW, Mantoura RFC, Wright SW (eds) Phytoplankton pigments in oceanography: Guidelines to modern methods. UNESCO, Paris, p 261
66. Zapata M, Garrido JL (1991) Chromatographia 31:589
67. Diehn B, Seely GR (1968) Biochim Biophys Acta 153:862
68. Latasa M, van Lenning K, Garrido JL, Scharek R, Estrada M, Rodríguez F, Zapata M (2001) Chromatographia 53:385
69. Mantoura RFC, Llewellyn CA (1983) Analytica Chim Acta 151:297
70. Gieskes WWC, Kraay GW (1983) Mar Biol 75:179
71. Gieskes WWC, Kraay GW (1986) Mar Biol 92:45
72. Wright SW, Shearer JD (1984) J Chromatogr 294:281
73. Roy S (1987) J Chromatogr 391:19
74. Zapata M, Ayala AM, Franco JM, Garrido JL (1987) Chromatographia 23:26
75. Garrido JL, Zapata M (1997) Chromatographia 44:43
76. Van Heukelem L, Thomas C (2001) J Chromatogr 910:31
77. Wright SW, Jeffrey SW, Mantoura RFC, Llewellyn CA, Welschmeyer N, Bjørnland T, Repeta D (1991) Mar Ecol Prog Ser 78:183
78. Zapata M, Rodríguez F, Garrido JL (2000) Mar Ecol Prog Ser 195:29
79. Jeffrey SW, Mantoura RFC (1997) Minimum criteria for identifying phytoplankton pigments. In: Jeffrey SW, Mantoura RFC, Wright SW (eds) Phytoplankton pigments in oceanography: Guidelines to modern methods. UNESCO, Paris, p 631
80. Airs RL, Atkinson JE, Keely BJ (2001) J Chromatogr A 917:167
81. Wright SW, Mantoura RFC (1997) Guidelines for selecting and setting up an HPLC system and laboratory. In: Jeffrey SW, Mantoura RFC, Wright SW (eds) Phytoplankton pigments in oceanography: Guidelines to modern methods. UNESCO, Paris, p 383
82. Jeffrey SW, Wright SW (1997) Qualitative and quantitative HPLC analysis of SCOR reference algal cultures. In: Jeffrey SW, Mantoura RFC, Wright SW (eds) Phytoplankton pigments in oceanography: Guidelines to modern methods. UNESCO, Paris, p 343
83. Mantoura RFC, Repeta DJ (1997) Calibration methods for HPLC. In: Jeffrey SW, Mantoura RFC, Wright SW (eds) Phytoplankton pigments in oceanography: Guidelines to modern methods. UNESCO, Paris p 407
84. Jeffrey SW (1974) Mar Biol 26:101
85. Jeffrey SW, Hallegraeff GM (1987) Mar Ecol Prog Ser 35:649
86. Barrett J, Jeffrey SW (1964) Plant Physiol 39:44
87. Barrett J, Jeffrey SW (1971) J Exp Mar Biol Ecol 7:255
88. Latasa M, Bidigare RR (1998) Deep-Sea Res II 45:2133
89. Jeffrey SW, Haxo FT (1968) Biol Bull 135:149
90. Mackey MD, Mackey DJ, Higgins HW, Wright SW (1996) Mar Ecol Prog Ser 144:65
91. Claustre H, Hooker SB, Van Heukelem L, Berthon JF, Barlowe R, Rasa J, Sessions H, Targa C, Thomas CS, van der Linde D, Marty JC (2004) Mar Chem 85:41
92. Garibotti IA, Vernet M, Kozlowski WA, Ferrario ME (2003) Mar Ecol Prog Ser 247:27
93. Prézelin BB, Hofmann EE, Mengelt C, Klinck JM (2000) J Mar Res 58:165
94. Marty JC, Chiavérini J, Pizay MD, Avril B (2002) Deep-Sea Res II 49:1965
95. Everitt DA, Wright SW, Volkman JK, Thomas DP, Lindstrom E (1990) Deep Sea Res 37:975
96. Letelier RM, Bidigare RR, Hebel DV, Ondrusek M, Winn CD, Karl DM (1993) Limnol Oceanogr 38:1420
97. Bidigare RR, Ondrusek ME (1996) Deep Sea Res II 43:809

98. Vidussi F, Marty J, Chiavérini J (2000) Deep-Sea Res I 47:423
99. Wright SW, Thomas DP, Marchant HJ, Higgins HW, Mackey MD, Mackey DJ (1996) Mar Ecol Prog Ser 144:285
100. Havskum H, Schlüter L, Scharek R, Berdalet E, Jacquet S (2004) Mar Ecol Prog Ser 273:31
101. Mackey MD, Higgins HW, Mackey DJ, Holdsworth D (1998) Deep-Sea Res 45:1441
102. Wright SW, van den Enden RL (2000) Deep-Sea Res 47:2363
103. Carreto JI, Montoya NG, Benavides HR, Guerrero R, Carignan MO (2003) Mar Biol 143:1013
104. van Lenning K, Latasa M, Estrada M, Sáez AG, Medlin L, Probert I, Véron B, Young B (2003) J Phycol 39:379
105. Latasa M, Scharek R, Le Gall F, Guilliou L (2004) J Phycol 40:1149
106. Stoń J, Kosakowska A, Łotocka M, Łysiak-Pastuszak E (2002) Oceanologia 44:419
107. Andersen RA, Bidigare RR, Keller MD, Latasa M (1996) Deep-Sea Res 43:517
108. Rodriguez J, Jimenez-Gomez F, Blanco JM, Figueroa FL (2002) Deep Sea Res II 49:693
109. Rodriguez F, Varela M, Fernández E, Zapata M (2003) Mar Biol 143:995
110. Ansotegui A, Sarobe A, Trigueros JM, Urrutxurtu I, Orive E (2003) J Plankton Res 25:341
111. Schlüter L, Møhlenberg F (2003) J Appl Phycol 15:465
112. Llewellyn CA, Fishwick JR, Blackford JC (2005) J Plankton Res 27:103
113. Lund JWG, Kipling C, Le Cren ED (1958) Hydrobiol 11:143
114. Veldhuis MJW, Kraay GW (2004) Deep-Sea Res I 51:507
115. Staehr PA, Henriksen P, Markager S (2002) Mar Ecol Prog Ser 238:47
116. Vidussi F, Claustre H, Manca BB, Luchetta A, Marty J-C (2001) J Geophys Res 106:19939
117. Barlow RG, Aiken J, Holligan PM, Cummings DG, Maritorena S, Hooker S (2002) Deep-Sea Res I 47:637
118. Gibb SW, Barlow RG, Cummings DG, Rees NW, Trees CC, Holligan P, Suggett D (2000) Prog Oceanogr 45:339
119. Di Tullio GR, Geesey M, Jones DR, Daly KL, Campbell L, Smith WO (2003) Mar Ecol Prog Ser 255:55
120. Welschmeyer NA, Lorenzen CJ (1984) Limnol Oceanogr 29:135
121. Geider RJ (1987) New Phytol 106:1
122. Geider RJ, La Roche J, Greene R M, Olaizola M (1993) J Phycol 29:755
123. Taylor AH, Geidler RJ, Gilbert FJH (1997) Mar Ecol Prog Ser 152:51
124. Riemann B, Simonsen P, Stensgaard L (1989) J Plankton Res 11:1037
125. Geider RJ, MacIntyre HL, Kana TM (1997) Mar Ecol Prog Ser 148:187
126. Hillebrand H, Dürselen CD, Kirschtel D, Pollingher D, Zohary T (1999) J Phycol 35:403
127. Montagnes DJS, Berges JA, Harrison PJ, Taylor FJR (1994) Limnol Oceanogr 39:1044
128. Menden-Deuer S, Lessard EJ (2000) Limnol Oceanogr 45:569
129. Redalje DG, Laws EA (1981) Mar Biol 62:73
130. Gieskes WWC, Kraay GW (1989) Deep Sea Res 36:1127
131. Goericke R, Welschmeyer NA (1993) Limnol Oceanogr 38:80
132. Stramski D (1999) Deep-Sea Res 46:335
133. Schmid H, Bauer F, Stich HB (1998) J Plankton Res 20:1651
134. Wright SW (1997) Summary of terms and equations used to evaluate HPLC chromatograms. In: Jeffrey SW, Mantoura RFC, Wright SW (eds) Phytoplankton pigments in oceanography: Guidelines to modern methods. UNESCO, Paris, p 622

135. Egeland ES, Guillard RRL, Liaaen-Jensen S (1997) Phytochem 44:1087
136. Stauber JL, Jeffrey SW (1988) J Phycol 24:158
137. Rodríguez F, Pazos Y, Maneiro J, Fraga S, Zapata M (2000) Ninth Int Conf on Algal Blooms, Hobart, Australia
138. Fawley MW (1991) J Phycol 27:544
139. Kolber ZS, Plumley FG, Lang AS, Beatty JT, Blankenship RE, VanDover CL, Vetriani C, Koblizek M, Rathgeber C, Falkowski PG (2001) Science 292:2492
140. Goericke R (2002) Limnol Oceanogr 47:290
141. Claustre H (1994) Limnol Oceanogr 39:1206
142. Brunet C, Casotti R, Aronne B, Vantrepotte V (2003) J Plankton Res 25:1413
143. Demers S, Roy S, Gagnon R, Vignault C (1991) Mar Ecol Prog Ser 76:185
144. Evens TJ, Kirkpatrick GJ, Millie DF, Chapman DJ, Schofield OME (2001) J Plankton Res 23:1177
145. Kashino Y, Kudoh S (2003) Phycol Res 51:168
146. Abrahamsson K, Lorén A, Wulff A, Wängberg S-A (2004) Deep-Sea Res II 51:2789
147. Andersson M, Van Nieuwerburgh L, Snoeijs P (2003) Mar Ecol Prog Ser 254:213
148. Bergmann T, Richardson TL, Paerl HW, Pinckney JL, Schofield O (2002) J Plankton Res 24:923
149. Ediger D, Raine R, Weeks AR, Robinson IS, Sagan S (2001) J Plankton Res 23:893
150. Furuya K, Hayashi M, Yabushita Y, Ishikawa A (2003) Deep Sea Res 50:367
151. Gibb SW, Cummings DG, Irigoien X, Barlow RG, Mantoura RFC (2001) Deep Sea Res 48:795
152. Gin KYH, Zhang S, Lee YK (2003) J Plankton Res 25:1507
153. Li H-P, Gong G-C, Hsiung T-M (2002) Bot Bull Acad Sin 43:283
154. Mackey DJ, Blanchot J, Higgins HW, Neveux J (2002) Deep-Sea Res 49:2561
155. Riegman R, Kraay GW (2001) J Plankton Res 23:191
156. Steinberg DK, Carlson CA, Bates NR, Johnson RJ, Michaels AF, Knap AH (2001) Deep-Sea Res 48:1405
157. Suzuki K, Hinuma A, Saito H, Kiyosawa H, Liu H, Saino T, Tsuda A (2005) Prog Oceanogr 64:167

Molecular Tools for the Analysis of DNA in Marine Environments

R. Danovaro (✉) · C. Corinaldesi · G. M. Luna · A. Dell'Anno

Department of Marine Sciences, Polytechnic University of Marche, Via Brecce Bianche, 60131 Ancona, Italy
danovaro@univpm.it

1	DNA in Aquatic Ecosystems	106
2	Extraction and Quantification of Intracellular DNA in Water Samples	109
3	Isolation and Quantification of Dissolved DNA	110
4	Intracellular DNA in Sediment Samples	111
5	Extracellular DNA in Sediment Samples	112
6	Degradation and Turnover of Extracellular DNA	113
7	Molecular Tools for the Analysis of Nucleic Acids in Marine Environments	114
8	Analysis of Microbial Diversity	117
9	Concluding Remarks	120
	References	121

Abstract In the last decade, microbial ecologists have increasingly applied molecular techniques to investigate microorganisms in natural environments. The use of molecular tools has allowed the identification of new and uncultured microbial species, and has greatly advanced our knowledge on the diversity and functioning of microbial communities in aquatic ecosystems. At the same time, the discovery of large quantities of extracellular DNA in both seawater and sediments is opening up new questions on the role and significance of these components in biogeochemical cycles and, potentially, also in horizontal gene transfer. This chapter describes the most recent methods for the extraction, quantification and isolation of intracellular and extracellular DNA in water and sediment samples. An outline of methods currently used in marine molecular ecology and their limitations is presented. The application of various molecular tools for studying DNA associated with microorganisms and for investigating the extracellular fraction is critically discussed. Some recent discoveries and new perspectives for future research are highlighted.

Keywords Diversity · Extracellular DNA · Intracellular DNA · Molecular analysis · Sediment · Water column

1
DNA in Aquatic Ecosystems

In aquatic ecosystems, the organic carbon inventory is largely dominated by non-living materials (i.e. detrital carbon) present in both the dissolved and particulate states, whose cycling is primarily mediated by heterotrophic prokaryotes. The labile fraction of the organic carbon pool in the oceans is mainly composed of simple (i.e. monomeric) and combined biochemical compounds. Among the biochemical classes of organic compounds, DNA ranks fourth after carbohydrates, proteins and lipids [1]. In aquatic environments, DNA is present in different forms: (1) associated with living organisms (i.e. intracellular DNA); (2) encapsulated by proteins (i.e. viral DNA); (3) free (i.e. soluble DNA) and (4) adsorbed to detrital and/or mineral particles [2]. Since viruses are a group of biological entities with a genome [3], only the two latter forms can be considered genuine extracellular DNA. DNA associated with living biomass is the ultimate source of extracellular DNA because cell-free DNA synthesis is not known to occur. Potential pathways of extracellular DNA production include: (1) exudation and excretion from viable cells; (2) losses associated with grazing activities; (3) passive release following cell death and lysis; (4) release due to virus-induced cell lysis; and (5) desorption/adsorption of dissolved DNA from seston particles.

The presence of DNA as a constituent of the dissolved organic matter pool in aquatic systems has been known since the early 1970s [4, 5], but our current understanding of the dynamics and distribution of dissolved DNA in marine environments is largely due to the works of Paul and co-workers [6–9] and DeFlaun and co-workers [10, 11]. Studies carried out on a regional scale in the Gulf of Mexico indicated that dissolved DNA concentrations are highest in estuarine environments (from 5 to 44 mg m^{-3}) and decrease with increasing distance from land (2–15 and 1–5 mg m^{-3} in coastal and offshore oceanic environments, respectively) and with increasing water depth (up to < 1 mg m^{-3} at bathyal depths). In addition, these studies revealed that the molecular size of the dissolved DNA pool ranges from < 0.5 to > 23.0 kbp, with DNA in offshore environments at the lower end of this range.

The recent discovery by microbial ecologists of high viral abundances in seawater has raised scepticism about the existence of a large pool of genuine extracellular DNA [12]. However, although common procedures for quantifying dissolved DNA do not distinguish soluble DNA from encapsulated DNA (i.e. viral DNA), several authors have demonstrated that viral DNA accounts generally for less than 20% of dissolved DNA pools [9, 13–16]. Quantitative estimates of the dissolved DNA pool in the water column thus reflect, to a large degree, extracellular DNA concentrations.

Besides quantitative estimates, other studies have specifically addressed questions about the production, degradation and cycling of extracellular DNA in aquatic ecosystems. For instance, by using radioactive precursors, Paul et al. [17] showed that heterotrophic bacterioplankton was the major source of dissolved DNA, while actively photosynthesizing phytoplankton did not contribute to this pool. However, subsequent studies by means of dot-blot hybridization allowed researchers to identify the presence of phytoplankton genes (i.e. ribulose biphosphate large subunit gene, rbcL) in the dissolved DNA fraction of freshwater and seawater samples [18].

Turnover times of extracellular DNA, calculated on the basis of estimates of degradation rates, range from 6.5 to 25 h [6, 8], indicating that in a pelagic environment extracellular DNA is a highly reactive macromolecule in the dissolved organic matter pool. Although Paul and co-workers showed that extracellular DNA is mainly a source of exogenous nucleotides, recycled by bacteria for the synthesis of new DNA, other studies suggested that extracellular DNA can be an important source of organic N and P for bacterioplankton metabolism [19–22]. Dissolved extracellular DNA alone may supply about 50% of the daily P requirements and about 10% of the daily N requirements of bacterioplankton [19, 20], and can play an even more important role in P-depleted ecosystems [23, 24]. Extracellular DNA can also have implications in horizontal gene transfer [25–29].

As far as particulate DNA is concerned, conceptual models indicate that more than 70% of the total particulate DNA pool in surface oceanic waters is accounted for by DNA associated with prokaryotes in the pico-plankton fraction (i.e. 0.2–1.0 μm, *sensu* [13]). However, particulate DNA pools may also include an extracellular fraction of detrital DNA (i.e. DNA adsorbed onto detrital particles; Fig. 1).

This is particularly evident in marine sediments, where these proportions may be inverted, the detrital fraction being largely dominant over the en-

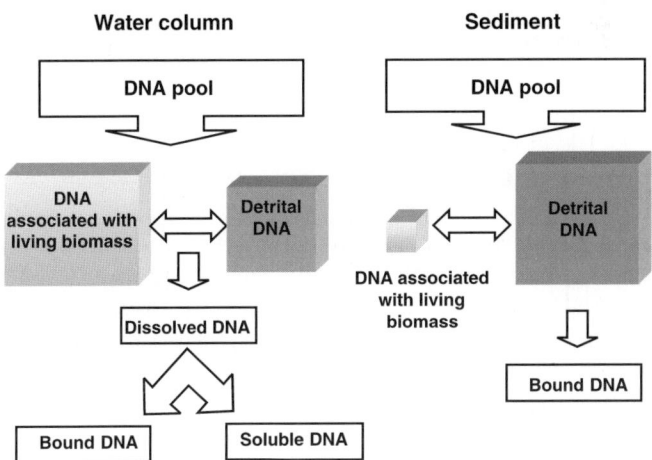

Fig. 1 Conceptual model of the different DNA pools in seawater (modified from [13]) and sediment (modified from [32, 60])

tire DNA pool. Particulate detrital DNA, estimated by the use of conversion factors, can account for a highly variable fraction of the total particulate DNA pool (0–93% [30, 31]). Estimates carried out on marine sediments highlighted that detrital DNA is the dominant component over the entire sedimentary DNA pool (up to 90% [1, 32, 33]). Concentrations of extracellular DNA in sediments from shallow depths down to the abyssal floor were 3 to 4 orders of magnitude higher than those in the water column. However, the use of conversion factors is questionable for providing accurate quantitative estimates of the relative importance of the extracellular DNA pool in different ecological compartments. Recently, Dell'Anno et al. [34] developed a new nuclease-based procedure for quantifying extracellular DNA concentrations in marine sediments. This procedure is highly specific for extracellular DNA, allowing one to obtain accurate quantitative estimates not biased by DNA contamination due to cell lysis or viral DNA. The results from this study clearly confirmed that extracellular DNA in marine sediments is the dominant fraction of the total sedimentary DNA pool.

The quantitative relevance of extracellular DNA in marine sediments is the result of complex interactions including DNA inputs from the photic layer through particle sedimentation, autochthonous DNA production, and degradation and/or utilization by heterotrophic organisms [35] (Fig. 2).

Extracellular DNA diagenesis in sediments is also influenced by DNA binding to complex refractory organic molecules and/or to inorganic particles, which protect DNA against nuclease degradation [36–38]. In this regard, Romanowski et al. [39] showed that DNA adsorbed on sand and clay becomes 100- to 1000-fold more resistant to DNase. Consequently, the half-life

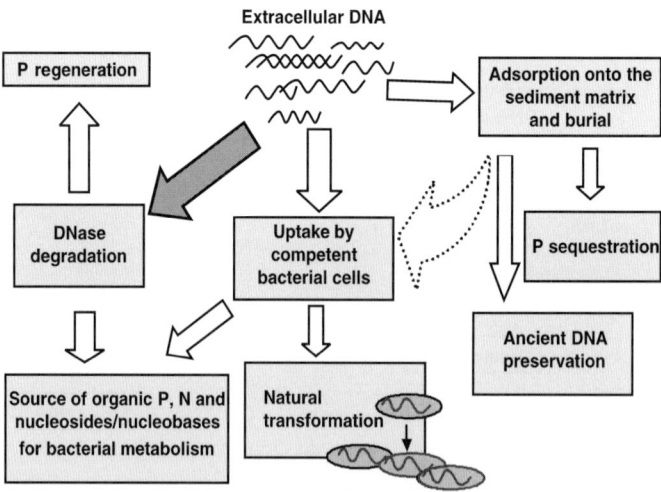

Fig. 2 Theoretical model of the fate and ecological significance of extracellular DNA in marine sediments

of extracellular DNA in sediments appears to be much longer than that in the water column [40]. The reduced degradability of extracellular DNA may explain why this molecule can also persist in deeper sediment layers (i.e. on geological timescales), thus representing a potential genetic marker of paleo-environments [35, 41, 42]. In fact, due to the higher resolving power of DNA sequences as compared to biomarkers, molecular characterization of ancient genetic material may improve the reconstruction of past communities and related paleo-environments. At the same time, since extracellular DNA production and accumulation in sediments represents a record of processes occurring in the pelagic and benthic domains at different temporal scales [32, 35, 41], analyses of sequences of structural and functional genes preserved in the extracellular DNA pool might provide new information about the ecological functioning of present-day ecosystems and paleo-ecosystems [34, 43].

Despite the overwhelming dominance of the detrital fraction in sedimentary DNA pools, in the last decade molecular techniques have been generally applied to the whole DNA pool, assuming that it is entirely associated with living biomass. The use of molecular techniques in the field of microbial ecology has greatly enhanced our understanding of the diversity and functioning of microorganisms in aquatic ecosystems [44], but questions related to potential biases due to the presence of extracellular DNA are still largely unsolved. Therefore, a crucial step for enhancing the accuracy of molecular tools is the development of reliable extraction methods able to separate intra- and extracellular DNA from a given environmental matrix.

2
Extraction and Quantification of Intracellular DNA in Water Samples

The standard approach utilized by most investigators for isolating particulate DNA is to concentrate cells on micropore membranes (0.2-μm pore size filters), after pre-filtration to avoid sample contamination with larger material, and then to lyse the cells retained on the filters [45, 46]. For example, Fuhrman et al. [45] described a method for DNA extraction based on filtration of marine and brackish waters (8–40 l) through 0.2-μm pore size filters. DNA was extracted directly from the filters in 1% sodium dodecyl sulphate (SDS) heated to 95–100 °C for 1.5–2 min. This procedure lyses essentially all bacterial cells and does not significantly denature the DNA, which is then purified by phenol extraction. DNA is quantified fluorometrically using Hoechst 33258 (i.e. a groove-binding DNA ligand that becomes brightly fluorescent when it binds to the double-strand form of the DNA).

Final yields are in the range of a few micrograms of DNA per litre and correspond, roughly, to 25–50% of the total bacterial DNA in the sample [45]. Although the bacterial community probably does not change during filtration, the large volumes of the water samples and the extended time of filtration required by this procedure could restrict the application of molecular tools in extensive surveys of aquatic systems and in experiments with multiple treatments or repeated sampling.

These problems could be circumvented by alternative techniques, such as the use of cylindrical filter membranes [46], tangential-flow filtration (TFF) [47] or vortex-flow filtration [48]. The first of these techniques is largely utilized for concentrating microbial cells [46]. This technique requires a high-capacity cylindrical filter, through which water is pumped and in which cell lysis is finally achieved [46]. Lysozyme, SDS and proteinase K are used in this step. DNA from the lysate can be purified by ethanol precipitation or buoyant-density centrifugation and utilized for molecular studies.

Recent studies on aquatic microbial communities used small volumes of water for recovering DNA [49–52]. Kirchman et al. [51] developed a method based on filtration of 10 ml (or less) of seawater through a polycarbonate filter which is sectioned, and a section is directly amplified by the polymerase chain reaction (this method has been designated as "filter PCR"). Molecular analyses revealed little difference when comparing the 16S rRNA amplicons obtained by other techniques and the "filter PCR" protocol.

3
Isolation and Quantification of Dissolved DNA

During the last 15 years, the isolation and quantification of dissolved DNA from water samples has been addressed by several studies using different approaches [6, 7, 9–11, 17, 24, 53–58]. Filtration of water samples through 0.2-μm pore size filters is the first step required for the isolation and quantification of dissolved DNA from microbial cells and other particulate material.

Several authors [6, 10, 11] have utilized ethanolic precipitation to concentrate dissolved DNA from freshwater and seawater samples. Concentrated DNA was quantified by the fluorescence of dye–DNA complexes (using Hoechst 33258 dye). To correct for the fluorescence not caused by DNA or caused by packaged phage DNA, samples treated with DNase were measured in parallel. As an example, De Flaun et al. [10] found that up to 76% of the fluorescence remained after DNase treatment of estuarine and oceanic water samples. The effectiveness of this procedure for concentrating dissolved DNA was demonstrated by the efficient (> 90%) recovery of internal standards. Further purification by chromatography, polyvinylpolypyrrolidone (PVPP) treatment and CsCl buoyant-density centrifugation gave preparations of sufficient purity for determination of the molecular weight of the extracted DNA and for the detection of specific genes by hybridization [7, 11, 18, 57].

Ethanolic precipitation is widely used in molecular ecology, but is time-consuming and is limited by the specificity of Hoechst 33258 towards double-stranded DNA. Karl and Bailiff [54] proposed an alternative technique based on addition of cetyltrimethylammonium bromide (CTAB) for concentrating dissolved DNA from water samples. The insoluble CTA-nucleic acid salts obtained are used to determine dissolved DNA concentrations by the fluorometric method using 3,5-diaminobenzoic acid (DABA). This procedure is compatible with rapid shipboard analyses, detects both single- and double-stranded DNA and, as opposed to ethanolic precipitation methods, proteins do not react with the precipitating agent.

Other compounds (such as diphenylamine, 4'-6-diamidino-2-phenylindole (DAPI), ethidium bromide and mithramycin) can, in theory, be used to quantify the dissolved DNA concentrated by the CTAB procedure. However, the investigation of Siuda and Güde [58] provided clear evidence that dissolved DNA concentrations were overestimated when determined by the CTAB–DAPI method in eutrophic freshwater samples. The CTAB technique, which causes the precipitation of DNA and other compounds [58], is not highly specific. This, together with the partial solubilization of various fluorescent components, might cause a significant alteration of the fluorescence during the assay [10, 58]. To avoid these problems, Siuda and Güde [58] estimated the DNA fraction that was hydrolysable by nucleases as the difference between the concentration of the DNA in samples with and without DNase treatment. The discrimination of the enzymatically hydrolysable DNA from the total dissolved DNA pool is important for a better understanding of the ecological role of extracellular DNA. In fact, the abundance of extracellular DNA in most aquatic systems makes it an important source of P and N, and/or nucleotides for aquatic microorganisms [59]. This discrimination can be achieved by using a nuclease-based procedure developed by Dell'Anno and Danovaro [60], in which extracellular DNA is cleaved into deoxynucleosides, which are then quantified fluorometrically by DABA or by HPLC. However, since this technique is based on nuclease hydrolysis of extracellular DNA, it does not allow the recovery of DNA for subsequent molecular studies, but only the quantification of the hydrolysis products (i.e. deoxynucleosides).

4
Intracellular DNA in Sediment Samples

Several protocols for DNA extraction from soils and sediments have been developed and improved in recent years [61–64], and a large effort has been devoted to purifying DNA and to enhancing DNA extraction efficiency [61, 62, 65, 66]. DNA extraction techniques usually involve a direct in situ lysis of cells and the subsequent release of DNA, and allow investigations of "community DNA" [67]. Other protocols involve the isolation of microbial cells from sediments prior to DNA extraction [68, 69]. Both these methods have advantages and disadvantages [66], but direct in situ lysis is more commonly used, mainly because of its faster times of extraction and much higher DNA yields [66].

In situ cell lysis from sediments can be achieved by means of physical (e.g. bead-mill homogenization, ultra-sonication and freeze–thawing) or chemical procedures (e.g. using SDS or Sarkosyl [61, 66]), or a combination of both. Freeze–thawing [70–72] and bead-mill homogenization [69, 73, 74] are commonly utilized, although bead-mill homogenization yields more DNA than the freeze–thaw procedure [70, 74–76]. However, a larger amount of contaminating humic acids are recovered by bead-mill homogenization [75–77] and, in certain cases, it can increase DNA shearing [75]. Chemical lysis can be obtained using solutions that contain SDS [45, 64, 73, 78–80] or Sarkosyl [81, 82]. The modifications of these chemical techniques also include high-temperature (from 60 °C to boiling) incubation [69, 70, 73], a phenol [72, 76, 83] or chloroform [84] extraction step, and the incorporation of chelating agents (EDTA and Chelex 100) to inhibit nucleases and disperse soil/sediment particles [85].

DNA extraction can also utilize enzymatic lysis, which involves the use of lysozyme [72, 73, 81, 83, 84], proteinase K [79, 86, 87], chromopeptidase [88] or pronase E [81]. All these enzymes have been employed to promote cell lysis, but due the lack of comparative studies it is not clear whether the addition of the enzymatic lysis step to other extraction protocols increases DNA yields.

The purification of DNA from sediments can be achieved by agarose gel electrophoresis [64, 74, 79, 80, 89], Sephadex G-200 column chromatography [70, 71, 83, 84, 90] and silica-based DNA binding [64, 73, 74], used individually or in combination. The purification efficiency is usually estimated by the amount of DNA recovered and by the effectiveness of the methods used to remove any contaminant that might inhibit PCR and/or other enzymes utilized for molecular analyses [91].

5
Extracellular DNA in Sediment Samples

Although in situ lysis is the most commonly utilized technique for DNA extraction from sediments, with this procedure extracellular DNA is co-extracted with nucleic acids released from the lysed cells, possibly leading to misinterpretation of the composition of the target community derived from molecular analysis [92]. Discrimination between intracellular and extracellular DNA in marine sediments is essential to carry out simultaneous molecular studies of these two DNA fractions. However, until recently, the isolation of extracellular DNA from sediments was an unsolved task, because the available procedures for the extraction of nucleic acids adsorbed on organic and inorganic particles disrupt living cells [40, 64, 91]. An attempt has been made to isolate extracellular DNA from aquatic sediments [77]. This protocol involves several washings of wet sediment samples with sodium phosphate, precipitation with ethanol and then purification by hydroxylapatite chromatography. However, this protocol has not been tested for possible contamination by intracellular DNA due to cell lysis during sediment handling. Moreover, the procedure of Ogram et al. [77] obtains extracellular DNA yields at least one order of magnitude lower than those obtained using the nuclease-based procedure of Dell'Anno and Danovaro [60]. Although extracellular DNA concentrations obtained by the nuclease-based procedure are not biased by DNA contamination due to cell lysis or by viral DNA [34, 60], this technique does not allow the recovery of DNA for subsequent molecular studies.

Recently, a new protocol has been developed to recover simultaneously DNA associated with microbial cells and extracellular DNA from the same sediment sample [93]. This protocol is an adaptation of the procedures extensively used for the isolation of microbial cells from sediments [67, 69, 94]. To date, this procedure has been applied only for extracting intact microbial cells from sediments, without considering the presence of extracellular DNA, which may be co-extracted. In order to recover simultaneously extracellular DNA and DNA associated with microbial cells, three washing steps of the samples using an isotonic solution of sodium phosphate buffer, supplemented with PVPP and low SDS concentrations (ten times lower than those generally used for in situ lysis), are required to improve the extraction efficiency of both extracellular and intracellular DNA pools. This protocol is suitable for molecular studies of extracellular DNA because it avoids any contamination by DNA released from cell lysis during handling and extraction, and provides adequate DNA yield and purity.

6
Degradation and Turnover of Extracellular DNA

Bacteria-mediated degradation of organic matter plays a key role in carbon cycling and nutrient regeneration in the world's oceans [39]. This process is largely mediated by extracellular enzymatic hydrolysis, which converts high molecular weight compounds into low molecular weight ones suitable for bacterial uptake [95–97]. The removal and cycling of extracellular DNA from marine environments occur through two basic mechanisms. The first is carried out by competent bacterial cells, which are able to internalize DNA fragments [40, 98]. The second is mediated by both cell-associated and free DNases, which are present in all aquatic environments and convert extracellular DNA into nucleosides and nucleobases [7, 57, 99]. This process is expected to be the main route for extracellular DNA cycling [40].

Nucleic acid turnover time can be defined as the ratio of the ambient (i.e. extracellular) DNA concentration and the velocity of its removal (i.e. degradation rates [6, 8]). The quantification of ambient extracellular DNA and the measurement of its degradation rates [34] are, therefore, indispensable in order to provide accurate estimates of extracellular DNA turnover rates in seawater and sediment samples. Previous studies carried out in marine environments estimated degradation rates of DNA in aquatic environments by analysing: (1) the decrease of acid-precipitable labelled DNA [8, 57, 100]; (2) the conversion of supercoiled into relaxed-circular or linear plasmid DNA [101]; and (3) the loss of hybridization signals of plasmid DNA in Southern transfer or dot blots [53]. However, these methods have mainly been applied for understanding the survival of specific DNA sequences [25, 40]. Moreover, being only a minor fraction of culturable marine bacteria [102], estimates of DNase activity based on isolates do not reflect the actual degradation of extracellular DNA in marine systems.

A new procedure for estimating extracellular DNA degradation rates in marine systems (seawater and sediment) is based on fluorometric detection of nuclease activity by means of a fluorescent DNA analogue [poly(dεA), polydeoxyribo-1-N_6ethenoadenylic acid (6 in apice)] [103]. This method, which was developed for in vitro studies, is based on the increase of the fluorescence of poly(dεA) due to degradation of polynucleotides [104] and it is highly specific for detecting exonuclease activity [105]. By this procedure it is possible to quantify the amount of nucleotides released from the degradation of the fluorescent DNA analogue. The conversion of extracellular DNA into nucleotides represents the key step for the subsequent bacterial uptake of nucleosides and nucleobases [7, 57]. The results of these studies indicate that poly(dεA) is effectively degraded into etheno-monomeric residues (i.e. dεAMP), and that this procedure can be routinely utilized for estimating extracellular DNA degradation rates in marine environments. In addition, since quantitative estimates of extracellular DNA do not necessarily reflect its actual bioavailability [13, 34], in order to calculate accurate turnover rates of extracellular DNA in seawater and sediment samples it has been suggested that the actual bioavailable fraction of extracellular DNA (i.e. hydrolysable by nucleases; [60, 103]) should be quantified. Turnover estimates of the bioavailable fraction of extracellular DNA are important for clarifying bacterial utilization pathways of extracellular DNA and provide new elements for a better comprehension of the mechanisms controlling DNA preservation in aquatic environments.

7
Molecular Tools for the Analysis of Nucleic Acids in Marine Environments

A wide variety of molecular techniques can be utilized for the analysis of nucleic acids (both DNA and RNA) in the marine environment. The choice of different molecular tools is clearly dependent upon the ecological objective. Basically there are two possible approaches for studying target genes or genomes of aquatic microorganisms: the first consists of the identification of target gene sequences within intact cells, without any extraction step of nucleic acids, while the second is based on cell lysis and nucleic acid isolation and recovery. If the target is a gene sequence within cells, it is possible to use probes (both fluorescent and radiolabelled) for its hybridization. The most common technique in marine microbial ecology is fluorescent in situ hybridization (FISH) [106], which utilizes fluorescent probes to hybridize complementary rRNA sequences (see [107] for details). For prokaryotes, the FISH technique allows one to identify and count, by epifluorescence microscopy, specific target genera or groups (e.g. *Cytophaga*, α-, β- and γ-Proteobacteria, Archaea [106]).

PCR is now a routinely used tool in marine molecular ecology. Once extracted and purified, the nucleic acid is amplified by PCR (polymerase chain reaction, in the case of DNA) or RT-PCR (reverse transcription-PCR in the case of RNA), thus allowing the production in vitro of large numbers of identical copies of a specific nucleic acid sequence [107]. For this purpose, it is necessary to know the sequences of the regions (primers: usually 15–20 nucleotides in length) flanking the two ends of the gene target. The PCR reaction is a very powerful tool as the reaction can theoretically proceed with just one single copy of the gene.

However, a review dealing with PCR amplification and the associated difficulties indicates that PCR is not completely free from pitfalls [108]. For example, PCR is strongly limited by the availability of known sequences in the database, which limits the design of gene-specific PCR primers: this means that primers may not include all the relevant naturally occurring genes, leading to an underestimation of gene diversity. Some authors [109] showed that biases can occur in the amplification step, caused by template annealing of mixtures of 16S rRNA genes, and showed how such biases were strongly dependent upon the number of cycles of replication. A potential bias can be introduced by the formation of chimeric products or artefacts. Another problem deals with the inhibition of the polymerase activity. The presence of inhibitors can often cause false-negative reactions (i.e. a target gene is present but not amplified). This is a frequent problem when working with marine sediment samples, where high concentrations of potential inhibitors (e.g. humic acids) are present [88]. However, in recent years several techniques have addressed and solved this problem [41, 42].

PCR generally utilizes primers for the 16S rRNA gene but other genes or gene families can also be employed for studying microbial diversity and phylogeny. Moreover, PCR allows investigation of the presence of mobile genetic elements, such as bacterial plasmids (in both the water column and marine sediments; [110–112]). Plasmid-encoded genes are a pool of mobile DNA, which is known to contribute significantly to genetic adaptation of natural microbial communities.

PCR does not allow recovery of intact genes, but only portions of them and only one single gene can be investigated each time. Different molecular tools must, therefore, be used if the target is a multiple gene or the entire genome. The recently proposed DNA microarray (or microchip) technology provides a platform for genome-wide hybridization experiments that can be utilized for identifying DNA sequences, for comparing different genomes and for monitoring gene expression [113–116].

DNA microarrays consist of thousands of unique DNA sequences connected to a small, solid surface (such as a glass slide [114]). Fluorescent or radioactivity-labelled mRNA or DNA derived from mRNA by RT-PCR or genomic DNA can bind to these sequences, thus producing a pattern indicative of nucleic acid sequences that can be qualitatively and quantitatively analysed [114]. Microarray technology has the advantage of rapid detection, automation [116, 117] and lower costs than conventional membrane-based hybridization (such as FISH; [107]). These characteristics make microarray technologies extremely useful for molecular characterization of mixed populations of microorganisms and their biological functions. Microarray technologies can theoretically be utilized for screening for the presence and expression of both prokaryotic and eukaryotic genes.

Another method for characterizing nucleic acids and their molecular diversity in the environment employs bacterial artificial chromosomes (BACs). BACs are capable of sustaining DNA inserts larger than 300 kbp [114] and allow construction of libraries of community DNA ("metagenomic libraries"). The "metagenome" is the whole genome pool within an environmental sample [118]. Subsequent sequencing or hybridization of BAC libraries allows the phylogenetic and genomic analysis of the entire microbial assemblage. To date, metagenomic libraries have been constructed from environmental DNA recovered from terrestrial soils [118, 119] and biofilms in drinking waters [120].

All these approaches have been utilized for studying a wide variety of genes associated with the DNA of living cells, but they could also be used for studying genes potentially present in the extracellular DNA pool. There is only molecular study that has addressed concomitantly, but separately, both intracellular and extracellular DNA in sediment [93]. This is particularly important in marine sediments, where extracellular DNA is characterized by high molecular weights, and is apparently also well preserved in subsurface sediment layers [34, 60].

Despite the advancement in molecular techniques, several questions still remain unanswered. What is the contribution of extracellular DNA to the metagenome? Does extracellular DNA contribute significantly to horizontal gene transfer through natural transformation in marine sediments? What is the fate of genes potentially released by genetically modified organisms (GMO) in the marine environment? What information is contained in the extracellular DNA preserved in the deeper sediment layers?

We do not know if the extracellular DNA pool contains functional sequences and we have no information on its origin. The extracellular DNA could represent a sort of "unexplored gene pool". Molecular studies specifically examining extracellular DNA from different environments could improve our comprehension of mechanisms controlling the persistence of nucleic acid molecules in the marine environment.

8
Analysis of Microbial Diversity

The origin of the molecular approach for identifying microorganisms can be traced back to the early work of Zuckerkandl and Pauling in the 1960s [121] and later to Woese's advances in microbial phylogeny [122]. Pace and coworkers [123, 124] are, however, considered to be the first to appreciate the power of molecular phylogeny for studying the diversity of microbial communities in the environment. Traditionally, the approaches for a taxonomic identification of bacteria were based on the identification of metabolic, phenotypic and physiological traits [124]. This required that bacteria had to be isolated on agar before being identified. The discovery that often less than 1% of the microbes in a sample can grow on agar plates, known as a central dogma in aquatic microbial ecology, demonstrated the need for new methodologies and approaches that were able to resolve the complex and diverse array of species making up the "black box" of marine microbial communities.

The most common approach for studying prokaryote diversity in the marine environment is based on the 16S rRNA gene. The rationale is to use a phylogenetic approach for establishing evolutionary relationships among microorganisms, and to use this as a framework for making inferences of community structure and biodiversity [125]. This target gene is particularly useful for studying microbial biodiversity [126] because it is present in all prokaryotes [127], and contains diagnostic variable regions (together with highly conserved regions), which are unique to specific populations or closely related groups [125]. Moreover, rRNA genes are thought to lack inter-specific horizontal gene transfer, in contrast to many other prokaryotic genes [127]. The methods based on 16S rRNA genes involve DNA extraction from the sample (sediment or seawater), followed by a PCR step using universal or specific primers (i.e. targeting all prokaryotic microorganisms or specific taxa), and then screening of the PCR products by means of one of the following techniques:

1. Cloning and sequencing [128, 129]: the PCR product is cloned into vectors and then randomly chosen clones are sequenced and their sequences aligned with those presented in databases. This permits the identification of prokaryotes, by assigning clones a phylogenetic identity.

2. DGGE (denaturing gradient gel electrophoresis [130]): an electrophoretic method allowing the separation of DNA fragments having the same length but different nucleotide sequences, which become visible as separate bands. These bands can be recovered for further cloning and sequencing. This method allows a molecular fingerprint of the microbial community in a sample to be obtained.
3. T-RFLP (terminal restriction fragment length polymorphisms [131]): a semi-quantitative technique allowing the estimation of the number and relative abundance of microbial ribotypes (often defined as OTU, operational taxonomic units) in a sample. In brief, the method involves a PCR amplification with fluorescently labelled primers, followed by digestion with restriction enzymes and screening of the number and types of restriction fragments. Screening and sizing of fragments can be performed on high-resolution (± 1 base) sequencing gels or on capillary electrophoresis systems providing digital outputs. Each fragment represents a single microbial ribotype. When compared with DGGE, the T-RFLP method has the advantage of being more sensitive and reproducible, but does not allow the recovery (and thus the cloning and sequencing) of the final product. Alternatively, using automated rRNA intergenic spacer analysis (ARISA) it is possible to track the presence and abundance of putative phylotypes over time, and compare community structures.
4. SSCP (single strand conformation polymorphisms [132]), ARDRA (amplified rDNA restriction analysis [133]) and heteroduplex mobility assay [134] are methods that are used less frequently.

In addition, real-time PCR allows the quantification of specific genes encountered in a sample. This methodology is based on the use of fluorescence reporters, which allow monitoring of the PCR reaction in a continuum.

Although the most utilized molecular marker for studying microbial diversity in the marine environment is the 16S rRNA gene for prokaryotes [135], 18S rRNA is increasingly used for eukaryotic microorganisms [136]. New genes have been proposed for studying microbial diversity in the marine environment, with special attention on microorganisms involved in specific biogeochemical processes [137–140]. For instance, much effort has been devoted to the study of denitrifying bacteria through functional genes, such as cd1-nir and Cu-nir genes, which encode for two forms of nitrite reductase [141].

The gene nosZ (encoding for nitrous oxide reductase) has been used as a molecular marker for studies of microbial diversity of benthic denitrifying bacteria [142]. Other genes such as nitrogenase reductase (NifH), cytochrome cd1-containing nitrite reductase (NirS), and Cu-containing nitrite reductase (NirK) have been used for studying the biodiversity of denitrifying and dinitrogen-fixing bacteria in terrestrial soils [143]. The conserved photosynthetic psbA gene (coding for the protein D1 of photosystem II reaction centre) has been utilized as a diversity indicator of marine oxygenic picophytoplankton, including cyanobacteria and eukaryotic algae [144]. The NH_3-monooxygenase subunit A gene (amoA) has been utilized for studying the diversity of ammonia-oxidizing bacteria [145]. Attempts to study the diversity of methanotrophic bacteria have been made by studying pmoA (a gene encoding the subunit of the particulate methane monooxygenase), mmoX (coding for subunits of soluble methane monooxygenase) and mxaF (methanol dehydrogenase; [146]).

The gene pufM (encoding the M subunit of the photosynthetic reaction centre) has been used for studying the diversity of anoxygenic phototrophs [147]. The diversity of autotrophic microorganisms can be assessed by studying genes encoding the ribulose-1,5-bisphosphate carboxylase/oxygenase (RuBisCO) enzyme, and attempts have been made using the ribulose bisphosphate carboxylase/oxygenase form I gene (rbcL; [148]) and the RuBisCO form II cbbM gene [149].

These new insights gained from the molecular approach are opening fresh ecological perspectives on microbial biodiversity in the marine environment and its interactions with ecosystem functioning, and not only for prokaryotes. Moreover, these techniques are providing important elements in the field of evolutionary biology [150].

The tree of life has been significantly revised after the recognition of Archaea as the third domain of life. A three-domain model, rather than one based on five kingdoms, has been proposed [151]. In a few years, the number of known major divisions within the two prokaryotic domains, Bacteria and Archaea, has doubled [152]. Indeed, 36 divisions have already been identified within the domain Bacteria, and 13 of them are known only from phylotypes [152].

Investigations based on 16S rRNA genes have revealed a previously unexpected microbial diversity in almost all aquatic ecosystems. Novel and yet-uncultured phylogenetic lineages have been discovered to be widely distributed in the marine environment [153]. For instance, among heterotrophic bacteria, those belonging to groups such as the Proteobacteria and the Cytophaga–Flavobacteria cluster have been shown to be extremely common in many oceanic habitats, accounting for as much as half of all bacteria identified with molecular microscopic techniques (i.e. FISH [154, 155]). Indeed, these bacteria have been demonstrated to be important consumers of dissolved organic matter in aquatic environments [156]. Molecular methods have shown how bacteria belonging to new and uncultured bacterial divisions, such as the SAR11 cluster (a phylogenetic group within the α-Proteobacteria) or W6, are often numerically important components of the marine picoplankton [129]. Members of the bacterial kingdom Acidobacterium, which has only one cultured member, has been found to be widespread, being present in most marine and freshwater sediments worldwide [157].

Archaea have been discovered only very recently to be common in marine ecosystems [158]. Initially known as Archebacteria [159], they represented until a decade ago a small group of atypical prokaryotes inhabiting unusual or extreme niches, such as those at high temperature, high salinity, extreme values of pH and/or in strictly anaerobic niches [160]. Recently, a wide number of studies, based on sequencing and comparison of 16S rRNA prokaryotic genes, have radically changed our view of the Archaea, revealing their ubiquitous distribution and their capability of thriving in aquatic and terrestrial temperate environments [160]. Studies of microbial diversity have revealed archaeal ribotypes to be a significant component of marine picoplankton assemblages [158, 161, 162], and even to dominate the mesopelagic prokaryotic communities of the north Pacific Ocean [163].

Culture-independent studies have shown that marine Archaea, which belong to the kingdom Crenarcheota (one of the three recognized kingdoms of the Archaeal domain), can be one of the most abundant cell types in the global ocean [163]. Archaea have been widely reported to also inhabit the benthic domain, including continental shelf anoxic sediments [164], freshwater sediments [165], and deep-sea and hydrothermal vent sediments [126, 160, 166]. Indeed, Archaea have been shown to possess important functional roles in marine carbon cycling. In this regard, Ouverney and Fuhrman [167] suggested that free-living plankton marine Archaea are involved in the heterotrophic uptake of dissolved amino acids, with activities comparable to those of their bacterial counterparts.

9
Concluding Remarks

Molecular tools will be increasingly useful in the future for gathering additional information on marine biodiversity and ecosystem functioning that traditional biogeochemical markers could not detect. For instance, lipid markers are useful tracers of organic matter sources [168]; in particular fatty acids having a great diversity of structures have been used as indicators of photosynthetic activity in surface water and phytoplankton taxonomic composition [169]. The relative abundances of individual fatty acids are also useful in evaluating the respective importance of inputs from bacteria, microalgae, marine fauna and continental higher plants [170]. Isotope analysis has been shown to be a powerful technique for distinguishing the sources of such compounds and tracing their metabolic pathways within organisms and food webs [171]. Pigment analysis by HPLC has also proven to be a valuable method for evaluating phytoplankton biomass, providing essential information regarding taxonomy, food-chain relationships, zooplankton grazing and detritus formation [172]. However, the fact that many fatty acid biomarkers and phytopigment products can originate from more than one source emphasizes the need for caution in assigning their biological origins based only on one single approach.

Analyses of nucleic acids through a variety of molecular techniques are a useful complement to all these biochemical approaches, and provide additional insights into the specific origin of DNA encountered in the ecosystem and into the functioning of microbes that play a key role in biogeochemical cycles. Moreover, molecular tools for the analysis of DNA in marine environments can be successfully applied to fossilized organic components, thus providing an archive of ancient aquatic microbial communities. Hence, they can be used to reconstruct variations in climate and their impacts on biodiversity. The combined stratigraphy on lipid and DNA analyses is opening the opportunity to reconstruct the paleo-microbiology and hence the paleo-ecosystem functioning with unprecedented detail [173].

References

1. Dell'Anno A, Marrale D, Pusceddu A, Danovaro R (1999) Mar Ecol Prog Ser 186:19
2. Maruyama A, Oda M, Higashihara T (1993) Appl Environ Microbiol 59:712
3. Weinbauer MG (2004) FEMS Microbiol Rev 28:127
4. Minear RA (1972) Environ Sci Technol 6:431
5. Pillai TNV, Ganguly AK (1972) J Mar Biol Assoc India 14:384
6. Paul JH, Jeffrey WH, DeFlaun MF (1987) Appl Environ Microbiol 53:170
7. Paul JH, DeFlaun MF, Jeffrey WH (1988) Appl Environ Microbiol 54:1682
8. Paul JH, David AW, DeFlaun MF, Cazares LH (1989) Appl Environ Microbiol 55:1823
9. Paul JH, Cazares LH, David AW, DeFlaun MF, Jeffrey WH (1991) Hydrobiologia 218:53
10. DeFlaun MF, Paul JH, Davis D (1986) Appl Environ Microbiol 52:654
11. DeFlaun MF, Jeffrey WH (1987) Mar Ecol Prog Ser 38:65
12. Wommack E, Colwell RR (2000) Microbiol Mol Biol Rev 64:69
13. Jiang SC, Paul JH (1995) Appl Environ Microbiol 61:317
14. Paul JH, Kellogg CA, Jiang SC (1994) Viruses and DNA in marine environments. In: Colwell RR, Simidu U, Ohwada K (eds) Microbial diversity in space and time. Plenum, New York, p 119
15. Weinbauer MG, Fuks D, Peduzzi P (1993) Appl Environ Microbiol 59:4074
16. Weinbauer MG, Peduzzi P (1995) J Plankton Res 17:1851
17. Paul JH, Jeffrey WH, Cannon JP (1990) Appl Environ Microbiol 56:2957
18. Paul JH, Cazares L, Thurmond JM (1990) Appl Environ Microbiol 56:1963
19. Jørgensen NOG, Kroer N, Coffin RB, Yang XH, Lee C (1993) Mar Ecol Prog Ser 98:135
20. Jørgensen NOG, Jacobsen CS (1996) Aquat Microb Ecol 11:263
21. Kroer N, Jørgensen NOG, Coffin RB (1994) Appl Environ Microbiol 60:4116
22. Turk V, Rehnstam AS, Lundberg E, Hagstrom A (1992) Appl Environ Microbiol 58:3744
23. Kolowith LC, Ingall ED, Benner R (2001) Limnol Oceanogr 46:309
24. Siuda W, Chrost RJ, Gude H (1998) Aquat Microb Ecol 15:89
25. Davison J (1999) Plasmid 42:73
26. Frischer ME, Stewart GJ, Paul JH (1994) FEMS Microbiol Ecol 15:127

27. Hermansson M, Linberg C (1994) FEMS Microbiol Ecol 15:47
28. Jeffrey WH, Paul JH, Stewart GJ (1990) Microb Ecol 19:259
29. Paul JH, Frischer ME, Thurmond JM (1991) Appl Environ Microbiol 57:1509
30. Bailiff MD, Karl DM (1991) Deep Sea Res 38:1077
31. Winn CD, Karl DM (1986) Limnol Oceanogr 31:637
32. Danovaro R, Dell'Anno A, Pusceddu A, Fabiano M (1999) Deep Sea Res I 46:1077
33. Dell'Anno A, Fabiano M, Duineveld GCA, Kok A, Danovaro R (1998) Appl Environ Microbiol 64:3238
34. Dell'Anno A, Bompadre S, Danovaro R (2002) Limnol Oceanogr 47:899
35. Dell'Anno A, Mei ML, Danovaro R (1999) Appl Environ Microbiol 65:4451
36. Crecchio C, Stotzky G (1998) Soil Biol Biochem 30:1061
37. Lorenz MG, Wackernagel W (1992) DNA binding to various minerals and retarded enzymatic degradation of DNA in a sand/clay microcosm. In: Gauthier MJ (ed) Gene transfer and environment. Springer, Berlin Heidelberg New York, p 103
38. Siuda W, Chrost RJ (2000) Aquat Microb Ecol 21:195
39. Romanowski G, Lorenz MG, Wackernagel W (1991) Appl Environ Microbiol 57:1057
40. Lorenz MG, Wackernagel W (1994) Microbiol Rev 58:563
41. Coolen MJ, Overmann J (1998) Appl Environ Microbiol 64:4513
42. Willerslev E, Hansen AJ, Binladen J, Brand TB, Thomas M, Gilbert P, Shapiro B, Bunce M, Wiuf C, Gilichinsky DA, Cooper A (2003) Science 300:791
43. Coolen MJ, Cypionka H, Sass AM, Sass H, Overmann J (2002) Science 296:2407
44. Pace NR (1997) Science 276:734
45. Fuhrman JA, Comeau DE, Hagstrom A, Chan AM (1988) Appl Environ Microbiol 54:1426
46. Sommerville CC, Knight IT, Straube WL, Colwell RR (1989) Appl Environ Microbiol 55:548
47. Barthel KG, Schneider G, Gradinger R, Lenz J (1989) J Plankton Res 11:1213
48. Jiang SC, Thurmond JM, Pichard SL, Paul JH (1992) Mar Ecol Prog Ser 80:101
49. Bernard L, Schaler H, Joux F, Courties C, Muyzer G, Lebaron P (2000) Aquat Microb Ecol 23:1
50. Fuchs BM, Zubkov MV, Sahm K, Burkill PH, Amann R (2000) Environ Microbiol 2:191
51. Kirchman DL, Yu L, Fuchs BN, Amann R (2001) Aquat Microb Ecol 26:13
52. Zubkov M, Fuchs BM, Archer SD, Kiene R, Amann R, Burkill PH (2001) Environ Microbiol 3:304
53. DeFlaun MF, Paul JH (1989) Microb Ecol 18:21
54. Karl DM, Bailiff MD (1989) Limnol Oceanogr 34:543
55. Paul JH, DeFlaun MF, Jeffrey WH, David AW (1988) Appl Environ Microbiol 54:718
56. Paul JH, David AW (1989) Appl Environ Microbiol 55:1865
57. Paul JH, Pichard SL (1989) Appl Environ Microbiol 55:2798
58. Siuda W, Gude H (1996) Aquat Microb Ecol 11:193
59. Dell'Anno A (1999) PhD thesis, University of Messina
60. Dell'Anno A, Danovaro R (2001) Nucleic acid turnover in aquatic environments. 1. Determination of total and extracellular DNA in marine sediments. In: Akkermans ADL, van Elsas JD, De Bruijn FJ (eds) Molecular microbial ecology manual. Kluwer, Dordrecht, p 1–9
61. Hurt RA, Qiu X, Wu L, Roh Y, Palumbo AV, Tiedje JM, Zhou J (2001) Appl Environ Microbiol 67:4495
62. Juniper SK, Cambon MA, Lesonger F, Barbier G (2001) Mar Geol 174:241
63. Rose-Amseleg CL, Garnier-Sillam E, Harry M (2001) Appl Soil Ecol 18:47

64. Zhou J, Bruns MA, Tiedje JM (1996) Appl Environ Microbiol 62:316
65. Martin-Laurent F, Philippot L, Hallet S, Chaussod R, Germon JC, Soulas G, Catroux G (2001) Appl Envir Microbiol 67:2354
66. Miller DN, Bryant JE, Madsen EL, Ghiorse WC (1999) Appl Environ Microbiol 65:4715
67. Atlas R (1993) Extraction of DNA from soils and sediments. In: Kemp P, Sherr B, Sherr E, Cole JJ (eds) Handbook of methods in aquatic microbial ecology. Lewis, Boca Raton, p 261
68. Riis V, Lorbeer H, Babel W (1998) Soil Biol Biochem 30:1573
69. Steffan RJ, Goksoyr J, Bej AK, Atlas RM (1988) Appl Environ Microbiol 54:2908
70. Kuske CR, Banton KL, Adorada DL, Stark PC, Hill KK, Jackson PJ (1998) Appl Environ Microbiol 64:2463
71. Rochelle PA, Fry JC, Parkes RJ, Weightman AJ (1992) FEMS Microbiol Lett 100:59
72. Tsai Y, Olson BH (1991) Appl Environ Microbiol 57:1070
73. Bruce KD, Horns WD, Hobman JL, Osborn AM, Strike P, Ritchie DA (1992) Appl Environ Microbiol 58:3413
74. Moré MI, Herrick JB, Silva MC, Ghiorse WC, Madsen EL (1994) Appl Environ Microbiol 60:1572
75. Leff LG, Dana JR, McArthur JV, Shimkets LJ (1995) Appl Environ Microbiol 61:1141
76. Smalla K, Cresswell N, Mendonca-Hagler LC, Wolters A, van Elsas JD (1993) J Appl Bacteriol 74:78
77. Ogram A, Sayler GS, Barkay T (1987) J Microbiol Methods 7:57
78. Holben WE (1997) Isolation and purification of bacterial community DNA from environmental samples. In: Hurst CJ, Knudsen GR, McInerney MJ, Stetzenbach LD, Walter MV (eds) Manual of methods in environmental microbiology. ASM, Washington, DC, p 431
79. MacGregor BJ, Moser DP, Wheeler Alm E, Nealson KH, Stahl DA (1997) Appl Environ Microbiol 63:1178
80. Ogram A (1998) Isolation of nucleic acids from environmental samples. In: Burlage RS, Atlas R, Stahl D, Geesey G, Sayler G (eds) Techniques in microbial ecology. Oxford University Press, New York, p 273
81. Holben WE (1994) Isolation and purification of bacterial DNA from soil. In: Weaver RW, Angle S, Bottomley P, Bezdicek D, Smith S, Tabatabai A, Wollum A (eds) Methods of soil analysis, part 2. Microbiological and biochemical properties. Soil Science Society of America, Madison, p 727
82. Trevors JT, Lee H, Cook S (1992) Microb Releases 1:111
83. Erb RW, Wagner-Döbler I (1993) Appl Environ Microbiol 59:4065
84. Herrick JB, Miller DN, Madsen EL, Ghiorse WC (1996) Extraction, purification, and amplification of microbial DNA from sediments and soils. In: Burke JF (ed) PCR: essential techniques. Wiley, New York, p 130
85. Jacobsen CS, Rasmussen OF (1992) Appl Environ Microbiol 58:2458
86. Malik M, Kain J, Pettigrew C, Ogram A (1994) J Microbiol Methods 20:183
87. Smith GB, Tiedje JM (1992) Appl Environ Microbiol 58:376
88. Tebbe CC, Vahjen W (1993) Appl Environ Microbiol 59:2657
89. DeGrange V, Bardin R (1995) Appl Environ Microbiol 61:2093
90. Young CC, Burghoff RL, Keim LG, Minak-Bernero V, Lute JR, Hinton SM (1993) Appl Environ Microbiol 59:1972
91. Tsai Y, Olson BH (1992) Appl Environ Microbiol 58:2292
92. Frostegard A, Courtois S, Ramisse V, Clerc S, Bernillon D, Le Gall F, Jeannin P, Nesme X, Simonet P (1999) Appl Environ Microbiol 65:5409

93. Corinaldesi C, Danovaro R, Dell'Anno A (2004) Appl Environ Microbiol 71:46
94. Holben WE, Jansson JK, Chelm BK, Tiedje JM (1988) Appl Environ Microbiol 54:703
95. Azam F (1998) Science 280:694
96. Hoppe HG (1991) Microbial extracellular enzyme activities: new key parameter in aquatic ecology. In: Chrost RJ (ed) Microbial enzymes in aquatic environments. Springer, Berlin Heidelberg New York, p 60
97. Keith SC, Arnosti C (2001) Aquat Microb Ecol 24:243
98. Dubnau D (1999) Annu Rev Microbiol 53:217
99. Ammerman JW, Azam F (1991) Limnol Oceanogr 36:1437
100. Novitsky JA (1986) Appl Environ Microbiol 52:504
101. Phillipps SJ, Dalgarn DS, Young SK (1989) J Water Pollut Control Fed 61:1588
102. Kaeberlein T, Lewis K, Epstein SS (2002) Science 296:1127
103. Dell'Anno A, Corinaldesi C (2004) Appl Environ Microbiol 70:4384
104. Chabbert M, Cazenave C, Helene C (1987) Biochemistry 26:2218
105. Takahashi M, Ling C (1991) Anal Biochem 198:246
106. Bouvier T, delGiorgio P (2002) FEMS Microbiol Ecol 1485:1
107. Amann RL, Ludwig W, Schleifer K (1995) Microbiol Rev 59:143
108. Mullis KB, Faloona F (1987) Methods Enzymol 155:335
109. Wilson IG (1997) Appl Environ Microbiol 63:3741
110. Suzuki MT, SJ Giovannoni (1996) Appl Environ Microbiol 62:625
111. Gotz A, Pukall R, Smit E, Tietze E, Prager R, Tschape H, van Elsas JD, Smalla K (1996) Appl Environ Microbiol 62:2621
112. Smalla K, Krogerrecklenfort E, Heuer H, Dejonghe W, Top E, Osborn M, Niewint J, Tebbe C, Barr M, Bailey M, Greated A, Thomas C, Turner S, Young P, Nikolakopoulou D, Karagouni A, Wolters A, van Elsas JD, Drønen K, Sandaa R, Borin S, Brabhu J, Grohmann E, Sobecky P (2000) Microbiology 146:1256
113. Sobecky PA, Mincer TJ, Chang MC, Helinski DR (1997) Appl Environ Microbiol 63:888
114. Ball KD, Trevors JT (2002) J Microbiol Methods 49:275
115. DeRisi JL, Iyer VR, Brown PO (1997) Science 278:680
116. Futcher B (2000) Curr Opin Cell Biol 12:710
117. Wu L, Thompson DK, Li G, Hurt RA, Tiedje JM, Zhou J (2001) Appl Environ Microbiol 67:5780
118. Rondon MR, August PR, Bettermann AD, Brady SF, Grossman TH, Liles MRL, Loiacono KA, Lynch BAL, MacNeil AI, Minor C, Tiong CL, Gilman M, Osburne MS, Clardy J, Handelsman J, Goodman RM (2000) Appl Environ Microbiol 66:2541
119. Voget S, Leggewie C, Uesbeck A, Raasch C, Jaeger KE, Streit WR (2003) Appl Environ Microbiol 69:6235
120. Schmeisser C, Stöckigt C, Raasch C, Wingender J, Timmis KN, Wenderoth DF, Flemming HC, Liesegang H, Schmitz RA, Jaeger KE, Streit WR (2003) Appl Environ Microbiol 69:7298
121. Zuckerkandl E, Pauling L (1965) J Theor Biol 8:357
122. Woese CR (1987) Microbiol Rev 51:221
123. Olsen GJ, Lane DJ, Giovannoni SJ, Stahl DA, Pace NR (1986) Annu Rev Microbiol 40:337
124. Pace NR, Stahl DA, Lane DJ, Olsen GJ (1986) Adv Microb Ecol 9:1
125. Kent AD, Triplett EW (2002) Annu Rev Microbiol 56:211
126. Moyer CL, Tiedje JM, Dobbs FC, Karl DM (1998) Deep Sea Res 45:303
127. Ward DM, Bateson MM, Weller R, Ruff-Roberts AL (1992) Adv Microb Ecol 12:219
128. Boucher Y, Doolittle WF (2002) Nature 417:27

129. Giovannoni SJ, Britschgi TB, Moyer CL, Field KG (1990) Nature 345:60
130. Muyzer G, de Waal EC, Uitterlinden AG (1993) Appl Environ Microbiol 59:695
131. Liu WT, Marsh TL, Cheng H, Forney LJ (1997) Appl Environ Microbiol 63:4516
132. Schwieger F, Tebbe CC (1998) Appl Environ Microbiol 64:4870
133. Massol-Deya A, Weller R, Rios-Hernandez L, Zhou JZ, Hickey RF, Tiedje JM (1997) Appl Environ Microbiol 63:270
134. Espejo RT, Feijoo CG, Romero J, Vasquez M (1998) Microbiology 144:1611
135. Head IM, Saunders JR, Pickup RW (1998) Microb Ecol 35:1
136. López-García P, Rodríguez-Valera F, Pedrós-Alió C, Moreira D (2001) Nature 409:603
137. Cary SC, Chisholm SW (2000) Ecological genomics: the application of genomic sciences to understanding the structure and function of marine ecosystems. Report of a workshop on marine microbial genomics to develop recommendations for the National Science Foundation, p 20
138. Paul JH, Pichard SL, Kang JB, Watson GMF, Tabita F (1999) Limnol Oceanogr 44:12
139. Paul JH, Alfreider A, Kang JB, Stokes RA, Griffin D, Campbell L, Ornolfsdottir E (2000) Mar Ecol Prog Ser 198:1
140. Paul JH, Alfreider A, Wawrik B (2000) Mar Ecol Prog Ser 198:9
141. Hallin S, Lindgren PE (1999) Appl Environ Microbiol 65:1652
142. Scala DJ, Kerkhof LJ (2000) Appl Environ Microbiol 66:1980
143. Rösch C, Mergel A, Bothe H (2002) Appl Environ Microbiol 68:3818
144. Zeidner G, Preston CM, Delong EF, Massana R, Post AF, Scanlan DJ, Béjà O (2003) Environ Microbiol 5:212
145. Mintie AT, Heichen RS, Cromack K Jr, Myrold DD, Bottomley PJ (2003) Appl Environ Microbiol 69:3129
146. Horz HP, Yimga MT, Liesack W (2001) Appl Environ Microbiol 67:4177
147. Achenbach LA, Carey J, Madigan MT (2001) Appl Environ Microbiol 67:2922
148. Pichard SL, Campbell L, Paul JH (1997) Appl Environ Microbiol 63:3600
149. Elsaied H, Naganuma T (2001) Appl Environ Microbiol 67:1751
150. Fuhrman JA, Lee SH, Masuchi Y, Davis AA, Wilcox RM (1994) Microb Ecol 28:133
151. Woese CR, Kandler O, Wheelis ML (1990) Proc Natl Acad Sci USA 87:4576
152. Hugenholtz P, Goebel BM, Pace NR (1998) J Bacteriol 180:4765
153. Schmidt TM, DeLong EF, Pace NR (1991) J Bacteriol 173:4371
154. Kirchman DL (2002) FEMS Microbiol Ecol 39:91
155. Kirchman DL, Yu L, Cottrell MT (2003) Appl Environ Microbiol 69:6587
156. Cottrell MT, Kirchman DL (2000) Appl Environ Microbiol 66:1692
157. Barnes SM, Takala SL, Kuske CR (1999) Appl Environ Microbiol 65:1731
158. DeLong EF (1992) Proc Natl Acad Sci USA 89:5685
159. Balch WE, Mangrum IJ, Fox GE, Wolfe RS, Woese CR (1977) J Mol Evol 9:305
160. Vetriani C, Jannasch HW, MacGregor BJ, Stahl DA, Reysenbach AL (1999) Appl Environ Microbiol 65:4375
161. DeLong EF, Ying Wu K, Prezelin BB, Jovine RVM (1994) Nature 371:695
162. Fuhrman JA, McCallum K, Davis AA (1992) Nature 356:148
163. Karner MB, DeLong EF, Karl DM (2001) Nature 409:507
164. Vetriani C, Reysenbach AL, Doré J (1998) FEMS Microbiol Lett 161:83
165. Hershberger KL, Barns SM, Reysenbach AL, Dowson SC, Pace NR (1996) Nature 384:420
166. Takai K, Horikoshi K (1999) Genetics 152:1285
167. Ouverney CC, Fuhrman JA (2000) Appl Environ Microbiol 66:4829
168. Wakeham S, Hedges JI, Lee C, Peterson ML, Hernes PJ (1997) Deep Sea Res II 44:2131

169. Shin KH, Hama T, Yoshie N, Noriki S, Tsunogai S (2000) Mar Chem 70:243
170. Sargent JR, Parkes RJ, Mueller-Harvey I, Henderson RJ (1987) Lipid biomarkers in marine ecology. In: Sleight MA (ed) Microbes in the sea. Wiley, New York, p 119–138
171. Hayes JM, Freeman KH, Popp BN, Hoham CH (1990) Org Geochem 16:1115
172. Claustre H, Marty JC, Cassiani L, Dagaut J (1989) Mar Microb Food Webs 3:51
173. Coolen MJL, Muyzer G, Rijpstra WIC, Schouten S, Volkman JK, Sinninghe Damsté JS (2004) Earth Planet Sci Lett 223:225

Biological Markers for Anoxia in the Photic Zone of the Water Column

J. S. Sinninghe Damsté (✉) · S. Schouten

Department of Marine Biogeochemistry and Toxicology, Royal Netherlands Institute for Sea Research, PO Box 59, 1790 AB Den Burg, The Netherlands
damste@nioz.nl, schouten@nioz.nl

1	Introduction	128
2	Taxonomy, Physiology, and Ecology of Anoxygenic Phototrophic Sulfur Bacteria	130
3	Specific Pigments of Green and Purple Sulfur Bacteria	132
4	Carbon Isotopic Fractionation of Anoxygenic Phototrophic Bacteria	136
5	Photic Zone Anoxia in Recent Depositional Environments and its Manifestation in the Sedimentary Record	137
5.1	Mahoney Lake	137
5.2	Ace Lake	139
5.3	Kyllaren Fjord	141
5.4	Black Sea	141
6	Ancient Sediments Containing Intact Pigments from Phototrophic Sulfur Bacteria	144
7	Dia- and Catagenesis of Carotenoid Markers for Photic Zone Anoxia	145
7.1	Hydrogenation	145
7.2	Cyclization and Aromatization	148
7.3	Aromatization of the Cyclohexenyl Moiety	149
7.4	Intramolecular Sulfurization	150
7.5	Incorporation into Macromolecular Organic Matter	150
7.6	Expulsion of Toluene and Xylene	151
7.7	C–C Bond Cleavage	152
8	Dia- and Catagenesis of Bacteriochlorophyll Markers for Photic Zone Anoxia	152
9	Evidence for Photic Zone Anoxia in Ancient Depositional Environments	153
10	The Fossil Record of Green and Purple Sulfur Bacteria	157
11	Conclusions	159
	References	159

Abstract In this chapter we review the current state of knowledge on the occurrence of anoxic conditions in the water column of past depositional systems using biomarkers. The recognition of such conditions is important for a better understanding of the deposition of petroleum source rocks and periods of widespread water column anoxia in the geological past (so-called oceanic anoxic events). The most reliable biomarkers for this purpose are specific pigments and bacteriochlorophylls derived from photosynthetic green and purple sulfur bacteria, which require both light and sulfide. These proxies for water column anoxia have been well tested in present-day lakes, fjords and stratified basins. With increasing burial in the sediment, the pigments and bacteriochlorophylls undergo a myriad of transformation reactions, but the products are in most cases still specific enough to use them as environmental indicators. In this way, the occurrence of photic zone euxinia has been demonstrated for a wide variety of settings (e.g. shelf seas, enclosed basins, open ocean systems) and during geological eras as old as the Precambrian.

Keywords Ancient depositional environments · Green sulfur bacteria · Molecular fossils · Sediments

Abbreviations

Fmo	Fenna–Matthews–Olsen protein
OM	organic matter
OC	organic carbon
GSB	green sulfur bacteria
PSB	purple sulfur bacteria
TCA	tricarboxylic acid
ATP	adenosine triphosphate
$\Delta\delta^{13}C$	$\delta^{13}C$ of the cell material relative to the $\delta^{13}C$ of CO_2
ε_p	isotopic fractionation factor
HPLC-PDA-MS	high performance liquid chromatography-photo diode array-mass spectroscopy
NMR	nuclear magnetic resonance
TMB	1,2,3,4-tetramethylbenzene
Me,Et maleimide	methyl ethyl maleimide
n-Pr	normal-propyl
i-Bu	iso-butyl
OAE	oceanic anoxic event
ODP	Ocean Drilling Project

1
Introduction

Vast amounts of organic matter (OM) have been buried in sediments over geological time, but only a small fraction has been transformed into petroleum. Today's human population relies critically on this geological transformation product of sedimentary OM for their ever-increasing energy demand. In the present-day oceans hot-spots for accumulation of OM sediments are restricted to those highly productive shelf areas that have a large input of nutrients, either by upwelling or input from rivers, or to anoxic basins such as the Black Sea [1]. It is believed that anoxic conditions of the water column play an important role in the preservation of oil-prone organic matter in sediments because aerobic degradation of OM is generally thought to be more efficient than anaerobic degradation [1]. Primary productivity [2] and sedimentation rate [3] may also exert an important control on OC accumulation rates, but in both cases there is a direct link to anoxia. Increased primary productivity increases the oxygen demand in the water column and pore waters of surface sediments and can ultimately induce anoxia. Sediments deposited in the oxygen minimum zone of the Arabian Sea have higher OC accumulation rates than those deposited above and below the oxygen minimum zone, even though the primary productivity in the photic zone is the same [4]. Hedges and co-workers [5, 6] showed that the time OM is exposed to oxygen (i.e. oxygen exposure time) is the primary control on the flux of OM to the deeper sediment and in this way explained the observed relationship with sedimentation rate.

There are indications that in the geological past the deposition of OC-rich sediments was more widespread during specific time periods [7] and also in settings (e.g. shelf seas and open marine basins) that are quite different from those in the present-day ocean where we find high OC accumulation rates. In these cases, it becomes more difficult to understand how petroleum source rocks were formed since present-day analogs do not exist. Geochemists, therefore, try to reconstruct the depositional conditions of these petroleum source rocks by using specific indicators (so-called proxies) for anoxic depositional conditions. For example, some trace metals (e.g. V, Ag and Cd) are selectively enriched relative to Al under anoxic conditions and can be used to trace anoxic conditions in past depositional systems [8]. Specific biomarkers (i.e. organic compounds found in sediments and oils which can unambiguously be related to specific lipids derived from characteristic organisms or groups of organisms) derived from organisms thriving under anoxic conditions can also be used [9].

It is often difficult to discriminate between environments having anoxic bottom waters and those in which only the pore waters of the sediment are anoxic since the microorganisms and processes in anoxic bottom waters and anoxic shallow sediments are often the same. Therefore, geochemists are interested in proxies that can be unambiguously related to water column anoxia. These are now available from the proxies derived from phototrophic anoxygenic bacteria which thrive in anoxic waters receiving light. In shallow aquatic settings such communities can be found at the water/sediment interface, but in most settings light does not reach the bottom and, consequently, the presence of photosynthetic anoxygenic bacteria indicates an almost complete anoxic water column.

Although cellular remains of photosynthetic anoxygenic bacteria are usually not found in sediments, specific organic components (i.e. carotenoids and bacteriochlorophylls) are, albeit often in a diagenetically altered form, and thus provide information on the redox state of the environment at the time of deposition. These biomarkers (the diagenetic products of characteristic natural products) have been found in sediments as old as the Paleoproterozoic and can even be traced back in petroleum. Here, we review the current state of knowledge with respect to biomarkers for photic zone anoxia in the water column. In this review we will limit ourselves to two important microbial groups: the green and purple sulfur bacteria and, specifically, to those that contain characteristic organic components that can be retrieved from the sedimentary record.

2
Taxonomy, Physiology, and Ecology of Anoxygenic Phototrophic Sulfur Bacteria

Phototrophic sulfur bacteria develop in anoxic environments in which light and reduced sulfur compound occur simultaneously. For this review two groups are important: the green sulfur bacteria (GSB; the family *Chlorobiaceae*) and the purple sulfur bacteria (PSB). This latter group is comprised of two families: the Chromatiaceae and the Ectothiorhodospiraceae. This latter family, however, does not contain characteristic organic compounds which can be traced back in the geological record and will not be considered here. There exists a variety of reviews on the microbiology of these groups [10–13]. Here we will only briefly describe some relevant aspects.

The GSB represent a phylogenetically isolated group within the domain Bacteria. They are presently divided into six genera: *Chlorobium, Ancalochloris, Chlorobaculum, Chloroherpeton, Pelodictyon*, and *Prosthecochloris* [14]. This classification was until recently based on morphology, motility and the presence or absence of gas vesicles in the cell [11]. Recent phylogenetic characterization using the 16S ribosomal DNA gene and fmo (Fenna–Matthews–Olsen protein) genes, however, indicated that the conventional taxonomy was not congruent with the phylogenetic tree: species from different genera were phylogenetically closely related, whereas some species from the same genus varied substantially in the 16S rDNA and fmo gene composition, requiring a substantial revision of the conventional taxonomy [15].

The Chromatiaceae family of PSB is presently comprised of 24 genera [14]. Together they form a well-defined phylogenetic cluster in the gamma proteobacteria group [10] that is quite closely related to the other family of PSB, the *Ectorhodospiraceae*. Within the Chromatiaceae a major phylogenetic branch contains only marine and halophilic species.

GSB and PSB require both light and a reduced form of sulfur, either sulfide or thiosulfate for growth. The GSB are non-motile, strictly anaerobic, and obligately phototrophic bacteria. Some GSB contain gas vacuoles which can be used to control buoyancy. They are autotrophic as they use carbon dioxide as their sole carbon source. Some simple organic components (e.g. acetate) can also be photoassimilated. Sulfide is used as an electron donor and is oxidized to sulfate. Depending on the sulfide concentration and light intensity, elemental sulfur may be formed as an intermediate product. The *Chromatiaceae* family of PSB form elemental sulfur as the oxidation product from sulfide or thiosulfate and store this internally in their cell. Because they are able to oxidize this further photoautotrophically to sulfate, the presence of elemental sulfur inside their cell provides the cell with a reservoir of photosynthetic electrons. Two major physiological groups are discriminated within the PSB. One group is comprised of strict anaerobes which are inhibited by oxygen, are obligately phototrophic and which require sulfide. Some simple organics (acetate and pyruvate) can be photoassimilated if sulfide and CO_2 are present. The other group is metabolically much more versatile and can photoassimilate organic substrates and use organic substrates as electron donors in the absence of sulfide.

GSB and PSB form dense layers in a variety of aquatic ecosystems such as lakes, lagoons, stratified marine basins and sediments receiving light. They are predominantly mesophilic although some thermophilic species (e.g. *Chlorobaculum tepidum, Chromatium tepidum*) are known. In an overview, van Gemerden and Mas [16] reported cell densities for over 60 different lakes and 7 sediment ecosystems containing phototrophic sulfur bacteria. Densities between 10^4 and 10^7 cells ml^{-1} and bacteriochlorophyll concentrations of $10-1000 \mu g L^{-1}$ are common in these settings. *Chromatiaceae* are typically found in freshwater and marine environments but not in hypersaline waters, where their sister group, the *Ectothiorhodospiraceae*, are dominant. Because of their metabolic requirements (see above), the phototrophic sulfur bacteria are limited to those environments where light reaches anoxic, sulfide-containing bottom layers. Because light and sulfide usually occur in opposing gradients, a niche of PSB and GSB only exists if there is a zone of overlap and this is only possible if the chemical gradient of sulfide is stabilized against vertical mixing.

In pelagic environments such as lakes, lagoons, fjords and stratified marine basins, chemical gradients are often stabilized by density differences between the oxic and anoxic water layers. Such density differences are either the result of thermal (and thus seasonal) stratification or substantial differences in the salt concentration of the surface and bottom waters. In these stratified systems, bottom waters become oxygen-depleted due to the remineralization of descending organic matter-containing particles by heterotrophic bacteria. If all the oxygen is consumed, sulfate-reducing bacteria produce sulfide, resulting in a plate of phototrophic sulfur bacteria in the water column of lakes and stratified marine basins which can extend over vertical distances of 10 cm [16, 17] up to 30 m [18], depending on the steepness of the chemical and light gradients. Although much more common in present-day stratified lakes and fjords, layers of phototrophic sulfur bacteria can also develop in stratified marine basins such as the Black Sea [18]. As we will discuss below, such systems may have been much more common in the geological past.

3
Specific Pigments of Green and Purple Sulfur Bacteria

The characteristic organic components found in GSB and PSB are directly related to their distinct physiology, i.e. anoxygenic photosynthesis. To absorb light they use chlorophylls and accessory pigments (carotenoids). Since the GSB and PSB occupy distinct ecological niches and receive light in much lower intensities and with a different electromagnetic spectrum than at the surface, they biosynthesize different pigments from those in oxygenic phototrophs, i.e. algae and cyanobacteria living in the upper part of the photic zone.

The PSB are characterized by the presence of bacteriochlorophyll-*a* or, in some cases, bacteriochlorophyll-*b* (see Scheme 1 for chlorophyll structures) and a number of characteristic carotenoids (Table 1) of which okenone (1, see Scheme 2 for carotenoid structures), an aromatic carotenoid, is of special geochemical interest since it is highly specific for PSB and can be traced through the geological record. This carotenoid has been reported in at least eight different genera of PSB (Table 1) and occurs in species from various branches in the phylogenetic tree of the *Chromatiaceae*. From the analysis of sediments of Lake Cadagno, a Swiss alpine stratified lake, Schaeffer et al. [19] suggested two other aromatic carotenoids (2 and 3) with the 2,3,4-trimethyl aromatic substitution pattern[1] also found in okenone. Their carbon isotopic composition suggests that they derive from *Chromatiaceae*, thriving at the shallow chemocline of Lake Cadagno. However, these pigments have not been identified so far in cultures of PSB.

The GSB contain a distinct suite of pigments: bacteriochlorophyll-*c*, -*d* or -*e* in combination with the aromatic carotenoids chlorobactene (5) and isorenieratene (7) (Table 1), sometimes in combination with smaller amounts of β-isorenieratene (6) and β-carotene. All these aromatic carotenoids are characterized by the 2,3,6-trimethyl substitution pattern of the aromatic ring. The distinct pigment composition of GSB results from their ecological niche which differs from that of the PSB; GSB can grow at lower light intensities than the PSB as can be clearly seen in anoxic, stratified lakes where often a purple bacterial "plate" (representing the PSB) overlays a green plate (representing the GSB) at the chemocline. Consequently, GSB also receive light

---

[1] Throughout this review we will use, for reasons of convenience, this numbering system and apply it also for diagenetic products of aromatic carotenoids even if this results in nomenclature not in agreement with general IUPAC rules.

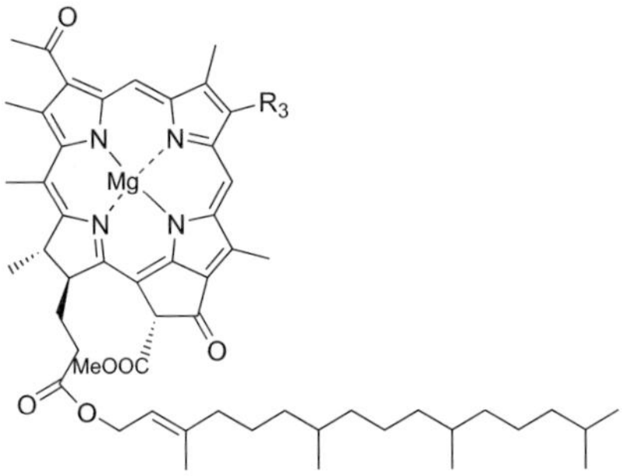

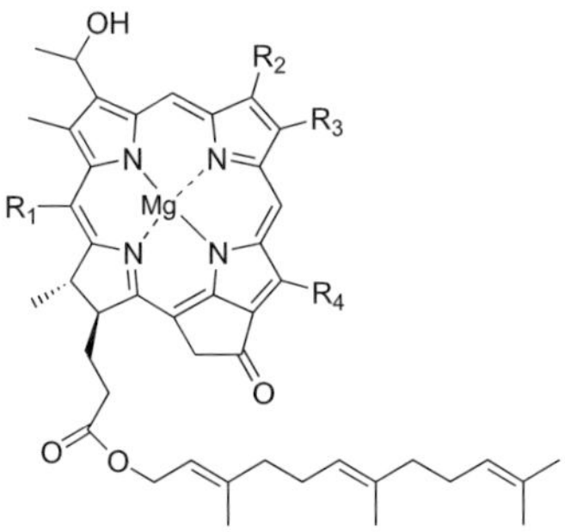

Biological Markers for Anoxia in the Photic Zone of the Water Column

bacteriochlorophyll a R_3 = ethyl
bacteriochlorophyll b R_3 = vinyl

bacteriochlorophyll c R_1 = Me, R_2 = Me, R_3 = Et, Pr, i-Bu, R_4 = Me, Et
bacteriochlorophyll d R_1 = H, R_2 = Me, R_3 = Et, Pr, i-Bu, R_4 = Me, Et
bacteriochlorophyll e R_1 = Me, R_2 = Me, R_3 = CHO, Pr, i-Bu, R_4 = Et

Scheme 1 Bacteriochlorophyll's of GSB and PSB

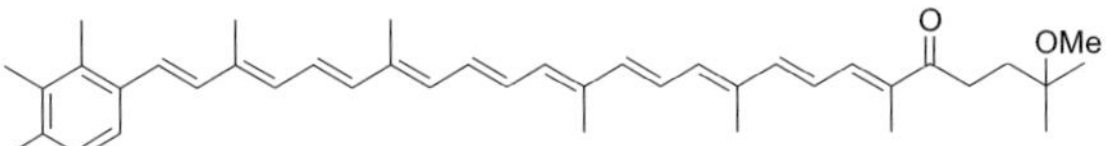

1. Okenone
2. 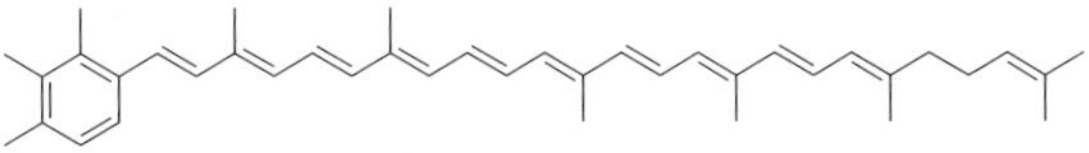
3. 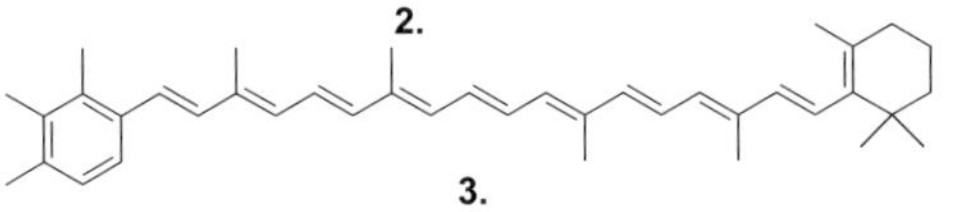
4. Renierapurpurin 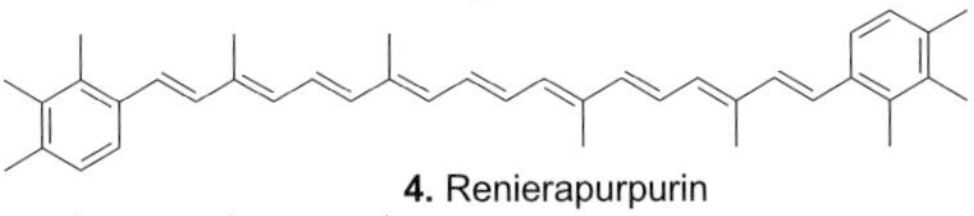
5. Chlorobactene 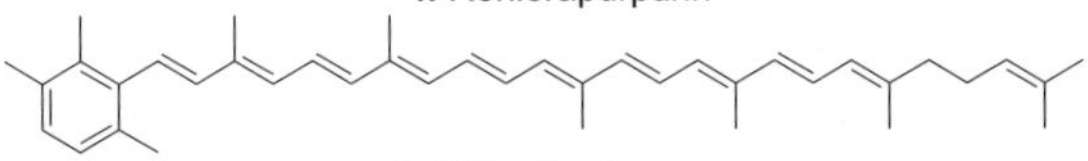
6. β-isorenieratene 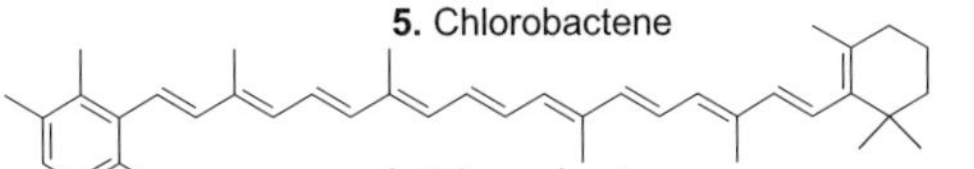
7. isorenieratene 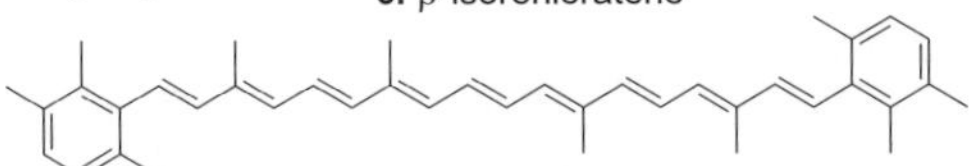
8. renieratene 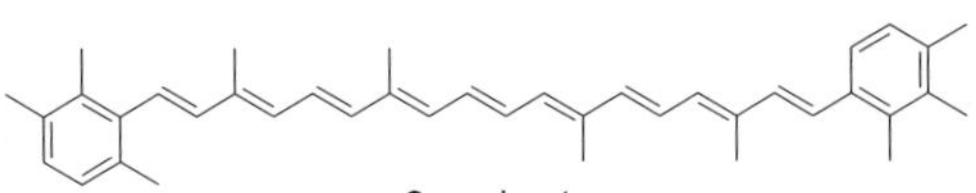
9. 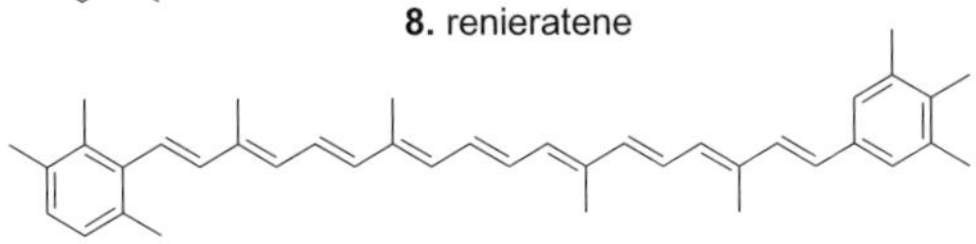

Scheme 2 Aromatic carotenoids

Table 1 Pigment composition of the major species[a] of PSB

Genus[a]	Species	Bacterio-chlorophyll	Major carotenoids[b]	Refs
Chromatium	C. okenii	a	Ok	[10]
	C. weissei	a	Ok	[10]
Allochromatium	A. vinosum	a	Sp, Ly, Rh	[10, 99]
	A. minutissimum	a	Sp, Ly, Rh	[10, 99]
	A. warmingii	a	Rl	[10, 99]
Halochromatium	H. salexigens	a	Sp, Ly, Rh	[10, 99]
	H. glycolicum	a	Sp	[10]
Isochromatium	I. buderi	a	Rl	[10]
Lamprobacter	L. modestohalophilus	a	Ok	[10]
Lamprocystis	L. roseopersicina	a	Rl	[10]
	L. purpurea	a	Ok	[99]
Marichromatium	M. gracile	a	Sp, Ly, Rh	[10, 99]
	M. purpuratum	a	Ok	[10, 99]
Rhabdochromatium	R. marinum	a	Ly	[10]
Thermochromatium	T. tepidum	a	Sp, Ly, Rh, Rv	[10, 99]
Thioalkalicoccus	T. limnaeus	a	Ts	[10]
Thiobaca	T. trueperi	a	Ly	[10]
Thiocapsa	T. roseopersicina	a	Ok, Sp	[10, 99]
	T. rosea	a	Sp	[10, 99]
	T. pendens	a	Sp	[10]
	T. litoralis	a	Sp	[10]
	T. marina	a	Ok	[100]
	T. sp.	a	Sp	[99]
Thiococcus	T. pfennigii	a	Ts	[10]
Thiocystis	T. gelatinosa	a	Ok	[10, 99]
	T. minor	a	Ok	[10, 99]
	T. violacea	a	Rl	[10, 99]
	T. violascens	a	Rl	[10]
Thiodictyon	T. elegans	a	Rl	[10]
	T. bacillosum	a	Rl	[10]
Thioflavicoccus	T. mobilis	a	Ts	[10]
Thiohalocapsa	T. halophila	a	Ok	[10, 99]
Thiolamprovum	T. pedioforme	a	Sp	[10]
Thiopedia	T. rosea	a	Ok	[10]
Thiorhodococcus	T. minor	a	Sp	[10]
Thiorhodovibrio	T. winogradskyi	a	Sp	[10]
Thiospirillum	T. jenense	a	Rh, Ly	[10]

[a] Classification according to Bergeys 2004;
[b] Key: Ok = okenone (1), Sp = spirilloxanthin, Ly = lycopene, Rh = rhodopin, Rv = rhodovibrin, Rl = rhodopinal, Ts = tetrahydro-spirilloxanthin [10, 99, 100]

with a different electromagnetic spectrum and have a different pigment composition accordingly. Both the specific bacteriochlorophylls (c, d and e) as well as the aromatic carotenoids are able to absorb light in the maxima of the electromagnetic spectrum of light in water at greater depth.

Within the GSB, there is a profound distinction in pigment composition between the green and brown colored GSB (Table 2). The green colored strains contain bacteriochlorophyll-c or -d and chlorobactene as the major pigments, whereas the brown colored species contain bacteriochlorophyll-e and isorenieratene. This distinction does not follow the molecular phylogeny and taxonomy of the GSB; i.e. some single species have both green and brown strains (Table 2). However, there is a clear relationship with the ecology of the strains; brown-colored GSB are often found in niches receiving even lower light intensities than those of the green colored GSB and, consequently, the specific pigment composition of the brown colored strains is interpreted as a further adaptation to reduced light intensity and alteration of the electromagnetic spectrum. With this specific set of pigments a strain of *Chlorobium phaeobacteroides* is capable of performing photosynthesis at a depth of ca. 80 m in the water column of the Black Sea [20–22]. GSB can also be part of so-called phototrophic con-

Table 2 Pigment composition of the major species[a] of GSB

Genus[a]	Species	Color	Bacterio-chlorophyll	Major carotenoids[b]	Refs
Chlorobium	C. limicola	Green	c or d	Cb	[11, 15, 99]
		Brown	e	Iso	[15]
	C. clathratiforme	Green	c or d	Cb	[11]
		Brown	?	Iso	[15]
	C. luteolum	Green	c or d	Cb	[11, 15]
	C. phaeobacteriodes	Brown	e	Iso	[11, 15, 99]
	C. phaeovibrioides	Brown	e	Iso	[11, 15]
Ancalochloris	A. perfilievii				
Chlorobaculum	C. tepidum	Green	c	Cb	[15]
	C. chlorovibrioides	Green	c or d	Cb	[11]
	C. limnaeum	Green	c	Cb	[15]
	C. parvum	Green	d	Cb	[15]
	C. thiosulfatophilum	Green	c	Cb	[15]
	"Clathrochloris sulfurica"	Green	c or d	Cb	[11]
Chloroherpeton	C. thalassium	Green	c or d	Ga	[11]
Pelodictyon	Pelodictyon phaeum	Brown	e	Iso	[11]
Prosthecochloris	P. aestuarii	Green	c or d	Cb	[11, 15]
	P. vibrioformis	Green	c or d	Cb	[11, 15, 99]
	P. phaeoaestuarii	Brown	e	Iso	[11]

[a] Classification according to Bergeys [14];
[b] Key: Cb = chlorobactene (5), Iso = Isorenieratene (7), Ga = γ carotene [11, 15, 99]

sortia. Glaeser and Overmann [23] showed that the phototrophic consortium "*Pelochromatium roseum*", which constituted 88% of the GSB (belonging to the brown colored strains) at the chemocline, did contain bacteriochlorophyll-*e* but, surprisingly, no isorenieratene or chlorobactene.

There are two other diaromatic carotenoids known to occur in nature; renieratene (**8**) and renierapurpurin (**4**). Although their structures resemble those of the aromatic carotenoids found in phototrophic sulfur bacteria, they have never been found in cultures to the best of our knowledge. They are thought to derive from sponges or sponge symbionts.

4
Carbon Isotopic Fractionation of Anoxygenic Phototrophic Bacteria

Both the GSB and the PSB are chemoautotrophic bacteria that use CO_2 as their carbon source. In *Chromatiaceae*, CO_2 is assimilated using the reductive pentose phosphate or Calvin cycle in which the first step is the fixation of CO_2 with the enzyme RuBisCo. Culture experiments with different species of *Chromatiaceae* have consistently shown a depletion of 20–23‰ for the cell material relative to CO_2 [24–27], in line with fractionations observed for autotrophic algae using the Calvin cycle. Carotenoids of PSB are approximately 3–5‰ depleted relative to the cell material [24, 27]. The lipids in a culture of *Thiocapsa roseopersicina* were found by van der Meer et al. [28] to be slightly enriched relative to the cell material and attributed this to the operation of alternative (i.e. non-Calvin) biochemical pathways.

GSB fix CO_2 through a completely different pathway: the reversed (or reductive) tricarboxylic acid (TCA) cycle. Compared with the Calvin cycle operated by the *Chromatiaceae* only slightly more than half of the amount of ATP is required per molecule of CO_2 in the reverse TCA cycle. Together with their specific pigment composition, this explains the low light adaptation of the GSB. The enzymes involved in the CO_2 fixation in the reverse TCA cycle only discriminate modestly against ^{13}C. Consequently, the cell material of GSB is isotopically heavy. Using cultures, Quandt et al. [25] and Sirevag et al. [26] reported $\Delta\delta^{13}C$ values ($\delta^{13}C$ of the cell material relative to the ^{13}C content of the CO_2) of 2.5 to 12.2‰.

Van der Meer et al. [28] determined the $\delta^{13}C$ of carotenoids, the esterifying alcohols of its bacteriochlorophyll-*c* and -*a* (farnesol and phytol, respectively) and fatty acids in a culture of *Chlorobium limicola*. The isoprenoid lipids (chlorobactene, farnesol and phytol) were ca. 2–3‰ enriched, whereas the straight-chain lipids were strongly enriched (11–16‰) relative to the total cell material. This trend is opposite that commonly observed where straight-chain lipids are ^{13}C-depleted compared to isoprenoid lipids, which in turn are ^{13}C-depleted compared to total cell material [29], which is a consequence of the biochemical pathway of lipid biosynthesis. This trend, however, can be explained by the specific biochemical pathways of the reverse TCA cycle. Recently, van Breugel et al. [30] determined ε_p (the isotopic fractionation factor) between the pigment isorenieratene and CO_2 in a natural population of GSB residing at the chemocline at ca. 4 m water depth of a stratified Norwegian fjord. These authors found a consistent ε_p of 4 ± 1‰, which was independent of the CO_2 and isorenieratene concentration. These data fit well with those of culture experiments [25, 26] and the ε_p of a natural population of GSB in the Black Sea. Glaeser and Overmann [23] reported a difference of 7‰ between CO_2 and farnesol, the esterifying alcohol of bacteriochlorophyll-e from a dense layer of the photoautotrophic consortium "*Pelochromatium roseum*", which was partially composed of brown-colored GSB, at the chemocline of a temperate holomictic lake (Lake Dagow, Brandenburg, Germany). These data indicate that pigments derived from GSB are generally enriched in ^{13}C relative to organic components derived from most other microorganisms and this has been very useful in the characterization of diagenetic products of pigments of GSB (see below).

Because both PSB and GSB reside at the chemocline of stratified basins, the CO_2 they fix is generally not completely derived from CO_2 from the atmosphere with a $\delta^{13}C$ of ca. -8‰, but it is also partly derived from remineralization of organic matter below the chemocline. This so-called respired CO_2 is much more depleted in ^{13}C and, consequently, $\delta^{13}C$ of CO_2 at the chemocline might be as depleted as -19‰ [31–35] in lakes and fjords. Although GSB fractionate to a minor extent (see above), such depleted $\delta^{13}C$ values for the CO_2 may still result in relatively light values for GSB pigments. Indeed, $\delta^{13}C$ values of up to -30‰ have been reported for isorenieratene [19, 30, 36]. For PSB, which use the Calvin cycle, the presence of respired CO_2 at the chemocline may result in quite depleted $\delta^{13}C$ values; values as depleted as -45‰ have been reported for okenone [19, 36].

5
Photic Zone Anoxia in Recent Depositional Environments and its Manifestation in the Sedimentary Record

The occurrence of phototrophic sulfur bacteria in environments which are characterized by an overlap of the photic and anoxic zones has been documented for a variety of settings (see for a review [16]). Here we will illustrate their ecological occurrence in four different settings for which also paleoecological information is available.

5.1
Mahoney Lake

Mahoney Lake is a small (800×400 m) and shallow (max. depth 18 m) meromictic lake in British Columbia (Canada, 49° N, 119° W) with a steep chemocline at 6.6 m (Fig. 1a). PSB form a dense, 10-cm-thick layer at the chemocline [17]. The dominant (98%) species, *Lamprocystis purpurea* formerly known as *Amoebobacter purpureus*, reached a maximum cell num-

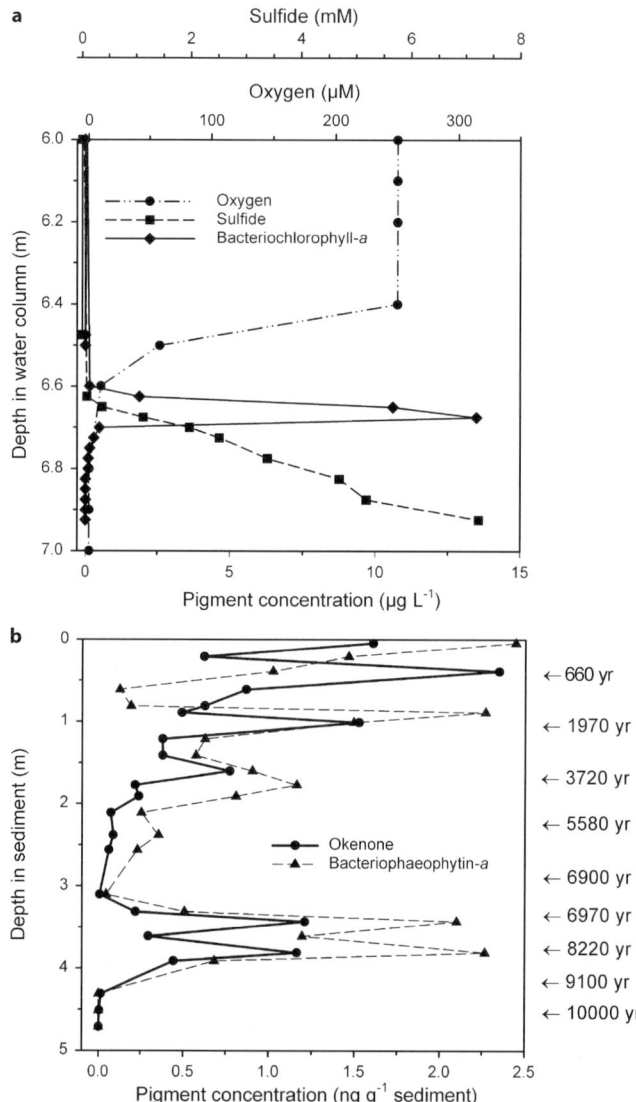

Fig. 1 **a** Vertical distribution of oxygen, sulfide, and bacteriochlorophyll-*a* derived from PSB in the present-day water column of Mahoney Lake in September 1989. Data are from [17]. **b** Concentration profiles of the PSB pigments okenone (**1**) and bacteriophaeophytin-*a*, a diagenetic product of bacteriochlorophyll-*a*, in the sediments of Mahoney Lake. Radiocarbon dates (in yr) are indicated. Data are from [37]

ber of 4×10^8 cells ml^{-1}. Extremely high concentrations of up to 21 mg bacteriochlorophyll-*a* L^{-1} were found in this layer and okenone (**1**) was detected as the dominant carotenoid [17]. Anoxygenic photosynthesis was limited by sulfide at the top of the layer, whereas the substantially reduced irradiance resulted in light limitation for most of the cell population below. GSB are not present in this lake and are probably inhibited by the high sulfide concentrations. Pigment analysis revealed that these specific conditions were common during the Holocene development of the lake; both fossil bacteriochlorophyll-*a* as well as okenone were identified in various strata (Fig. 1b) [37]. This was supported by fossil DNA data which revealed four different 16S rRNA gene sequences falling in the group of *Chromatiaceae*. The sequence from sediments 9100-year-old was 99.2% identical to the 16S rRNA gene sequence of *Lamprocystis purpurea* isolated from the present-day chemocline of the lake. At other sediment intervals other dominant 16S rRNA gene sequences were found that were related to *Marichromatium* and *Thiorhodovibrio* species, indicating that the PSB population varied substantially during the Holocene.

5.2
Ace Lake

Post-glacial Ace Lake is a small (200 × 300 m) and shallow (max. depth 25 m), meromictic, saline lake in the Vestfold Hills of eastern Antarctica (68° S, 78° E) and usually covered by ice for about 11 months of the year. It contains anoxic, sulfidic, sulfate-depleted, and methane-saturated bottom waters (Fig. 2a) [38]. The euxinic chemocline at 11.7 m depth is colonized by obligate anoxygenic photolithotrophic GSB [39] with the brown-colored *Chlorobium phaeovibrioides* DSMZ 269^T as their closest relative (99.6% sequence similarity) [40]. The concentration of a characteristic carotenoid for GSB, chlorobactene (**5**), peaked at the chemocline with concentrations of 4 mg l^{-1} (Fig. 2a). Smaller amounts of isorenieratene (**7**) and β-isorenieratene (**6**) were also found. This is somewhat surprising since *C. phaeovibrioides* is a brown strain of GSB and hence would be expected to have isorenieratene as its most abundant carotenoid (Table 2). However, some species of GSB are now known to have both green and brown colored strains (Table 2). PSB were not encountered in this lake, probably due to the relatively low light intensity at the chemocline. Analysis of fossil carotenoids in a sediment core covering the last 10.5 ka revealed that almost immediately after marine waters entered the paleo freshwater lake as a result of post-glacial sea-level rise, Ace Lake became meromictic with the formation of sulfidic bottom waters and a chemocline colonized by GSB (Fig. 2b). The carbon isotopic composition of fossil chlorobactene (ca. − 18‰) confirmed its origin from GSB [41]. Analysis of fossil 16S rRNA genes revealed that the biological source of chlorobactene throughout the Holocene was identical to that residing at the present-day chemocline.

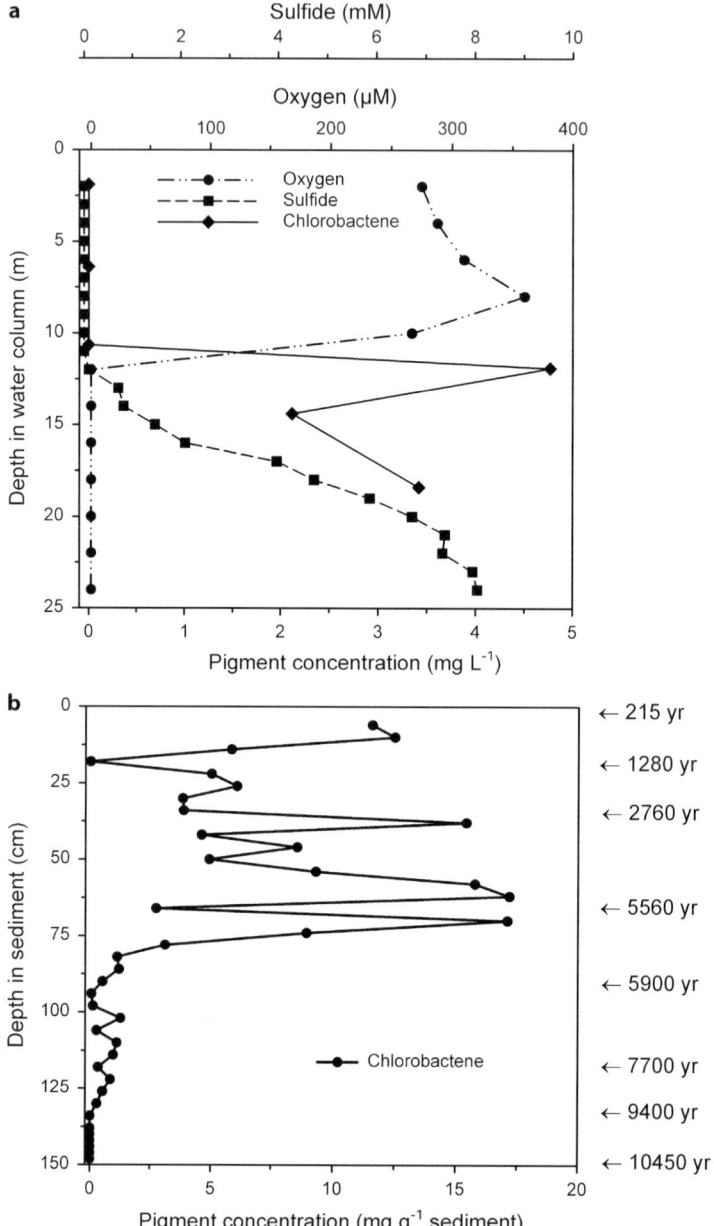

Fig. 2 **a** Vertical distribution of oxygen, sulfide, and chlorobactene derived from GSB in the present-day water column of Ace Lake (November 2000). Data are from [38, 40]. **b** Concentration profiles of the GSB pigments chlorobactene (5) in the sediments of Ace Lake. Radiocarbon dates (in yr) are indicated. Data are from [40]

5.3
Kyllaren Fjord

Kyllaren is a small (ca. 0.35 km^2), 29 meter deep fjord on the west coast of Norway (62° S, 5° E). It is connected to the Norwegian Sea through a shallow inlet which consists of a tidal flat and a 1–2 m deep water channel, depending on the tide. Consequently, bottom water exchange is restricted, resulting in a permanent halocline/chemocline at 2–5 m water depth, at which a population of GSB resides. GSB carotenoids detected in the water column of Kyllaren fjord are dominated throughout the year by isorenieratene and β-isorenieratene, peaking at the chemocline (Fig. 3a), with concentrations up to 9.2 and 3.0 µg L^{-1}, respectively [30]. The stable carbon isotope composition of isorenieratene and β-isorenieratene vary from −24 to −30‰, which, in combination with the strongly depleted δ^{13}C values of CO$_2$ at the chemocline (−20 to −27‰), confirms their origin from GSB. The presence of fossil carotenoids in the upper 20 cm of sediments (Fig. 3b), indicate that GSB and, surprisingly, also PSB, as indicated by the presence of okenone 1, thrived at the chemocline for the last 200 yr [42]. Their isotopic compositions (−25‰ for isorenieratene and −39‰ for okenone) are in agreement with these assignments of the biological sources for these fossil carotenoids [42], considering that these are relatively light due to the prominence of the respired CO$_2$ recycling process [31]. During the last 200 years there has been a tendency of GSB to become more abundant than PSB (Fig. 3b) in this fjord. This may result from the increased primary productivity in the fjord, induced by an increase in nutrient load [42]. The increase in algal biomass is thought to have affected the light intensity and light quality at the chemocline, inducing a shift in the composition of the phototrophic bacteria.

5.4
Black Sea

The Black Sea is the world's present-day largest anoxic basin. It has an area of 422 000 km^2 and its maximum depth is ca. 2200 m. It is a restricted basin because of the shallow connection to the Mediterranean Sea (Strait of Bosporus), which substantially limits the mixing of low salinity surface water with the high salinity (and therefore more dense) bottom water. This results in a strong density stratification with an associated chemocline. In 1989, Repeta and co-workers [22] showed that, although the chemocline is much deeper than in stratified lakes and fjords, the Black Sea's deep chemocline (at ca. 80 m) still provided a niche for GSB. This was revealed by the analysis of the characteristic pigments, bacteriochlorophyll-*e*, isorenieratene, and β-isorenieratene in the water column (Fig. 4a), which all peaked at ca. 80 m depth. Overmann et al. [43] showed by cultivation that these GSB are brown-colored *Chlorobium phaeobacteroides* strains, which are adapted to

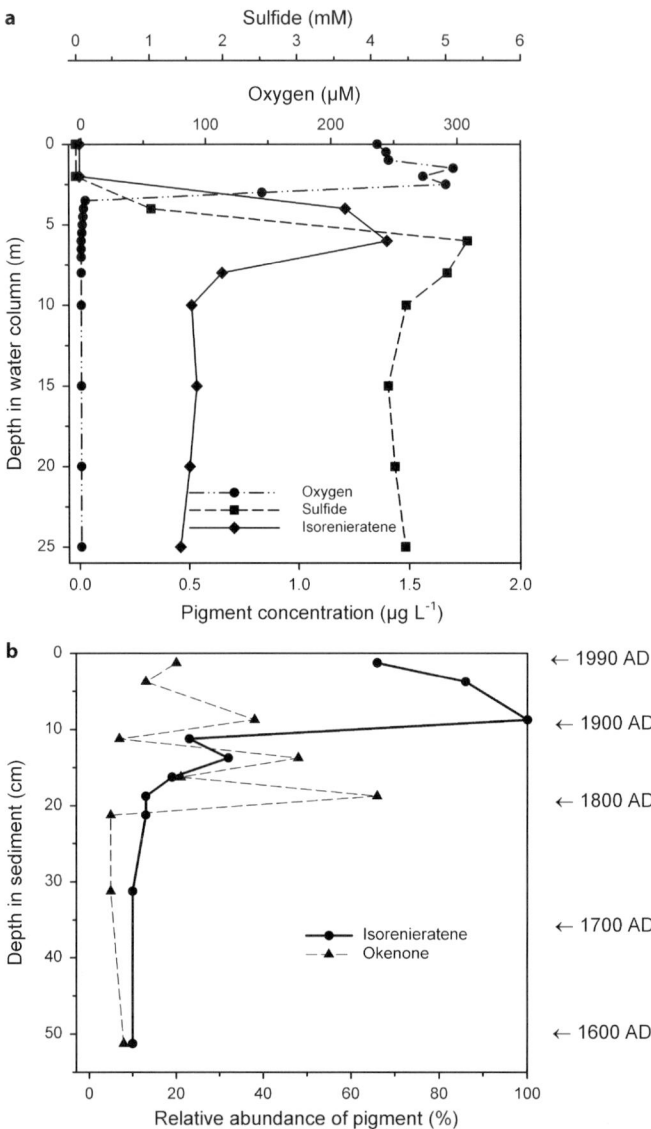

Fig. 3 **a** Vertical distribution of oxygen, sulfide, and isorenieratene derived from GSB in the present-day water column of Kyllaren fjord (August 2002). Data are from [30, 37]. **b** Concentration profiles (in relative abundance) of the GSB and PSB pigments isorenieratene (**7**) and okenone (**1**) in the sediments of Kyllaren fjord. Estimated dates of sediment intervals (in yr), based on ^{210}Pb, are indicated. Data are from [42]

the very low light intensities (ca. 0.003 µmol Quanta m^{-2} s^{-1}) at the chemocline. At these low light intensities, it is estimated that each bacteriochlorophyll molecule will absorb a photon only once every 8 hours. Consequently,

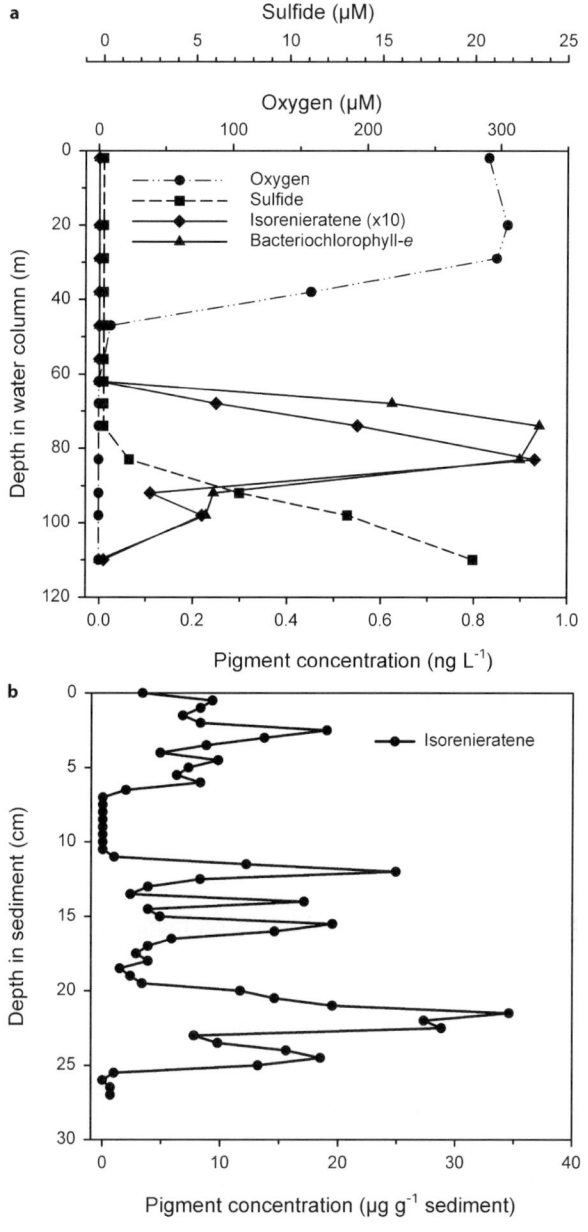

Fig. 4 a Vertical distribution of oxygen, sulfide and the GSB pigments bacteriochlorophyll-*e* and isorenieratene in the present-day water column of Black Sea (May 1988). Data were obtained from Station BS2-2 of the RV Knorr cruise 134, Leg 9 and are from [22]. **b** Concentration profiles of the GSB pigments isorenieratene (7) in the unit-1 sediments of the Black Sea from the same station. The core reflects ca. 1500 yr of sedimentation. Data are from [44]

these GSB have to be extremely efficient with respect to energy. Indeed, in comparison with the type strain of *C. phaeobacteroides*, the GSB isolated from the Black Sea chemocline have a very low maintenance energy requirement and an increased concentration of light-harvesting pigments [43]. Repeta [44] and Sinninghe Damsté et al. [45] showed that the present overlap of the photic and sulfidic zone has not always been the case in the last 8200 years. Repeta [44] provided high-resolution data (Fig. 4b) of intact isorenieratene in the sediments, which revealed substantial fluctuations related to the depth of the chemocline. Sinninghe Damsté et al. [45] showed that fossil isorenieratene is indeed derived from GSB since its δ^{13}C values are relatively enriched, consistent with the use of the reverse TCA cycle of the GSB.

These four cases provide insight in the type of environments in which the GSB and PSB can thrive and indicate that the pigments of these types of bacteria can become part of the sediments and, thus, can provide insight into past communities of phototrophic bacteria. Since these bacteria require both light and sulfide, this provides important information on past water column anoxia.

6
Ancient Sediments Containing Intact Pigments from Phototrophic Sulfur Bacteria

In the previous section we have seen that intact pigments of GSB and PSB are preserved in sediments up to 9 ka. This is already surprising since both carotenoids and bacteriochlorophylls are labile components that undergo rapid transformation reactions during burial (so-called dia- and catagenesis) [44, 46]. In the next section, we will describe these pathways in detail, but first we will discuss the occurrence of intact pigments in ancient sediments.

The oldest well-documented occurrence of intact pigments from GSB and PSB is in the Upper Miocene Gessosso-Solfifera Formation in Italy. Maxwell and co-workers [47, 48] identified intact isorenieratene by HPLC-PDA-MS in the marl layers of this 6 Ma year-old formation. It was accompanied by smaller amounts of *cis*-isorenieratene, obviously formed by *trans-cis* isomerization of the double bond(s) of isorenieratene, which are all *trans*. Intact isorenieratene was also identified in Pliocene sapropels ca. 3 Ma years old, deposited in the Eastern Mediterranean Sea [49]. The amount of intact isorenieratene (summed all-*trans* and *cis* isomers) ranged from non-detectable at the base and top of a sapropel up to 140 µg g^{-1} sediment in the central part of the sapropels.

Isorenieratene accumulation rates at a central and western site in the Eastern Mediterranean Basin are remarkably similar and increase sharply to levels of up to 3.0 mg m^{-2} yr^{-1} in the central part of the sapropel and then drop to low levels. Since Ba/Al ratios indicate enhanced paleoproductivity during sapropel formation, these data support previously proposed models, according to which increased productivity is the driving force for the generation of euxinic conditions [50]. Apart from intact isorenieratene, these sapropels, which are buried 50–100 m in the subsurface, also contain components that have been interpreted as diagenetic derivatives of isorenieratene [51, 52]. They are formed by cyclization reactions involving the polyene system of double bonds in these carotenoids. The following section will discuss some major diagenetic products of aromatic carotenoids that have been identified so far.

7
Dia- and Catagenesis of Carotenoid Markers for Photic Zone Anoxia

The first indication that carotenoid carbon skeletons can be preserved in ancient sediments came from the identification of three perhydro diaromatic carotenoids (**16**, **13**, and **17**; see Scheme 3 for structures of perhydro aromatic carotenoids) in the Schistes Carton of the Paris Basin [53]. Summons and Powell [54, 55] subsequently identified two series of aryl isoprenoids (**19** and **20**; see Scheme 4 for diagenetic products of aromatic carotenoids) in sediments and petroleum. These series of components had the 2,3,6- and 2,3,4-trimethyl substitution pattern of the aromatic moiety, which suggested a relationship with isorenieratene (**7**) or chlorobactene (**5**) and okenone (**1**) or renierapurpurin (**4**), respectively, through hydrogenation and C–C bond cleavage. Since the series of 2,3,6-trimethyl aryl isoprenoids extended to C_{30} (and not to C_{40} as would be expected for a chlorobactene origin), it was likely that these products were derived from isorenieratene. Summons and Powell [54, 55] also showed that the $\delta^{13}C$ of the 2,3,6-trimethyl aryl isoprenoids was relatively enriched, lending further support to their derivation from isorenieratene produced by GSB using the reverse TCA cycle (see above).

Subsequently, a multitude of dia- and catagenetic products of aromatic carotenoids has been identified (see for a review [46]) and the processes by which they are formed are now reasonably well understood. Isolation of key components and elucidation of their structure by high-field NMR techniques and the determination of their ^{13}C content has been essential in this respect. For proper assessment of photic zone anoxia in past depositional settings, it is important to understand these dia- and catagenetic pathways as discussed below.

7.1
Hydrogenation

The first indication that intact aromatic carotenoid carbon skeletons can be preserved in ancient sediments came from the identification of the

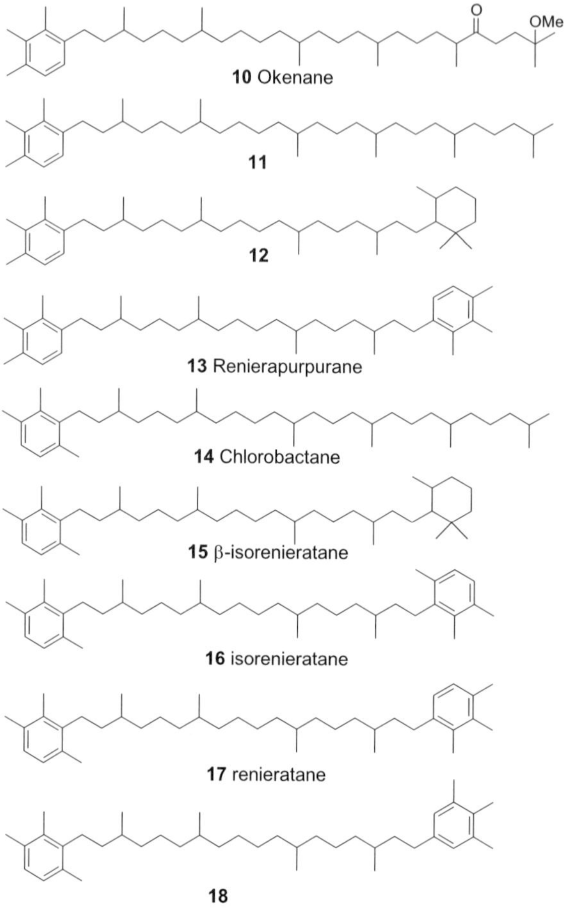

Scheme 3 Perhydro aromatic carotenoids

"perhydro" derivatives of the diaromatic carotenoids, isorenieratene, renieratene, and renierapurpurin (**13**, **16**, **17** [53]), β-isorenieratene (**15** [56]), and chlorobactene (**14** [57]). Subsequently, **13**, **15**, **16**, and **17** have also been identified in crude oils [58–62]. These findings strongly suggested that these molecular fossils were formed through hydrogenation of the polyene chain of the preserved carotenoids. Schaeffer et al. [63] recently identified some partially hydrogenated derivatives of isorenieratene and okenone (e.g. **21**) in surface sediments of Lake Cadagno, indicating that hydrogenation may occur during the first phases of diagenesis.

The analysis of fossil "perhydro" carotenoids may also provide clues as to the structure of carotenoids that once existed but have become extinct during evolution. In a suite of samples from the Williston Basin the "perhydro" carotenoids were dominated by a diaryl isoprenoid **18** possessing

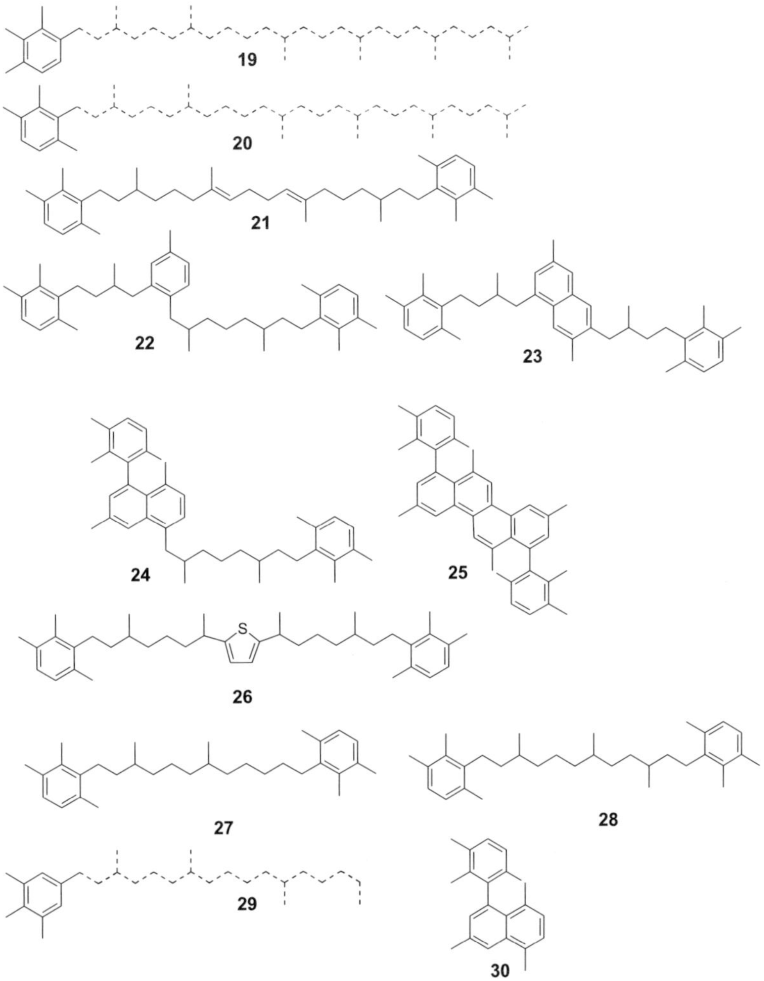

Scheme 4 Examples of dia- and catagenetic products of aromatic carotenoids

the unprecedented 3,4,5-trimethyl substitution pattern for the aromatic ring in combination with the more common 2,3,6-trimethyl substitution pattern [59–61]. Since shifts of methyl groups of aromatic rings during diagenesis were deemed highly unlikely, this suggested a novel carotenoid **9** (Scheme 2) as precursor. Since this carotenoid has not been reported to occur in nature and diaryl isoprenoid **18** is distinct from other molecular fossils in its carbon isotopic composition, it was speculated that this may be derived from a now extinct species of GSB.

7.2
Cyclization and Aromatization

A large suite of diagenetic products of isorenieratene (e.g. 22-25) and other aromatic carotenoids has been identified with structures containing one to four additional, in most cases aromatic, rings [52, 64-67]. Most of these components have only been characterized by mass spectrometry, but in specific cases key structures (22-23) have been isolated from sediment extracts and fully characterized by NMR [52, 64]. The identical carbon isotopic compositions of isorenieratane and these C_{40} isorenieratene derivatives confirm their diagenetic relationship (Fig. 5). The structures of these diagenetic products strongly suggest that they have been formed by a sequence of cyclization, aromatization, and hydrogenation reactions occurring within the polyene system (Fig. 6). Initial cyclization seems to occur mainly after appropriate *trans-cis* isomerization at two specific sites, which favors a Diels-Alder reaction [65].

These pathways are supported by calculations using molecular dynamics simulation [68] which allowed the entire cyclization pathway to be simulated, including stable intermediates as well as energy barriers related to transition states. The simulations indicated that the formation of tetracyclic isorenieratene derivatives is likely to occur via an A-ring initiated reaction mechanism,

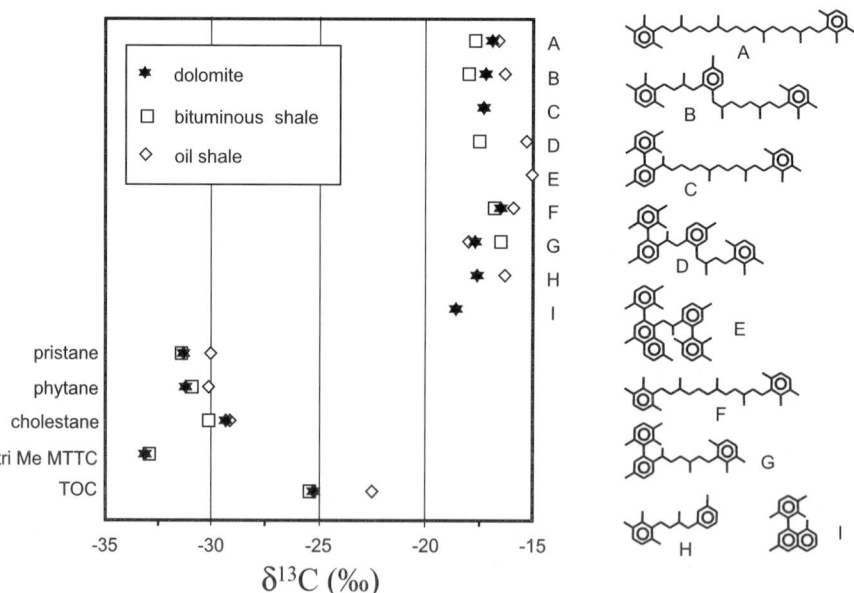

Fig. 5 Stable carbon isotopic composition of isorenieratene derivatives and some algal biomarkers present in three sedimentary facies of the Lower Kimmeridge Clay Fm. [118]

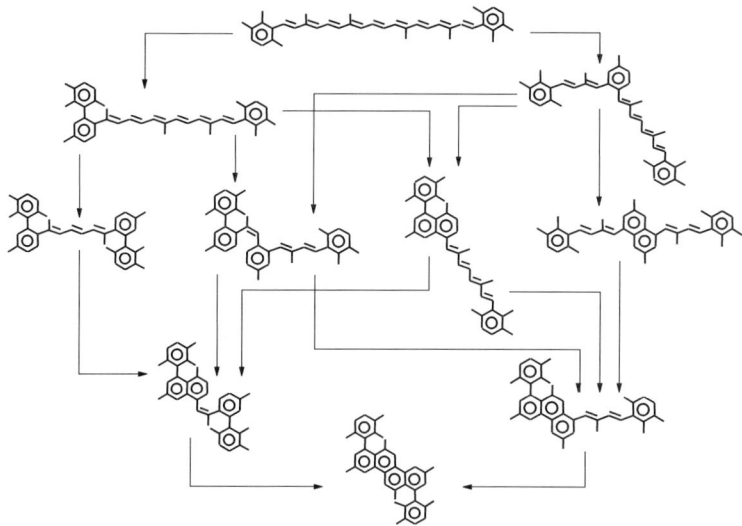

Fig. 6 Cyclization and aromatization of isorenieratene resulting in the formation of diagenetic derivatives [49]

as the reaction product resulting from A-ring closure is more stable than that derived from B-ring closure. Furthermore, the A-ring initiated tetracyclization pathways contain one fewer high-energy hydrogen shift step than their B-ring initiated counterparts, indicating that B-ring initiated cyclization is more likely to result in the formation of monoaromatic compounds. These observations are in excellent agreement with observed distributions of isorenieratene derivatives in sediments.

7.3
Aromatization of the Cyclohexenyl Moiety

In addition to hydrogenation of the cyclohexenyl moieties of carotenoids, as in β-isorenieratene, aromatization with concomitant transfer of methyl groups is also thought to occur. A 1,2-shift of a geminal methyl group of β-carotene may occur during aromatization as has been simulated in the laboratory [62]. In this way, β-isorenieratane (**15**) present in a North Sea crude oil is thought to be derived from aromatization and hydrogenation of β-carotene and not from hydrogenation of β-isorenieratene [52, 62]. This was corroborated by its stable carbon isotopic composition which was identical to that of β-carotane in the same oil, but 15‰ depleted relative to that of isorenieratane derived from green sulfur bacteria, which are also known to biosynthesize β-isorenieratene. In this way, aryl isoprenoids may also partly derive from β-carotene and, indeed, the aryl isoprenoids in the North Sea crude oil studied by Koopmans et al. [62] had a carbon isotopic composition indicating a mixed origin from both β-carotene and isorenieratene. Therefore, aryl isoprenoids should be used with caution and we recommend that their δ^{13}C value should be determined to be assured of an origin from *Chlorobiaceae*.

7.4
Intramolecular Sulfurization

Various sedimentary organic sulfur compounds possessing the isorenieratane skeleton have been identified by mass spectrometry and Raney Ni desulfurization [65]. They occur as thiophenes (e.g. **26**), dithiophenes, bithiophenes, benzo(*b*)thiophenes, and cyclic (poly)sulfides (thianes and dithianes). Sedimentary organic sulfur compounds result from a reaction of reduced inorganic sulfur species, formed by sulfate reducing bacteria, and functionalized lipids during early diagenesis ("natural sulfurization"; see for a review [69]). Obviously, the polyene system of carotenoids is prone to natural sulfurization as well and, after the subsequent reduction of double bonds, leads to the formation of sulfur compounds with aromatic carotenoid skeletons.

7.5
Incorporation into Macromolecular Organic Matter

The largest fraction (> 90%) of sedimentary organic matter is composed of "kerogen", macromolecular organic matter which is insoluble in water and common organic solvents. In addition, the extractable organic matter also contains macromolecular aggregates. These fractions often contain sequestered molecular fossils which can be released by chemical and thermal degradation methods. The polyene system of aromatic carotenoids has been shown to be prone to reactions leading to sequestration. The most common of these reactions is natural sulfurization but, in this case, operating in an intermolecular fashion and leading to cross-linking of carbon skeletons through (poly)sulfide linkages. A variety of "perhydro" aromatic carotenoids, β-chlorobactane **14**, isorenieratane **15**, isorenieratane **16**, renieratane **17**, and diaryl isoprenoid **18** have been released from soluble macromolecular aggregates [45, 56, 59, 65, 70–80] and from kerogen [81] by various desulfurization methods. Isorenieratane was also released by selective cleavage of polysulfide bonds, indicating that this carotenoid skeleton may also be bound only via polysulfide linkages [65, 71, 82]. However, the amounts released in this way are much lower since carotenoid skeletons are likely to be bound via several S-linkages and the chance that these are only polysulfide linkages is small. In addition, diagenetic products of isorenieratene formed by cyclization/aromatization (e.g. **22**) have also been found upon desulfurization of geomacromolecules [65, 77].

Thermal degradation has also been used to analyze aromatic carotenoids sequestered in high-molecular-weight material. The high abundance of 1,2,3,4-tetramethylbenzene (TMB) in flash pyrolysates of kerogens [58, 59, 61, 83–85] has been explained by its formation through β-cleavage of a benzene ring of aromatic carotenoids incorporated into the kerogen. TMB can be derived from almost all mono- and diaromatic carotenoids depicted in Scheme 2. However, a distinction can be made based on the distribution of γ-cleavage products in the flash pyrolysates and the ^{13}C content of the β- and γ-cleavage products. 1-Ethyl-2,3,6-trimethylbenzene is often a dominant C_5 alkylbenzene in pyrolysates of kerogens [59, 84] consistent with a derivation from isorenieratene or chlorobactene. TMB and 1-ethyl-2,3,6-trimethylbenzene released after off-line pyrolysis of a kerogen isolated from the Duvernay Formation also had anomalously high ^{13}C contents consistent with this explanation [59, 61]. Pedentchouk et al. [86] showed that 1,2,4-trimethylbenzene in pyrolysates may also be (partially) derived from kerogen-bound isorenieratene.

In contrast, the ^{13}C content of TMB in a pyrolysate of a kerogen isolated from Indian Ocean surface sediments is identical to that of algal lipids, excluding an origin from isorenieratene [87], indicating that there are multiple origins for TMB. The anomalously high concentration of TMB and 1-ethyl-2,3,6-trimethylbenzene in petroleums from the Williston Basin [59] and their high ^{13}C contents indicates that these products are also formed from macromolecularly bound aromatic carotenoids during cracking in the subsurface. Similar observations have been made for products derived from diaromatic carotenoid **9** (i.e. 1,3,4,5-tetramethylbenzene and 1-ethyl-3,4,5-trimethylbenzene [59]).

Although most studies have shown that sulfur bonding is important for sequestration of carotenoids in macromolecular fractions, a combined chemical and thermal degradation study of diaromatic carotenoid skeletons in a sediment of the Duvernay Formation has indicated that only a minor fraction is bound solely via S-linkages [59], indicating that alternative modes of binding such as oxygen or carbon linkages are involved as well. Quantitative analysis of all organic matter fractions from this sediment revealed that the pool of diaromatic carotenoids initially biosynthesized was almost completely (ca. 99%) incorporated into macromolecular fractions [59], demonstrating the importance of aromatic carotenoid sequestration during diagenesis.

7.6
Expulsion of Toluene and Xylene

C_{32} (**27**) and C_{33} (**28**) pseudohomologs of isorenieratane have been reported both as free components or in a S-bound form [65, 66, 77]. Mild heating of β-carotene produces C_{32} and C_{33} "carotenoids" and *m*-xylene and toluene, respectively (e.g. [88]) by a pericyclic reaction via an eight-membered ring transition state of the polyene chain. Since this mechanism only involves the polyene system of the carotenoid, it is postulated that isorenieratene and other aromatic carotenoids can undergo similar reactions, resulting in the formation of C_{32} and C_{33} diaromatic "carotenoids" [65, 67]. It is thought that these expulsion reactions also occur in immature sediments and that subsequent hydrogenation leads to compounds **27–28**. The cyclization, aromatization, and sulfurization reactions that lead to the C_{40} diagenetic products of isorenieratene also occur with C_{32} and C_{33} "carotenoids".

7.7
C – C Bond Cleavage

A wide variety of putative products from C – C bond cleavage of fossil aromatic carotenoid derivatives have been identified [65]. These include the pseudo-homologous series of aryl isoprenoids **19**, **20**, and **29**, whose structures are sub-units of the "perhydro" aromatic carotenoids [54, 55, 59, 89]. Other products (e.g. **30**) are formed from carotenoid units which have undergone cyclization and aromatization, expulsion, or sulfurization reactions. These products are probably not directly formed from fossil carotenoids but rather form after initial incorporation of carotenoids in high-molecular-weight fractions and subsequent C – C bond cleavage during increasing thermal stress due to burial in the subsurface [58, 65].

8
Dia- and Catagenesis of Bacteriochlorophyll Markers for Photic Zone Anoxia

Bacteriochlorophylls are subjected to substantial alterations during sediment dia- and catagenesis, ultimately resulting in the formation of porphyrins which can still be found in petroleum. This has been the subject of various reviews (e.g. [90]) and thus will not be covered here. Members of the bacteriochlorophylls *c*, *d*, and *e* are characterized by the presence of additional carbon atoms at position C-8 (leading to e.g. propyl and *i*-butyl substitution), C-12 and C-20 of the chlorophyll skeleton (Scheme 1) in comparison with chlorophyll *a*. This higher and characteristic degree of alkylation can be found in diagenetic products of the bacteriochlorophylls (e.g. [91, 92]) and in this way photic zone anoxia in past depositional systems can be assessed.

Oxidative degradation productions of bacteriochlorophylls have also been recognized in the fossil record. Grice et al. [93] identified maleimides (1*H*-pyrrole-2,5-diones) as degradation products of photosynthetic tetrapyrrole pigments for the first time in ancient marine sediments. The maleimides had a relatively simple distribution, dominated by Me,Et maleimide, which is predominantly derived from chlorophyll of planktonic origin. Me,*n*-Pr and Me,*i*-Bu maleimides, present in low abundance, were thought on structural grounds to be derived from the bacteriochlorophylls *c*, *d*, or *e* of *Chlorobiaceae*. This was confirmed by their enrichment in ^{13}C resulting from the reversed TCA cycle employed by the GSB. The structurally more specific Me,*i*-Bu maleimide was slightly more enriched in ^{13}C than Me,*n*-Pr maleimide, suggesting that the latter is partly derived from phytoplanktonic chlorophyll. These results provide evidence for the existence in both depositional settings of microbial communities containing *Chlorobiaceae*. It remains enigmatic when the maleimides are formed: by oxidation of the bacteriochlorophylls at the time of deposition, during weathering or during sample processing.

9
Evidence for Photic Zone Anoxia in Ancient Depositional Environments

The dia- and catagenetic products of the pigments of GSB have been used to assess photic zone anoxia in a wide variety of settings. Table 3 provides a literature overview of the depositional environments for which molecular paleontological evidence exists for an overlap of the photic and sulfidic zone. It also specifies whether this information is based on fossil carotenoid derivatives or on fossil bacteriochlorophyll derivatives or both. In most cases, the evidence is based on the presence of carotenoid derivatives, probably because they are easier to analyze than the bacteriochlorophyll derivatives. In most instances where intact aromatic carotenoid derivatives were found, products derived from isorenieratene and not from chlorobactene or *β*-isorenieratene were encountered. On the basis of the relatively high abundance of isorenieratene in the brown strains of the GSB (Table 2), this indicates a relatively deep chemocline in the photic zone.

Chlorobactene derivatives were only encountered in a few settings: during OAE-2 in the Tarfaya and Cape Verde (DSDP Site 367) basins and during the Toarcian anoxic event in the Paris Basin (Schistes Carton Formation) and SW Germany (Posidonia Shale) and the 1.64 Ga Barney Creek Formation [94]. In these latter deposits the hydrogenated form of okenone, okenane, derived from PSB was also found. This would indicate even shallower anoxic conditions in the photic zone. Dia- and catagenetic products of the hypothetical diaromatic carotenoid (9) with the 3,4,5-trimethyl substitution pattern of one of the rings is only found in sediments of 450 to 310 Ma in age, but were not present in the Paleoproterozoic Barney Creek Formation.

Most of the settings for which we have evidence for photic zone anoxia were apparently highly productive shelf seas with paleo water depths of 100–300 m (the so-called continental or epeiric seas). The stratification in these systems is believed to be seasonally controlled [95] and photic zone anoxia is probably highly dynamic. The Oxford Clay is probably a typical example of this type of setting since in the same stratum both delicately preserved fossils of benthic organisms requiring oxygen and isorenieratene

Table 3 Molecular paleontological evidence for photic zone anoxia in ancient depositional systems

Black shale	Geological age	Aromatic carotenoid[a] derivatives	Bacteriochlorophyll derivatives[b]	Refs
Sediments				
Eastern Mediterranean sapropels ODP 964, 967, 969 leg 160	Pliocene	7		[49–52]
Sdom Fm, Dead Sea (Israel)	Miocene/Pliocene	7		[101]
Gessoso-Solfifera Fm. (Italy)	Messinian	7	P	[47, 48, 62, 72, 76, 77]
Gibellina Marl (Sicily, Italy)	Messinian	7, 5		[70]
Menilite Fm. (Poland)	Oligocene	7		[65, 102]
Mulhouse Basin (France)	Oligocene		P	[92, 103]
Qianjiang Fm., Jianghan Basin (China)	Middle/Late Eocene	7, 6, 5		[104]
Messel (Germany)	Eocene		P	[91]
Xingouzhui Fm, Jianghan Basin, (China, 113°E)	Paleocene-Early Eocene	7		[104]
Deep Ivorian Basin (ODP 959)	Coniacan/Santonian	7, 5		[105]
Canje Fm. (British Guyana)	Late Turonian	7		[65]
OAE-2				
Tarfaya (Morocco)	Cenomanian-Turonian	7, 5		[56, 79, 80, 106]
Cape Verde Basin, North Atlantic (DSDP 367)	Cenomanian-Turonian	7, 5	P, Ma	[56, 79, 107]
Cape Verde Rise, North Atlantic (DSDP 368)	Cenomanian-Turonian		P, Ma	[107]
Demerara Rise, North Atlantic (DSDP 144)	Upper Cenomanian	7, 5		[56, 79]
Cape Hatteras, North Atlantic (DSDP 603B)	Cenomanian-Turonian	7	Ma	[107, 108]
Western Interior Seaway (USA)	Cenomanian-Turonian	7		[109]
Bahloul Fm. (Tunesia)	Cenomanian-Turonian	7	P, Ma	[107]
Livello Bonarelli (Italy)	Cenomanian-Turonian		P, Ma	[107]
Exmouth Plateau, Indian Ocean (ODP 763C)	Cenomanian-Turonian		Ma	[107]

Table 3 Continued

Black shale	Geological age	Aromatic carotenoid[a] derivatives	Bacterio-chlorophyll derivatives[b]	Refs
OAE-1a				
Livello Selli (Italy)	Early Aptian	7	P, Ma	[107, 110]
Maculungo Shale, Kwanza Basin (West Africa, 10°S, 14°E)	Barremian	7		[86]
Sunniland Limestone Fm (USA)	Early Cretaceous	AI		[111]
Calcaires en Plaquettes Fm. (France)	Late Jurassic	7, 5		[57, 65, 97]
Kimmeridge Clay Fm (UK)	Late Jurassic	7, 4, 8		[52, 65, 112]
Toarcian OAE				
Schistes Cartons (France)	Toarcian	7		[53, 65, 107]
Posidonia Shale (Germany)	Toarcian	7, 5, 6	P	[107, 113, 114]
Whitby Fm. (UK)	Toarcian	AI	P	[115]
Marche-Umbria Basin (Italy)	Toarcian		P, Ma	[107]
Oxford Clay Fm (UK)	Middle Callovian	7		[96]
Allgäu Fm. (Germany)	Early Jurassic (Toarcian)	7		[65, 116]
Hauptdolomit (Germany)	Late Triassic	7		[65]
Kössen Marl (Hungary)	Late Triassic	7		[65]
Sepiano Shale (Switzerland)	Middle Triassic		P, Ma	[117, 118]
Kupferschiefer (Germany)	Late Permian	7	P	[66, 93, 118–122]
Minnelusa Fm (USA)	Late Carboniferous	7		[65]
Lodegepole Fm (Western Canada Basin)	Early Carboniferous	AI		[58]
Exshaw Fm (Western Canada Basin)	Early Carboniferous	9, 7		[58, 65]
Bakken Fm (Williston Basin, Canada)	Early Carboniferous	AI		[58]
Holy Cross Mountains (Poland)	Late Devonian	9, 7		[123]

Table 3 Continued

Black shale	Geological age	Aromatic carotenoid[a] derivatives	Bacterio-chlorophyll derivatives[b]	Refs
Duvernay Fm. (Canada)	Late Devonian	9, 7, 8		[58–60]
Ratner Shale (Western Canada Basin)	Middle Devonian	AI		[58]
Keg River Fm. (Canada)	Middle Devonian	9, 7		[98, 124]
Salina, Michigan Basin (Canada)	Silurian	AI		[54, 55]
Maquoketa, Illinois Basin (USA)	Late Ordovician	AI		[125]
Boas Oil Shale (Canada)	Late Ordovician	9, 7		[126]
Decorah Fm. (USA)	Middle Ordovician	7		[127]
Womble Shale (USA)	Middle Ordovician	7, 9		[126]
Barney Creek Fm. (Australia)	Paleoproterozoic	4, 5, 1, 8, 7		[94]
Oils				
Sunniland Limestone Fm (USA)	Early Cretaceous	AI		[111]
Lodegepole Fm (Western Canada Basin)	Carboniferous	AI		[58]
Exshaw Fm (Western Canada Basin)	Carboniferous	AI		[58]
Bakken Fm (Williston Basin, Canada)	Carboniferous	AI		[58]
Duvernay (Western Canada Basin)	Upper Devonian	9, 7		[58–60]
Ratner Shale (Western Canada Basin)	Middle Devonian	AI		[58]
Keg River (Western Canada Basin)	Middle Devonian	AI		[54, 55]
Pripyat River (Belarussia)	Devonian	9, 7		[67]
Salina (Michigan Basin, Canada)	Silurian	AI		[54, 55]

[a] numbers refer to carotenoids indicated in Scheme 1, AI = aryl isoprenoids;
[b] P = prophyrins, Ma = maleimides

derivatives occur. This has been explained by the concept of intermittent anoxia [96, 97]. The problem with the understanding of these settings is that we have no contemporary analogue for these systems.

Some of the settings where photic zone anoxia occurred in the geological past represent deep water settings (e.g. Mediterranean Sea, Atlantic Ocean during OAE-2, and the Ivorian Basin) and these cases likely represent more or less stagnant basins for which the present-day Black Sea can serve as a modern analogue. The Eastern Mediterranean basin is probably the most comparable to the present-day Black Sea as it is also a land-locked basin with a shallow sill.

10
The Fossil Record of Green and Purple Sulfur Bacteria

The occurrence of diagenetic products derived from the pigments of GSB also provides an insight into the evolution of these bacteria. The oldest reported occurrence of GSB is from the Paleoproterozoic Barney Creek Formation. Both isorenieratane and chlorobactane were reported in these 1.64 Ga deposits, testifying to the existence of an adaptation to different light regimes already present at that time [94]. Unfortunately, due to the high background of organic constituents in these very old sediments, it has been impossible to measure the δ^{13}C values of individual carotenoid derivatives and, in this way, unambiguously confirm their origin from GSB. However, since their structures are so specific and their occurrence fits so well to the assumed anoxicity of the ocean at that time [94], there remains little doubt that GSB indeed thrived in the Paleoproterozoic.

A large gap in the fossil record falls from 1.64 Ga to ca. 0.45 Ga with no reported occurrences of fossil molecules derived from GSB. However, from the Ordovician up to the present-day numerous occurrences of GSB have been documented (Table 3) and in many cases it has been shown that the carotenoid derivatives bear the ^{13}C signature of the reverse TCA cycle (Fig. 7). This indicates that for at least 450 Ma the GSB biosynthesized isorenieratene through a specific biochemical pathway. This shows that lipid biochemistry can be very conservative and, consequently, that lipid biomarkers can indeed be used for tracing back specific organisms on time scales of hundreds to even thousands of millions of years.

An interesting phenomenon in the evolution of the GSB is the existence of the hypothetical carotenoid 9 with the 3,4,5-trimethyl substitution pattern of one of the rings. The structural resemblance of its diagenetic derivatives to isorenieratene (7) in combination with its distinct isotopic signature [59, 67] confirms its origin from GSB. However, diagenetic products from this carotenoid have only been found in sediments and oils from the Ordovician to the Devonian (Table 3) and have not been found in any samples

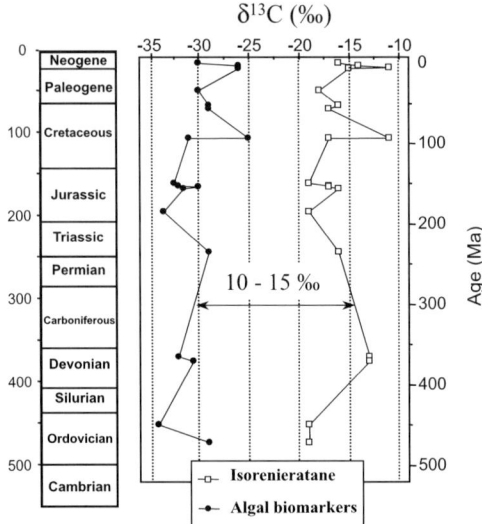

Fig. 7 Stable carbon isotopic composition of isorenieratene derivatives through time

of younger age. This suggests that the group of GSB capable of biosynthesizing this carotenoid had become extinct during evolution of the GSB. We can only speculate why. Perhaps the carotenoid with its odd substitution pattern has slightly different light absorption characteristics. It is known that the light intensity of the sun reaching the Earth's surface has increased over geological time and, perhaps, the specific carotenoid helped GSB to cope with this in their specific ecological niches. However, it is peculiar that derivatives of this carotenoid have not been found in the much older Paleoproterozoic sediments (Summons, per commun.), although isorenieratene derivatives are only minor components, due to the assumed shallow anoxic conditions [94]. Behrens et al. [98] noted that in sediments from the lower Keg River Formation derivatives from carotenoid **9** were predominant, whereas in sediment from the Upper Keg River only derivatives of isorenieratene **7** occurred. They attributed this to a change in the population of *Chlorobiaceae* resulting from a change in salinity or light availability.

The sedimentary record of PSB is much more sketchy. Derivatives of okenone are much less common, although aryl isoprenoids with the 2,3,4-trimethyl aryl substitution pattern have been widely reported. The perhydro derivative of okenone has, however, been found in the Paleoproterozoic Barney Creek Fm., indicating that ancestors of this group within the proteobacteria existed at that time and probably had the same physiology as the present day organisms.

11
Conclusions

Both intact pigments and bacteriochlorophyll of green and purple sulfur bacteria and their diagenetic and catagenetic derivatives can provide information on the spatial overlap of the photic and euxinic zone in past depositional systems. This is important in understanding the processes and conditions associated with burial of organic matter in sediments. In contrast to present day oceanic settings, photic zone anoxia occurred quite commonly during certain periods in the geological past and this often resulted in the deposition of black shales, the precursor of petroleum.

References

1. Demaison GJ, Moore GT (1980) AAPG Bull 64:1179
2. Suess E (1980) Nature 288:260
3. Henrichs SM (1992) Mar Chem 39:119
4. Sinninghe Damsté JS, Rijpstra WIC, Reichart GJ (2002) Geochim Cosmochim Acta 66:2737
5. Hartnett HE, Keil RG, Hedges JI, Devol AH (1998) Nature 391:572
6. Hedges JI, Sheng Hu F, Devol AH, Hartnett HE, Tsamakis E, Keil RG (1999) Am J Sci 299:529
7. Klemme HD, Ulmishek GF (1991) AAPG Bull 75:1809
8. Nijenhuis IA, Bosch H-J, Sinninghe Damsté JS, Brumsack HJ, de Lange GJ (1999) Earth Planet Sci Lett 169:277
9. Peters KE, Moldowan JM (1993) The biomarker guide. Interpreting molecular fossils in petroleum and ancient sediments. Prentice-Hall, Englewood Cliffs, New Jersey
10. Imhoff JF (2000) The Chromatiaceae. The Prokaryotes. http://141.150.157.117:8080/prokPUB/chaprender/jsp/showchap.jsp?chapnum=472
11. Overmann J (2000) The family Chlorobiaceae. The Prokaryotes. http://141.150.157.117:8080/prokPUB/chaprender/jsp/showchap.jsp?chapnum=323
12. Overmann J, van Gemerden H (2000) FEMS Microbiol Rev 24:591
13. Pfennig N (1989) Ecology of phototrophic purple and green sulfur bacteria. In: Schlegel HG, Bowien B (eds) Autotrophic Bacteria. Springer, Berlin Heidelberg New York, p 97
14. Bergey (2004) Manual of systematic bacteriology, second edition. Taxonomic outline of the prokaryotes—on-line edition. http://141.150.157.80/bergeysoutline/main.htm
15. Imhoff JF (2003) Int J Syst Evol Microbiol 53:941
16. van Gemerden H, Mas J (1995) Ecology of phototrophic sulfur bacteria. In: Blankenship RE, Madigan MT, Bauer CE (eds) Advances in photosynthesis, Vol 2: Anoxygenix photosynthetic bacteria. Kluwer, Dordrecht, p 49
17. Overmann J, Beatty JT, Hall KJ, Pfennig N, Northcote TG (1991) Limnol Oceanogr 36:846
18. Repeta DJ (1989) Geochim Cosmochim Acta 53:699
19. Schaeffer P, Adam P, Wehrung P, Albrecht P (1997) Tetrahedron Lett 38:8413
20. Repeta DJ, Simpson DJ (1991) Deep-Sea Res Part A 38:S969

21. Overmann J, Cypionka H, Pfennig N (1992) Limnol Oceanogr 37:150
22. Repeta DJ, Simpson DJ, Jorgensen BB, Jannasch HW (1989) Nature 342:69
23. Glaeser J, Overmann J (2003) Appl Environ Microbiol 69:3739
24. Wong W, Sackett WM, Benedict CR (1975) Plant Physiol 55:475
25. Quandt I, Gottschalk G, Ziegler H, Stichler W (1977) FEMS Microbiol Lett 1:125
26. Sirevag R, Buchanan BB, Berry JA, Troughton JH (1977) Arch Microbiol 112:35
27. Madigan MT, Takigiku R, Lee RG, Gest H, Hayes JM (1989) Appl Environ Microbiol 55:639
28. van der Meer MTJ, Schouten S, Sinninghe Damsté JS (1998) Org Geochem 28:527
29. Schouten S, Klein Breteler W, Blokker P, Schogt N, Rijpstra WIC, Grice K, Baas M, Sinninghe Damsté JS (1998) Geochim Cosmochim Acta 62:1406
30. van Breugel Y, Schouten S, Paetzel M, Ossebaar J, Sinninghe Damsté JS (2005) Earth Planet Sci Lett 235:427
31. Van Breugel Y, Schouten S, Paetzel M, Nordeide R, Sinninghe Damsté JS (2005) Org Geochem 36:1763
32. Wachniew P, Rozanski K (1997) Geochim Cosmochim Acta 61:2453
33. Velinsky DJ, Fogel ML (1999) Mar Chem 67:161
34. Rau GH (1978) Science 201:901
35. Fry B, Jannasch HW, Molyneaux SJ, Wirsen CO, Muramoto JA, King S (1991) Deep-Sea Res 38:1003
36. Hartgers WA, Schouten S, Lopez JF, Sinninghe Damsté JS, Grimalt JO (2000) Org Geochem 31:777
37. Coolen MJL, Overmann J (1998) Appl Environ Microbiol 64:4513
38. Rankin LM, Gibson JAE, Franzmann PD, Burton HR (1999) Polarforschung 66:33
39. Burke CM, Burton HR (1988) Hydrobiologia 165:13
40. Coolen MJL, Muyzer G, Schouten S, Volkman JK, Sinninghe Damsté JS (2005) Sulfur and methane cycling during the Holocene in Ace Lake (Antarctica) revealed by lipid and DNA stratigraphy. NATO Conference Proceedings (in press)
41. Schouten S, Rijpstra WIC, Kok M, Hopmans EC, Summons RE, Volkman JK, Sinninghe Damsté JS (2001) Geochim Cosmochim Acta 65:1629
42. Smittenberg RH, Pancost RD, Hopmans EC, Paetzel M, Sinninghe Damsté JS (2004) Palaeogeogr Palaeoclimat Palaeoecol 202:331
43. Overmann J, Sandmann G, Hall KJ, Northcote TG (1993) Aquatic Sci 55:31
44. Repeta DJ (1993) Geochim Cosmochim Acta 57:4337
45. Sinninghe Damsté JS, Wakeham SG, Kohnen MEL, Hayes JM, de Leeuw JW (1993) Nature 362:827
46. Sinninghe Damsté JS, Koopmans MP (1997) Pure Appl Chem 69:2067
47. Putschew A, Schaeffer P, Schaeffer-Reiss C, Maxwell JR (1998) Org Geochem 29:1849
48. Keely BJ, Blake SR, Schaeffer P, Maxwell JR (1995) Org Geochem 23:527
49. Menzel D, Hopmans EC, Van Bergen PF, de Leeuw JW, Sinninghe Damsté JS (2002) Mar Geol 189:215
50. Passier HF, Bosch H-J, Nijenhuis IA, Lourens LJ, Bottcher ME, Leenders A, Sinninghe Damsté JS, de Lange GJ, de Leeuw JW (1999) Nature 397:146
51. Bosch H-J, Sinninghe Damsté JS, de Leeuw JW (1998) Proc ODP, Sci Res 160:285
52. Sinninghe Damsté JS, Schouten S, van Duin ACT (2001) Geochim Cosmochim Acta 65:1557
53. Schaefle J, Ludwig B, Albrecht P, Ourisson G (1977) Tetrahedron Lett 41:3673
54. Summons RE, Powell TG (1986) Nature 319:763
55. Summons RE, Powell TG (1987) Geochim Cosmochim Acta 51:557
56. Sinninghe Damsté JS, Köster J (1998) Earth Planet Sci Lett 158:165

57. Van Kaam-Peters HME, Sinninghe Damsté JS (1997) Org Geochem 27:371
58. Requejo AG, Allan J, Creaney S, Gray NR, Cole KS (1992) Org Geochem 19:245
59. Hartgers WA, Sinninghe Damsté JS, Requejo AG, Allan J, Hayes JM, Ling Y, Xie TM, Primack J, de Leeuw JW (1994) Org Geochem 22:703
60. Hartgers WA, Sinninghe Damsté JS, Koopmans MP, de Leeuw JW (1993) J Chem Soc-Chem Commun 1715
61. Hartgers WA, Sinninghe Damsté JS, Requejo AG, Allan J, Hayes JM, de Leeuw JW (1994) Nature 369:224
62. Koopmans MP, Schouten S, Kohnen MEL, Sinninghe Damsté JS (1996) Geochim Cosmochim Acta 60:4873
63. Schaeffer P, Schmitt G, Behrens A, Adam P, Hebting Y, Bernasconi S, Albrecht P (2001) Abstract book, 20th International Meeting on Organic Geochemistry, Nancy 2:365
64. Sinninghe Damsté JS, Köster J, Baas M, Koopmans MP, van de Graaf B, Geenevasen JAJ, Kruk C (1995) J Chem Soc Chem Commun 187
65. Koopmans MP, Köster J, van Kaam-Peters HME, Kenig F, Schouten S, Hartgers WA, de Leeuw JW, Sinninghe Damsté JS (1996) Geochim Cosmochim Acta 60:4467
66. Grice K, Schaeffer P, Schwark L, Maxwell JR (1997) Org Geochem 26:677
67. Clifford DJ, Clayton JL, Sinninghe Damsté JS (1998) Org Geochem 29:1253
68. van Duin ACT, Sinninghe Damsté JS (2003) Org Geochem 34:515
69. Sinninghe Damsté JS, de Leeuw JW (1990) Org Geochem 16:1077
70. Schaeffer P, Reiss C, Albrecht P (1995) Org Geochem 23:567
71. Kohnen MEL, Sinninghe Damsté JS, Kock-van Dalen AC, de Leeuw JW (1991) Geochim Cosmochim Acta 55:1375
72. Kohnen MEL, Schouten S, Sinninghe Damsté JS, de Leeuw JW, Merritt DA, Hayes JM (1992) Science 256:358
73. Hofmann IC, Hutchison J, Robson JN, Chicarelli MI, Maxwell JR (1992) Org Geochem 19:371
74. Adam P, Schmid JC, Mycke B, Strazielle C, Connan J, Huc A, Riva A, Albrecht P (1993) Geochim Cosmochim Acta 57:3395
75. Schouten S, Pavlovic D, Sinninghe Damsté JS, de Leeuw JW (1993) Org Geochem 20:901
76. Sinninghe Damsté JS, Frewin NL, Kenig F, de Leeuw JW (1995) Org Geochem 23:471
77. Kenig F, Sinninghe Damsté JS, Frewin NL, Hayes JM, de Leeuw JW (1995) Org Geochem 23:485
78. Wakeham SG, Sinninghe Damsté JS, Kohnen MEL, de Leeuw JW (1995) Geochim Cosmochim Acta 59:521
79. Kuypers MMM, Pancost RD, Nijenhuis IA, Sinninghe Damsté JS (2002) Paleoceanogr 17:10.1029/2000PA000569
80. Kolonic S, Lüning S, Akmal M, Wagner T, Chelai H, Tsikos H, Kuhnt W, Sinninghe Damsté JS (2005) Paleoceanogr 20:PA1006
81. Schaeffer P, Harrison WN, Keely BJ, Maxwell JR (1995) Org Geochem 23:541
82. Schouten S, Sinninghe Damsté JS, Baas M, Kock-van Dalen AC, Kohnen MEL, de Leeuw JW (1995) Org Geochem 23:765
83. Douglas AG, Sinninghe Damsté JS, Fowler MG, Eglinton TI, de Leeuw JW (1991) Geochim Cosmochim Acta 55:275
84. Hartgers WA, Sinninghe Damsté JS, de Leeuw JW (1994) Geochim Cosmochim Acta 58:1759
85. Gelin F, Sinninghe Damsté JS, Harrison WN, Maxwell JR, de Leeuw JW (1995) Org Geochem 23:555

86. Pedentchouk N, Freeman KH, Harris NB, Clifford DJ, Grice K (2004) Org Geochem 35:33
87. Hoefs MJL, Sinninghe Damsté JS, De Leeuw JW (1995) Org Geochem 23:263
88. Byers JD Erdman JG (1983) Low temperature degradation of carotenoids as a model for early diagenesis in recent sediments. In: Bjorøy M, Albrecht P, Cornford C, De Groot K, Eglinton G, Galimov E, Leythaeuser D, Pelet R, Rullkötter J, Speers G (eds) Advances in Organic Geochemistry 1981. Wiley, Chichester, p 725
89. Ostroukhov SB, Aref'yev OA, Makuina VM, Zabrodina MN, Petrov AlA (1982) Neftekhimiya (in Russian) 22:723
90. Callot HJ, Ocampo R (2000) In: Kadish KM, Smith KM, Guillard R (eds) The Porphyrin Handbook. Academic Press, London, p 350
91. Ocampo R, Callot HJ, Albrecht P (1985) J Chem Soc Chem Commun 200
92. Keely BJ, Maxwell JR (1993) Org Geochem 20:1217
93. Grice K, Gibbison R, Atkinson JE, Schwark L, Eckardt CB, Maxwell JR (1996) Geochim Cosmochim Acta 60:3913
94. Brocks JJ, Love GD, Summons RE, Logan GA (2004) Geochim Cosmochim Acta 68:796
95. Tyson RV, Pearson TH (eds) (1991) Modern and ancient continental shelf anoxia: an overview. Geol Soc Lond Spec Publ London 58:1
96. Kenig F, Hudson JD, Sinninghe Damsté JS, Popp BN (2004) Geology 32:421
97. Van Kaam-Peters HME, Rijpstra WIC, de Leeuw JW, Sinninghe Damsté JS (1998) Org Geochem 28:151
98. Behrens A, Wilkes H, Schaeffer P, Clegg H, Albrecht P (1998) Org Geochem 29:1905
99. Casamayor EO, Calderon-Paz JI, Mas J, Pedros-Alio C (1998) Arch Microbiol 170:269
100. Caumette P, Guyoneaud R, Imhoff JF, Suling J, Gorenko V (2004) Int J Syst Evol Microbiol 54:1031
101. Grice K, Schouten S, Nissenbaum A, Charrach J, Sinninghe Damsté JS (1998) Org Geochem 28:195
102. Köster J, Rospondek M, Schouten S, Kotarba M, Zubrzycki A, Sinninghe Damsté JS (1998) Org Geochem 29:649
103. Mawson DH, Walker JS, Keely BJ (2004) Org Geochem 35:1229
104. Grice K, Schouten S, Peters KE, Sinninghe Damsté JS (1998) Org Geochem 29:1745
105. Wagner T, Sinninghe Damsté JS, Hofmann P, Beckmann B (2004) Paleoceanogr 19:PA409910.1029/2004PA001087
106. Tsikos H, Jenkyns HC, Walsworth-Bell B, Petrizzo MR, Forster A, Kolonic S, Erba E, Silva IP, Baas M, Wagner T, Sinninghe Damsté JS (2004) J Geol Soc Lond 161:711
107. Pancost RD, Crawford N, Magness S, Turner A, Jenkyns HC, Maxwell JR (2004) J Geol Soc Lond 161:353
108. Kuypers MMM, Lourens L, Rijpstra WIC, Pancost RD, Nijenhuis IA, Sinninghe Damsté JS (2004) Earth Planet Sci Lett 228:465
109. Simons DJH, Kenig F, Schroder-Adams CJ (2003) Org Geochem 34:1177
110. van Breugel et al. (2004) unpublished results
111. Xinke Y, Pu F, Philp RP (1990) Org Geochem 15:433
112. van Kaam-Peters HME, Schouten S, Köster J, Sinninghe Damsté JS (1998) Geochim Cosmochim Acta 62:3259
113. Schouten S, van Kaam-Peters HME, Rijpstra WIC, Schoell M, Sinninghe Damsté JS (2000) Am J Sci 300:1
114. Schwark L, Frimmel A (2004) Chem Geol 206:231
115. Saelen G, Tyson RV, Telnæs N, Talbot MR (2000) Palaeogeog Palaeoclimat Palaeoecol 163:163

116. Köster J, Schouten S, Sinninghe Damsté JS, de Leeuw JW (1995) Reconstruction of the depositional environment of Toarcian marlstones (Allgäu Formation, Tirol/Austria) using biomarkers and compound specific carbon isotope analyses. In: Grimalt JO, Dorronsoro C (eds) Organic geochemistry: developments and applications of energy, climate, environment and human history. A.I.G.O.A, San Sebastian, Spain, p 76
117. Rosell-Mele A, Carter JF, Maxwell JR (1999) Rapid Commun Mass Spectrom 13:568
118. Grice K, Schaeffer P, Schwark L, Maxwell JR (1996) Org Geochem 25:131
119. Schwark L, Puttmann W (1990) Org Geochem 16:749
120. Gibbison R, Peakman TM, Maxwell JR (1995) Tetrahedron Lett 36:9057
121. Bechtel A, Püttmann W (1997) Palaeogeog Palaeoclimat Palaeoecol 136:331
122. Pancost RD, Crawford N, Maxwell JR (2002) Chem Geol 188:21763
123. Joachimski MM, Ostertag-Henning C, Pancost RD, Strauss H, Freeman KH, Littke R, Sinninghe Damsté JS, Racki G (2001) Chem Geol 175:109
124. Clegg H, Horsfield B, Stasiuk L, Fowler M, Vliex M (1997) Org Geochem 26:627
125. Guthrie JM (1996) Org Geochem 25:439
126. Koopmans MP, de Leeuw JW, Lewan MD, Sinninghe Damsté JS (1996) Org Geochem 25:391
127. Pancost RD, Freeman KH, Patzkowsky ME, Wavrek DA, Collister JW (1998) Org Geochem 29:1649

Atmospheric Transport of Terrestrial Organic Matter to the Sea

Bernd R. T. Simoneit

Environmental and Petroleum Geochemistry Group, College of Oceanic and Atmospheric Sciences, Oregon State University, Corvallis, OR 97331, USA
simoneit@coas.oregonstate.edu

1	Introduction	166
1.1	Global Dust and Wind Systems	166
1.2	Inorganic Tracers	168
1.3	Urban–Global Mixing and Transport	169
1.4	Dust Input to the Ocean	169
1.5	Definitions of Terms Used	170
2	**Terrestrial Organic Matter in the Atmosphere**	171
2.1	Volatile Compounds	172
2.2	Organic Matter of Atmospheric Particles	173
2.2.1	Organic Compounds	173
2.2.2	Urban Emissions	173
2.2.3	Burning of Biomass	174
2.2.4	Aliphatic Homologous Compounds	175
2.2.5	Terpenoid and Steroid Biomarkers	179
2.2.6	Compound-Specific Isotope Analysis	180
2.2.7	Polar Compounds	180
2.2.8	PAH and Oxy-PAH	181
2.2.9	Humic and Fulvic Acids	183
2.3	Elemental Carbon	183
2.4	Source Signatures	183
2.4.1	Ambient Vegetation	184
2.4.2	Fossil Fuel Utilization	184
2.4.3	Biomass Burning	186
2.4.4	Soil/Sand Resuspension	188
2.5	Secondary Reactions	190
2.6	Fallout and Washout	191
2.7	Modeling	191
2.8	Water-Soluble Organic Matter	192
3	**Sampling and Analytical Methods**	192
3.1	Sampling	192
3.2	Extraction and Fractionation	193
3.3	Instrumental Analyses	194
3.4	Data Modeling	195

4	Overview	196
4.1	Major Terrestrial Organic Tracers and Sources	196
4.2	Major Secondary Products	196
5	Conclusions	198
	References	199

Abstract The transport of atmospheric particles with associated terrestrial organic matter to the oceans is an important process affecting various global concerns such as climate change and environmental and human health. Aerosol particulate matter over the oceans is derived from autochthonous emissions admixed with varying amounts of continental effluents. It is important to be able to assess the sources and fate (receptors) of aerosols, and both inorganic and organic tracers are of utility. The organic compounds of atmospheric particles from marine and terrestrial regions can be characterized by specific tracers for (1) natural emissions (marine lipids, vegetation waxes, terpenes), (2) fossil fuels utilization (internal combustion engine emissions, coal burning), (3) biomass burning (taxon specific, wildfires), (4) anthropogenic emissions (industry, urban activity), and (5) soil and desert sand resuspension (agriculture, wind erosion). The precursor–product chemistry can then be used to assess the secondary reactions (thermal or atmospheric) and the fate of aerosol organic matter.

Keywords Aerosols · Biomarkers · Lipids · Fossil fuels · Saccharides

1
Introduction

Organic matter can be transported to the sea by river, atmospheric, and ice rafting processes. For all these processes the continental material must first be broken down into smaller particles, i.e., eroded, before removal. This erosion is mainly by the action of flowing water, but air erodes the finer particles from areas where no moisture binds the soil or minerals together [1]. Rivers deposit large amounts of mainly mineral detritus near their deltas in the ocean. Atmospheric fallout and washout deposit lesser amounts of mineral and associated organic matter over larger areas of the oceans [2–4]. This chapter deals with the atmospheric transport of the particle-associated organic matter to the sea.

1.1
Global Dust and Wind Systems

The earliest observations on eolian dust fallout in the Atlantic Ocean were reported by Darwin [5] and Ehrenberg [6], who concluded that Africa was the most likely source of the dust on the basis of the fresh-water diatom fossil content. The identification of eolian material in marine sediments was initially reported by Radczewski [7, 8], using iron oxide coated quartz particles

(*Wüstenquarz*) as the diagnostic marker for desert dust derived from central North Africa. The presence of *Melosira granulata* in eolian dust collected at Barbados [9], and the report that this same fresh-water diatom species was present in deep-sea sediments of the tropical belt of the Atlantic Ocean [10], confirmed the wind transport of these diatoms across the sea from Africa. They originated in lakes, rivers, and swamps of central Africa, where, after the dessication during the dry season, fine dust from the bottom muds (often with plant detritus and ash) was taken up by the trade winds and blown out to sea [9]. Rex and Goldberg [11] surveyed wind-transport mechanisms and summarized three generic classes for particulate matter: (1) extraterrestrial particles, (2) solids of biological origins (including anthropogenic emissions)

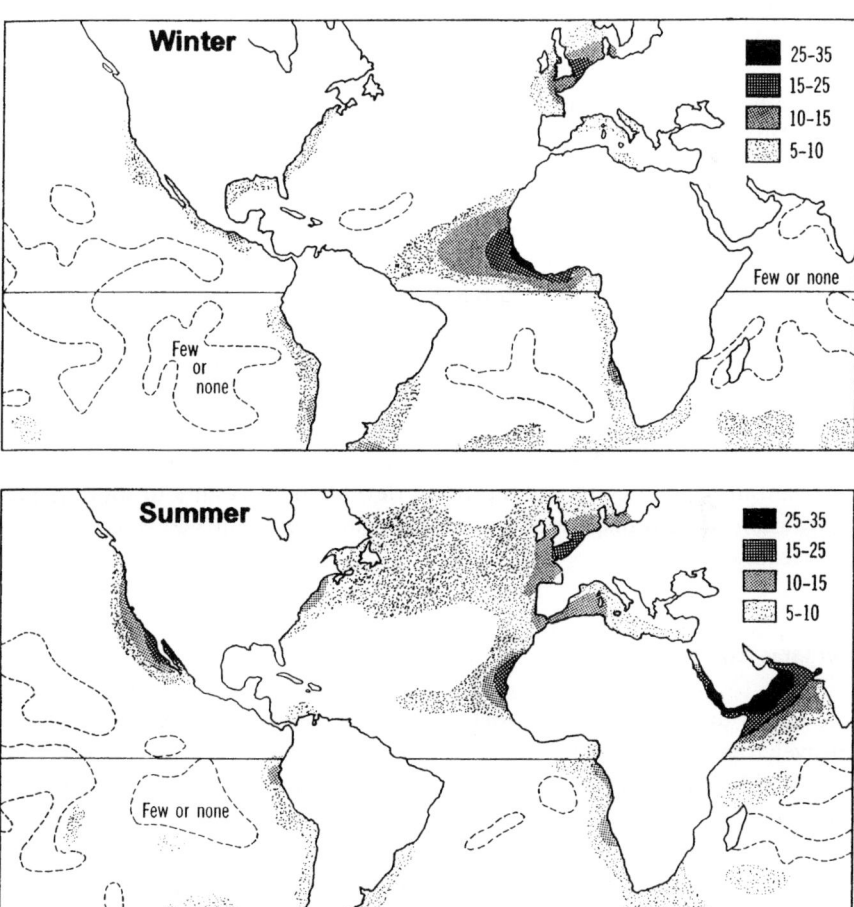

Fig. 1 Average frequency of haze over the ocean during the northern winter (*upper map*) and summer (*lower map*). Frequencies are given in percent of total number of observations. (Redrawn from Arrhenius [233]. Reproduced with permission, J. Wiley & Sons)

from the continents, and (3) solids of inorganic origins from the lithosphere, including debris from volcanic activity.

Generic class 2 is the major contributor of organic matter to eolian particulate matter and includes man's activities. Most atmospheric fall-out and washout originates from the lower troposphere (altitude below 15 km [12]). The major wind patterns of the troposphere are the trade winds and westerlies. The haze observed over certain parts of the world ocean (Fig. 1) is due mainly to atmospheric particulate matter. Parts of the Atlantic Ocean (15°N–50°N, westerlies driven by the jet stream; 15°S–15°N, trade winds; 15°S–50°S, westerlies driven by the jet stream), the Indian Ocean (15°S–20°N, trade winds; 20°S–50°S, westerlies), and some areas of the Pacific Ocean (15°N–50°N, westerlies; 15°N–15°S, trade winds; 15°S–50°S, westerlies) are the major regions of the world ocean where atmospheric input to the sediments is presently significant [13].

Aerosols influence the radiative balance of the Earth through absorption and scattering of solar radiation back to space, known as the direct effect, and by increasing the amount of solar radiation that clouds reflect, the indirect effect [14]. The direct radiative effect of aerosols in strongly influenced by particle size and composition. Radiative effects of anthropogenic aerosols are relatively large compared with their mass contribution because they are in the size range which is radiatively most active [15–17]. Particles emitted from biological organic matter are a source that contributes significantly to the total particle burden in the atmosphere.

Another major concern is the potential detrimental human health effects of fine aerosol particulate matter with its associated toxic organic compounds (e.g., polycyclic aromatic hydrocarbons, PAHs [18, 19]). These health effects have been addressed mainly in urban areas; however, increasing global industrialization with concomitant urban development results in more global distributions and transport of the anthropogenic toxic emissions, even across the oceans.

1.2
Inorganic Tracers

The eolian dust veil over the North Atlantic has been investigated more extensively than any other marine area, especially in terms of inorganic chemistry [4, 20–28]. A dust fall south of the Cape Verde Islands on 17 January 1965 [25, 26], when examined under the microscope, consisted of mostly quartz and calcite in the coarse fraction (silt). Most of the remainder was phytoliths (about 5%), fresh-water diatoms (about 3%), and fungal spores (about 2%). Insect scales, plant tissue, and a few opaque spherules (urban emissions) were also observed, but pollen was not found. The phytoliths were associated with the fresh-water diatoms, indicating the same areal source, such as grasslands around intermittent lakes [25].

The major terrigenous marker minerals composing eolian dusts from various global areas are clays, quartz (usually coated with iron oxide), and feldspars [4, 29]. The clays consist of montmorillonite, illite, chlorite, and kaolinite and are all of a land-derived origin [24].

1.3
Urban–Global Mixing and Transport

The atmosphere–ocean interface is under intensive study to assess the transport and deposition of organic and inorganic chemical species from land to ocean, and ultimately to the marine sedimentary record [30, 31]. The atmospheric input of terrigenous organic carbon to the world ocean is about equivalent to the organic carbon washed out by the rivers, and is estimated to be 20 Tg per year [32]. Most of this carbon is of a contemporary biological origin, and is associated with transport of mineral dust by major wind systems that are seasonally active from the continents to the oceans [2, 3, 13, 27, 33–38]. The increasing atmospheric burden of urban particulate matter intermingles with the natural and anthropogenic emissions in the continental rural areas, and the total mixture is eventually transported to the atmosphere over the oceans [39, 40]. Major oxidative-photochemical reactions alter the organic matter composition during transport, forming derivative products with higher oxygen contents, i.e., greater polarity [41, 42]. Those secondary products form cloud condensation nuclei, complementing inorganic species such as SO_x and NO_x. Thus, molecular characterization and provision of mass balance estimates for aerosols near sources and in downwind regions will continue to be a major need.

A precautionary comment needs to be made regarding the analyses of low-level organic tracers in atmospheric particles transported over long distances. This was discussed with an example of a study from Puerto Rico which addressed organic aerosols of the trade winds in the Caribbean [43]. These authors reported primarily synthetic organic compounds found in suntan lotions and other lubricants, and proposed that those compounds were natural components of the trade wind particulate matter. All the compounds reported appear to be artifacts and contaminants from the sampling and experimental procedures, and the inferred results are not valid [44]. Successful contamination control for aerosol sampling on islands has also been extensively discussed for the SEAREX Program (Refs. [45, 166] and references therein).

1.4
Dust Input to the Ocean

The actual input of eolian dust to marine sediments occurs by two major pathways, namely, dry fallout and wet washout [45]. When the dust falls out or is washed out by rain onto the sea surface, it rapidly flocculates [46]. The floc formation appears to be aided by the presence of organic matter [47]. The settling speeds of such flocculated aggregates are many times greater that those of the constituent particles [46]. Marine filter feeders may also ingest some of the organic-rich flocs and thus aggregate the material into fecal pellets, which then also sink rapidly [9, 48, 49]

1.5
Definitions of Terms Used

Some of the terms and concepts used in this chapter need to be defined, because they are not necessarily the same as used by scientists in other disciplines. A brief selection follows:

Aerosol An atmospheric suspension of liquid and solid particles, including adsorbed water and volatile organic components on the solids.

Anthropogenic Designates derivation from human activity.

Biomarker Molecules that have definitive chemical structures, relatable either directly or indirectly through a set of diagenetic alterations to biogenic precursors (sources), and that cannot be synthesized by abiogenic processes.

Biomass Any biosynthetic matter, such as all vegetation (flora), its recent detritus (e.g., litter, humus, peat to lignite), all fauna (e.g., rendering, cooking), and all products produced by industry utilizing biomass as raw material (e.g., paper, rubber).

Bitumen The geological equivalent of lipids, consisting in the widest sense of any sedimentary hydrocarbon mixture ranging in state from tarry (asphalt), through viscous to liquid (petroleum).

Desert sand The fine fraction of wind-eroded particulate matter carried by major wind systems from continents to the sea (e.g., Kosa Asian dust, Sirocco).

Lipids A broad term that includes all oil-soluble, water-insoluble organic substances, such as fats, waxes, fatty acids, fatty alcohols, sterols, pigments, and terpenoids, all biosynthesized by contemporary biota.

Particles Can be solid or a liquid in air (gas). A mixture of such solid and liquid particles constitutes an aerosol. Typical dimensions are greater than 0.001 mm and particles can be sampled by filtration.

Smoke Small gasborne particles formed during incomplete combustion. It consists mainly of carbon and minor organic compounds and inorganic matter with typical dimensions greater than 0.01 µm.

Soil The fine particle fraction injected into the atmosphere during agricultural tilling and harvesting of fields, and by wind erosion of fallow fields.

2
Terrestrial Organic Matter in the Atmosphere

Atmospheric particulate matter contains organic tracers which are characteristic of (1) their sources (biological and geological), (2) the mode of formation, and (3) subsequent alteration during transport (secondary products) downwind. The major source categories and the production mechanisms for atmospheric particulate matter with the associated organic compound tracers are summarized in Table 1. The compound composition of urban, rural, and remote (including marine) aerosol samples will be described here briefly by using some illustrative examples. The major compound groups derived from emission sources of atmospheric particles are summarized in Table 2.

Table 1 Source categories and production mechanisms for atmospheric particulate matter (*PM*) and associated organic tracers

Source category, process	Interpretation/Application
Natural or uncontrolled processes	
1. Terrestrial	
(a) Vegetation (wind abrasion and electrostatic)	Natural-background PM (biogenic)
(b) Soil erosion/resuspension	Natural-background PM (biogenic/synthetic)
(c) Wild fires, biomass burning	Smoke/ash PM, biomarkers (biogenic)
2. Marine	
Autochthonous lipids (seaslick)	Natural-background PM (biogenic)
Anthropogenic activities	
1. Urban (industries, food preparation, home heating)	Smoke/dust, biomarkers, and synthetic compounds
2. Rural/remote (agriculture, controlled burning, soil resuspension)	Dust/smoke, biomarkers (biogenic)
3. Marine (fallout of ship emissions/ resuspension of seaslick)	Exhaust biomarkers (geologic)
Fossil fuels	
1. Petroleum (internal combustion engines)	Exhaust biomarkers (geologic)
2. Coal (burning)	Smoke biomarkers (geologic)

Table 2 Key source-specific tracers for organic components in aerosol particles

Compound or compound class		Major source [a]	Emission process [b]
n-Alkanes	$C_{15}-C_{20}$	Microbial (urban)	Direct/resuspension (vehicle exhaust)
	$C_{20}-C_{37}$	Plant waxes (urban)	Direct/biomass burning (vehicle exhaust)
n-Alkenes	$C_{15}-C_{37}$	Biomass (coal)	Burning
n-Alkan-2-ones	$C_{15}-C_{35}$	Biomass (coal)	Biodegradation (burning)
n-Alkanals	$C_{15}-C_{35}$	Biomass (coal)	Biodegradation (burning)
	C_3-C_{10}	Microbial (fossil fuel)	Photo-oxidation (combustion)
n-Alkanoic acids	$C_{12}-C_{19}$	Microbial (biomass)	Direct/resuspension (burning)
	$C_{20}-C_{36}$	Higher plants	Direct (burning)
n-Alkanols	$C_{14}-C_{36}$	Biomass	Direct (burning)
n-Alkyl amides	$C_{14}-C_{32}$	Biomass	Burning (cooking)
n-Alkyl nitriles	$C_{14}-C_{32}$	Biomass	Burning (cooking)
Alkanedioic acids	C_2-C_{28}	Various	Photo-oxidation (combustion)
Wax esters	$C_{28}-C_{58}$	Plant waxes	Biomass burning
Triterpenyl alkanoates		Tropical vegetation	Biomass burning
Triacylglycerides		Flora (fauna)	Biomass burning (cooking)
Methoxyphenols		Biomass with lignins	Burning
Levoglucosan (mannosan, galactosan)		Biomass with cellulose	Burning
Saccharides		Soil (sugar processing)	Direct resuspension (agricultural/industrial)
Cholesterol		Urban (algae)	Cooking (direct)
Phytosterols		Higher plants	Burning (direct)
Triterpenoids		Higher plants	Burning (direct)
Diterpenoids (resin acids)		Higher plants, i.e., conifers	Burning (direct)
Hopanes/steranes		Petroleum	Urban (vehicle exhaust, etc.)
UCM		Petroleum	Urban (vehicle exhaust, etc.)
Alkylpicenes/alkylchrysenes		Coals	Urban (burning/heating)
PAHs		Ubiquitous	All pyrogenic processes

UCM unresolved complex mixture, *PAHs* polycyclic aromatic hydrocarbons
[a] Secondary source in *parentheses*
[b] Listed in order of importance

2.1
Volatile Compounds

Volatile compounds in urban and rural airsheds are primarily hemi-, mono-, and sesquiterpenoids from vegetation and anthropogenic spillage and use of solvents, gasoline, etc. These compounds react quickly with OH, ozone, NO_x, or UV radiation, producing secondary aerosols [17, 50]; thus, the precursors and possibly the products are not expected to survive long-range transport. Volatile compounds are therefore not considered further in this review, although their secondary products may contribute to the low molecular weight polar compounds (e.g., oxalic and $C_3 - C_5$ dicarboxylic acids) in air parcels transported over long distances.

2.2
Organic Matter of Atmospheric Particles

The organic matter of atmospheric particles described here consists of extractable (solvent-soluble) organic compounds. It includes the fulvic and humic acid components that occur in some aerosols, but does not consider the biogenic particles such as bacteria, algae, pollen, and spores. The soluble components of biogenic particles are also extracted if they are not removed by size exclusion during sample acquisition.

2.2.1
Organic Compounds

The organic matter of atmospheric particles consists predominantly of lipids, soot, and humic and fulvic acids, and is now firmly established as a major fraction of aerosols [37, 38, 51–120]. Many of these reports present data for biological, fossil fuel, and other-source organic matter admixed in the aerosols. Furthermore, in comparison to the relatively extensive studies that have been carried out on hydrocarbons in aerosols of both natural and anthropogenic origins from urban and rural/remote regions, less molecular information is available on polar and water-soluble compounds [38–40, 45, 70, 82, 103, 105, 107, 108, 111, 121–126].

The oxidative breakdown of organic compounds in the atmosphere is another active research area. It produces, for example, short-chain-length monocarboxylic, dicarboxylic and ketocarboxylic acids (C_2–C_6), which are the secondary products (oxidation) from higher molecular weight organic compounds of both biological and fossil fuel origins, and not solely from unsaturated fatty acids [33, 34, 42, 127–129]. These carboxylic acids are found globally in the atmospheres of both urban and remote areas.

2.2.2
Urban Emissions

There have been numerous reports on organic compounds in urban atmospheric samples and most deal with the compounds related to human health effects (e.g., PAHs) and emissions from traffic. These will not be reviewed here and the reader is referred to earlier reviews [130–133], and to key summaries for the toxic PAHs [134–140]. Chemical fingerprints of emission sources have commonly been utilized to distinguish specific inputs from point sources, rather than correlation of an air parcel to its regional source [91, 141, 142]. The transport of urban emissions to remote and marine areas is an important component superimposed on the natural organic matter background and must therefore be considered in global modeling [39, 40, 143, 144].

2.2.3
Burning of Biomass

Burning of biomass is a ubiquitous process and comprises natural wildfires and numerous anthropogenic activities, such as controlled burning in agriculture and forestry, burning wood for heating or other purposes, food preparation such as grilling and frying, tobacco use, rendering industries, crematoria, and garbage disposal. Smoke, especially from wildfires, is transported globally. The fossil fuels, i.e., petroleum, coal, lignite, and peat, are derived from biological organic matter of past geological times and when burned emit compounds which are distinguishable from those derived from biomass burning. This topic and the associated organic tracers have been reviewed recently and the review also includes emissions from coal combustion [145].

Biomass burning is an important primary source of many organic compounds which are reactants in atmospheric chemistry, and of soot particulate matter that decreases visibility and absorbs incident radiation [146–148]. Thus, it is the smoke from burning processes which needs more source testing and characterization. The compound composition data of smoke particulate matter is important for understanding the contribution of organic components from biomass burning emissions to atmospheric chemistry and complements existing reports on the characterization of direct organic emissions (natural background) from biomass sources [37, 62, 81, 87, 90, 94, 99, 104, 105, 117, 125, 126, 140, 145, 149–161].

The varying temperature conditions during burning determine the molecular alteration and transformation of the organic compounds emitted from biomass fuels [125, 126]. The heat intensity, aeration, and duration of smoldering and flaming conditions determine the distributions and ratios of the natural versus altered compounds present in smoke [117, 126]. Thus, the directly emitted and the thermally altered molecular tracers in smoke provide a chemical fingerprint which is source-specific and useful for identifying contributions from single or multiple biomass fuels in samples of atmospheric particulate matter. Such data can complement the inorganic marker potassium from "potash" (water-soluble), which has been suggested as a tracer for wood smoke in receptor models [162]. However, K^+ is also emitted by other major sources in urban areas, such as meat cooking and refuse incineration [92, 163, 164], which must be taken into account in mass balance models.

2.2.4
Aliphatic Homologous Compounds

Aerosol particulate matter transported over the ocean contains mainly aliphatic lipids from terrestrial plant waxes admixed with varying amounts of urban emissions, autochthonous marine lipids, and secondary oxidation products [37, 38, 41, 45, 68, 69, 99, 100, 113, 165]. This was studied in detail during the SEAREX Program ([45, 166] and references therein) and is being confirmed under the auspices of the ACE-Asia Program [38, 40, 165, 167].

The distributions of the n-alkanes in ambient aerosol samples range from C_{16} to C_{40}, with a typical C_{max} at 29 or 31 and high carbon preference index (CPI) values [77, 105]. The presence of an unresolved complex mixture (UCM) of branched and cyclic hydrocarbons in urban samples (Fig. 2a, c), the n-alkanes from C_{16} to C_{22}, pristane, and phytane, indicates a direct input from petroleum fuel use (Fig. 2b, d; minor n-alkanes are better discernable in m/z 85 key ion plots of gas chromatography–mass spectrometry, GC-MS, data) [101, 103]. The higher molecular weight constituents (above C_{23}) represent the typical petroleum components in urban atmospheric particles with a minor amount of natural plant wax alkanes superimposed [82, 101, 103, 105, 115]. In remote and marine areas, the aerosols contain predominantly natural plant wax components as illustrated in Fig. 2c and d. The plant epicuticular wax signature is evident in the key ion plot (m/z 85) showing the strong odd-carbon-number predominance and C_{max} at 29. The GC-MS data for the total extract (silylated) of aerosol particles taken in Sapporo, Hokkaido, Japan, during the Asian dust event of 2001 are summarized in Fig. 3. The major components are polar compounds with minor aliphatic lipids. Key ion plots are used to define the homologous series (e.g., m/z 85 alkanes, m/z 117 alkanoic acids). The alkanes in the Sapporo aerosol can be apportioned into plant wax and fossil fuel emissions as drawn in Fig. 3b.

The n-alkanoic (fatty) acids are significant components in extractable aerosol lipids (especially marine samples: Figs. 2a, c, 3a, c) and generally range from C_{10} to C_{32}, with a strong even-carbon-number predominance (CPI > 5) and bimodal distribution with C_{max} at 16 and a secondary maxium at 24 or 26. They are interpreted to be of a biogenic origin. The homologs shorter than n-C_{20} may be derived in part from microbial/marine sources, although these acids are ubiquitous in biota [107]. The homologs longer than n-C_{22} are derived primarily from vascular plant wax [37, 168]. Long-chain wax esters (LCWE) are present in aerosols and in the smoke from burning many plant species, typically with compounds from 36 up to 60 total carbon numbers and a strong even-carbon-number predominance [105, 107, 169]. The LCWE series generally comprise mainly palmitic acid esterified with fatty alcohols ranging from C_{22} to C_{34} and minor amounts of stearic (C_{18}) and arachidic (C_{20}) acids esterified with the C_{32} and C_{34} alcohols. Wax esters occur in smoke aerosols at abundances comparable to those of the n-alkanes,

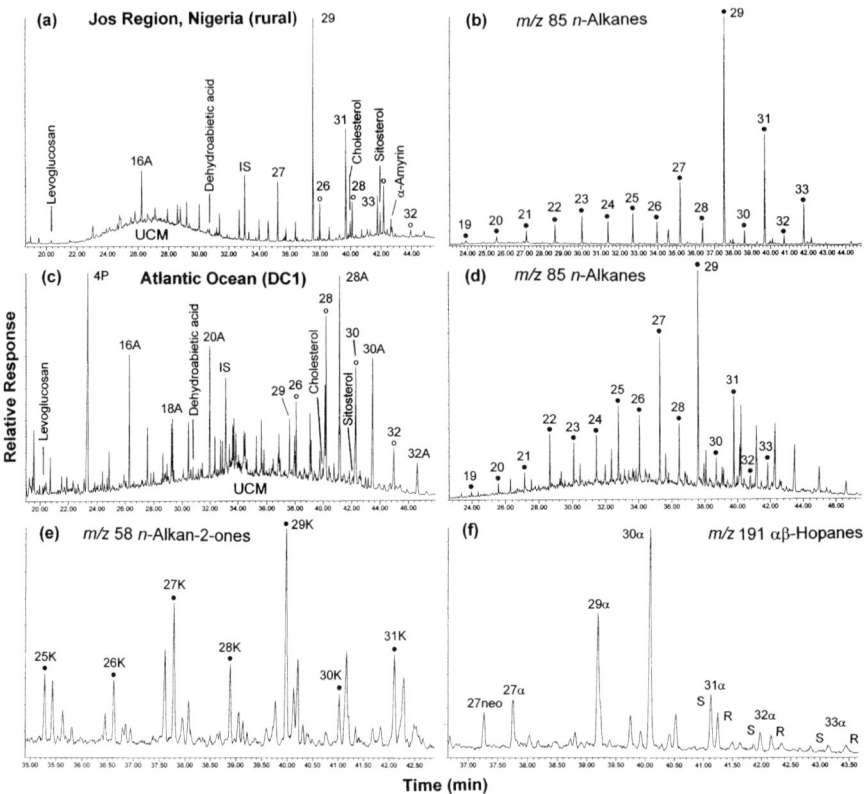

Fig. 2 Examples of gas chromatography–mass spectrometry (*GC-MS*) data for typical total extracts from aerosol particle samples (analyzed as silylated derivatives): **a** total ion current (*TIC*) trace for Jos region aerosol, Nigeria (March 1979), **b** m/z 85 fragmentogram (key ion for *n*-alkanes) for Jos aerosol showing the odd-carbon-number predominance of the wax alkanes, **c** TIC trace for Atlantic Ocean aerosol (DC2, December 1974), **d** m/z 85 fragmentogram, **e** m/z 58 fragmentogram (key ion for *n*-alkane-2-ones), and **f** m/z 191 fragmentogram (key ion for the hopane biomarkers) for DC2. The GC temperature program is different for the various samples. *UCM* unresolved complex mixture, *IS* internal standard ($C_{24}D_{50}$), *numbers* refer to carbon-chain length of *n*-alkanes, *iA n*-alkanoic acids, *open circles* alkanols (analyzed as trimethylsilyl ether or methyl esters)

providing further evidence for direct volatilization of very high molecular weight compounds into smoke and aerosols. They are found at trace levels in aerosols, indicating rapid and extensive alteration (hydrolysis) probably to the constituent fatty acids and alcohols.

Minor series of *n*-alkan-2-ones are often present in aerosol samples (Fig. 2e) and range from C_{25} to C_{35} with an odd-carbon-number predominance (CPI > 4). These compounds have been described for aerosols enriched in anthropogenic components [103, 104], and because they have essentially the same distribution and range as the wax alkanes indicate an origin from

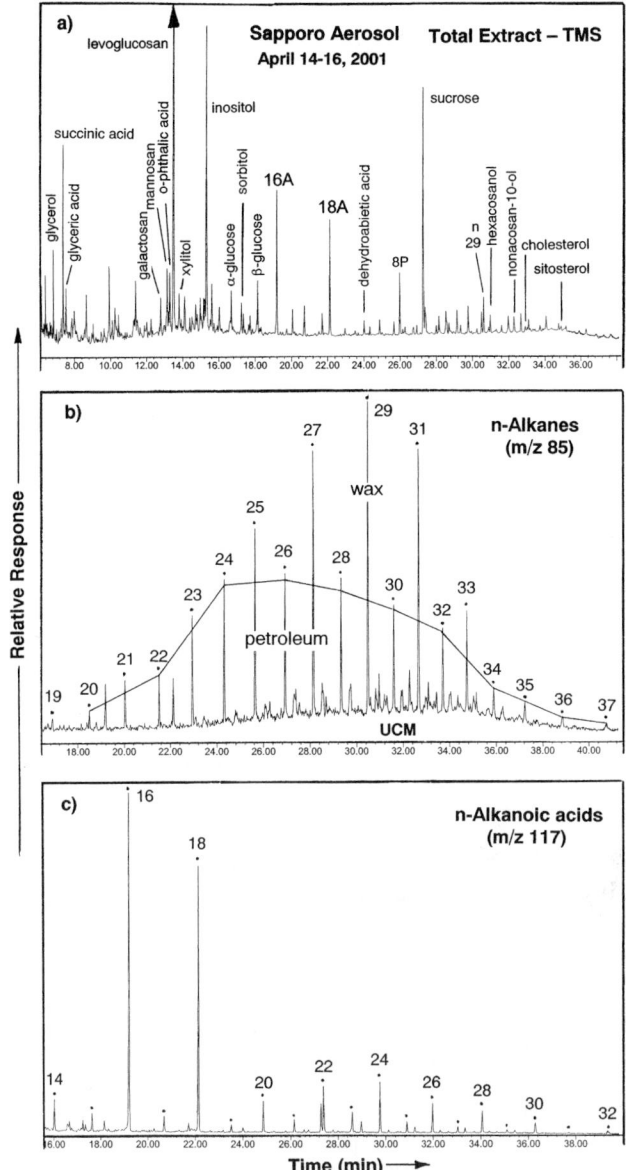

Fig. 3 GC-MS data for total extract of urban aerosol particles from Sapporo, Japan (April 2002, as silylated derivatives): **a** TIC trace, levoglucosan is shown at 0.4 of its total intensity, **b** m/z 85 fragmentogram (key ion for n-alkanes), and **c** m/z 117 fragmentogram (key ion for fatty acid TMS esters). Tracer compounds are labeled, *16A* and *18A* are n-alkanoic acids, *8P* is diethylhexyl phthalate, *n29* is nonacosane, *numbers over dots* indicate carbon-chain length, *UCM* is the unresolved complex mixture, *petroleum* is the n-alkane contribution under the envelope drawn from fossil fuel emissions and *wax* is the amount of superimposed n-alkanes derived from epicuticular plant wax

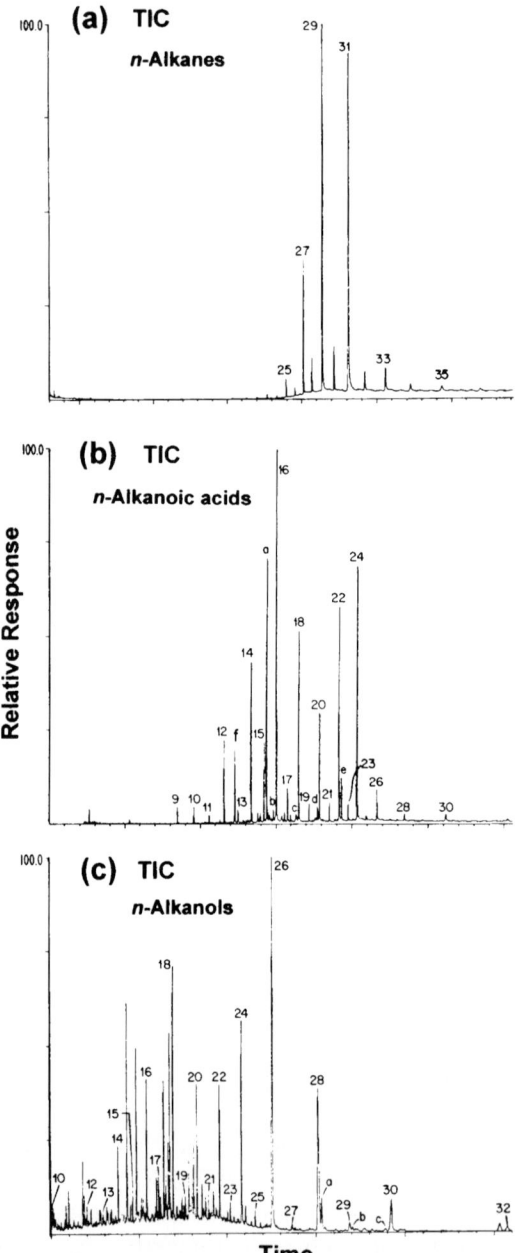

Fig. 4 TIC traces for the GC-MS analyses of extract fractions from a rural aerosol sample (Corvallis region, OR, USA): **a** total *n*-alkanes (F1), **b** alkanoic acids (analyzed as methyl esters) and *n*-alkan-2-ones (F2 and F3) (*a* phytone, *f* farnesone, *b–e n*-alkan-2-ones), and **c** alcohols (analyzed as trimethylsilyl ethers *TMS*, F4) (*a–e* sterols). *Numbers* refer to carbon-chain length

oxidation of vegetation wax alkanes, probably by burning [170] or microbial β-oxidation of alkanes.

Series of α,ω-alkanedioic acids are generally present, especially in marine areas, and range from C_5 to C_{29}, with C_{max} at 9 and 22 or 24 and CPI $\sim$ 1.0. These compounds are atmospheric oxidation products from biopolymers or other lipid components (e.g., ω-hydroxyalkanoic acids) or incomplete combustion products [30, 107, 149]. The short-chain (shorter than C_{10}) dicarboxylic acids are also found and represent ubiquitous photo-oxidation products from numerous organic compounds in the atmosphere [34].

Alkanols are significant components of the extractable lipids (Figs. 2a, c, 3a, 4c) and typically range from n-C_{12} to n-C_{34}, with a strong even-carbon-number predominance (CPI > 5) and C_{max} at 26 or 28. The homologs longer than n-C_{20} are characteristic of vascular plant wax and those shorter than n-C_{20} may be derived from microbial/marine sources, because they are not major constituents of plant waxes [37, 107, 171, 172].

2.2.5
Terpenoid and Steroid Biomarkers

The major natural product (i.e., biogenic) biomarkers detected in aerosols are sterols, triterpenols, triterpenones, and diterpenoid acids (Table 2, Figs. 2, 3). The sterols (I, see the "Appendix" for structures cited in the text) consist primarily of cholesterol (R is =H) and minor amounts of campesterol (R is α-CH_3) and sitosterol (R is β-C_2H_5). The origin of the significant amount of cholesterol in marine areas is from algae and zooplankton. The triterpenones consist of α-amyrone and β-amyrone (II, III, R is O, respectively), which are oxidative derivatives from α-amyrin and β-amyrin (i.e., II, III, R is OH, respectively), both terrestrial natural products. The precursors are dominant constituents of higher-plant (angiosperm) epicuticular wax and gums and are converted to the ketones after direct emission and/or by combustive processes [149]. The diterpenoid acids found are generally based on the abietane skeleton and consist of dehydroabietic acid (IV) and 3-oxodehydroabietic (V) and 7-oxodehydroabietic (VI) acids (Figs. 2a, c, 3a). The resin acid precursors or retene are generally not detectable in aerosols which have been exposed to photoreactions. The presence of the diterpenoid compounds indicates an input from conifer wood combustion [125, 161]. Conifers are not common in tropical regions [149] and thus additional imported wood may be used as fuel in such areas (Fig. 3a). Dehydroabietic acid is detectable over the ocean off Africa (Fig. 2c), indicating its long-range transport.

Specific biomarkers that are attributable to biomass burning sources are also present and even dominant in heavily impacted airsheds. Levoglucosan (VII), with lesser amounts of mannosan, galactosan, and 1,6-anhydro-β-D-glucofuranose are dominant in urban areas owing to wood burning and globally from wildfires (Figs. 2a, c, 3a). They are detectable in lower amounts in remote areas and over the ocean. Levoglucosan and the other anhydrosaccharides are derived exclusively from cellulose by burning [126]. Various phenols and methoxyphenols (e.g., vanillic acid, **VIII**; catechol), derived from lignin in biomass by burning [125], are also detectable close to their source, but they are readily degraded to secondary products.

Petroleum biomarkers, mentioned here for completeness, are specific indicator compounds consisting of mainly hydrocarbons which can be utilized to define both the fossil origin and the geological source of petroleum residues [101, 103]. These compounds are relatively stable in ambient atmospheric environments. An example of such a biomarker distribution pattern is shown in Fig. 2f. The $17\alpha(H), 21\beta(H)$-hopane series (**IX**) is present as the predominant biomarker group (steranes are present at trace levels). The hopanes generally range from C_{27} to C_{35} (usually with little or no C_{28}), and minor moretanes (**X**) are sometimes detectable. The presence of hopane biomarkers in aerosols confirms an input source from fossil fuel utilization (i.e., vehicular traffic). However, it should be noted that an unusual hopane distribution pattern observed for samples may be due to some blow-off from the filters, i.e., compounds shorter than C_{29} are retained less effectively by the filter owing to their volatility. Furthermore, coal burning emits a different suite of hopanes and moretanes [173] and burning of biomass, such as tropical forest litter, has yet another hopane/moretane composition [174].

2.2.6
Compound-Specific Isotope Analysis

Organic compounds of aerosols have been analyzed by GC combustion isotope ratio MS to determine their carbon isotopic compositions [57, 98, 175–183]. This method has been applied to alkanes, fatty acids, fatty alcohols, and PAHs to differentiate the carbon fixation biochemistry of the source vegetation of the wax lipids and to distinguish the pyrogenic origin of the PAH sources. Isotope analysis of bulk aerosol organic fractions is also of utility in complementing molecular source assessments [180, 184]. For example, the carbon isotope compositions of the total hydrocarbon fractions from African aerosols support the assignment of their origin from C_4 plants typical of savannah regions [180].

2.2.7
Polar Compounds

The polar compounds documented for aerosol particles are water-soluble and consist of short-chain dicarboxylic acids, anhydrosaccharides from biomass burning (e.g., levoglucosan), saccharides from soil resuspension and microbiota, and to some extent phenolics from biomass burning [125, 126, 185].

Oxalic acid, with the other short-chain homologs of dicarboxylic, hydroxydicarboxylic, aromatic dicarboxylic, oxocarboxylic, and hydroxycarboxylic acids, as mentioned before have been reported as major water-soluble components of aerosols [34, 41, 186]. Most of these carboxylic acids are secondary oxidation products of atmospheric organic compounds and are found in remote marine as well as continental rural to urban areas [39, 40, 165].

The tracers for emissions from biomass burning, primarily levoglucosan (VII) from cellulose decomposition, with lesser amounts of mannosan (XI), galactosan (XII), and 1,6-anhydroglucofuranose (XIII), are common aerosol components [39, 40, 126, 145, 185]. These anhydrosaccharides are completely soluble in water and have been reported globally in remote-to-urban airsheds [121, 157, 187–190]. Lignin, another major biopolymer of wood, yields methoxyphenolic tracers in the smoke upon burning with distributions characteristic of the fuel type [125, 153]. These tracers (e.g., vanillic acid) are not completely water-soluble, although phenolics are polar and hygroscopic.

Major amounts of saccharides have been reported to be present in aerosols of certain geographic regions such as Amazonia, Brazil [121], northwestern Pacific with Korea and Japan [39, 40], and Santiago, Chile [191, 192]. The dominant primary saccharides consist of α- and β-glucose (XIV), α- and β-fructose (XV), sucrose (disaccharide, XVI) and mycose (XVII, trehalose), with lesser amounts of inositols, α- and β-mannose, α- and β-xylose, and α- and β-galactose. In addition, saccharide polyols (alditols, i.e., reduced sugars) are also found and include sorbitol (XVIII, D-glucitol), xylitol, mannitol, arabitol, erythritol, and glycerol [40, 121]. All these saccharides are emitted directly from sources and not significantly by thermal stripping during burning. They are completely soluble in water. The primary and polyol saccharides have been shown to be source-specific tracers for resuspended soils derived from agricultural tilling, harvesting, husbandry, construction, engineering, traffic on unpaved roads, and natural wind erosion [185].

2.2.8
PAH and Oxy-PAH

PAHs have been studied extensively in urban airsheds and some long-range transport data have also been published [143, 144, 193–195]. A typical PAH composition is shown in Fig. 5 for a total extract from aerosol particles collected on Gosan Island, south of Korea [165]. These aerosols were derived from Asia during the 2001 dust transit episode [40]. The PAHs range from phenanthrene (3-ring) to coronene (6-ring) and their total concentrations are significant, ranging from 0.005 to 7.8 ng m^{-3} in Gosan and from 1.7 to 19.3 ng m^{-3} in Sapporo. Their distribution patterns are similar during the dust event, indicating common sources. 1,3,5-Triphenylbenzene is present at significant concentrations, indicating a major input source of PAHs

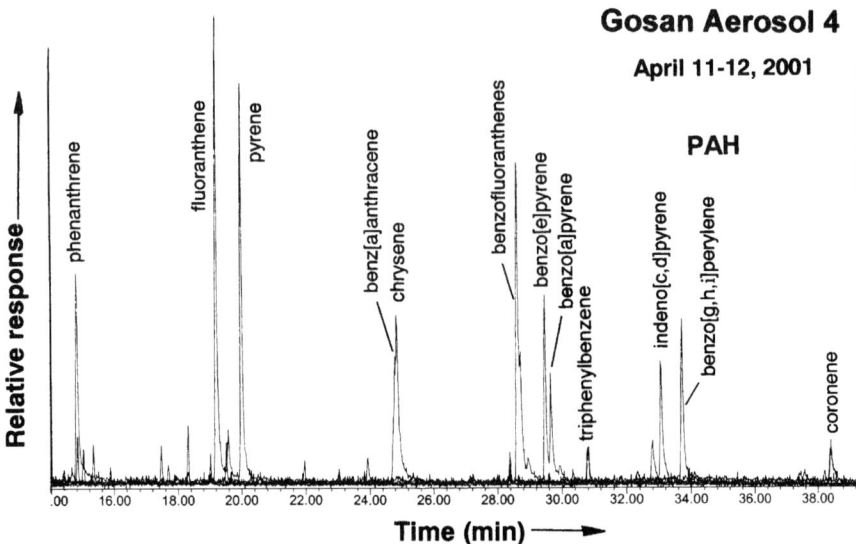

Fig. 5 Mass fragmentogram (m/z 178, 202, 228, 252, 276, 300 + 306) for typical and major polycyclic aromatic hydrocarbons (*PAHs*) in aerosol sampled during the ACE-Asia campaign on Gosan, Korea (major dust event from Asia)

from incineration and burning of refuse and plastics in urban areas [196]. This has also been reported for various cities in China [114, 197] with total PAH concentrations of 8–450 ng m^{-3}. Retene from burning conifer wood was not detectable in any of these samples. PAHs are a health concern because of their carcinogenicity, genotoxicity, and endocrine disrupting potential. Thus, the benzopyrenes are generally analyzed in environmental monitoring studies. The benzopyrenes and cyclopenta[c, d]pyrene are found at high concentrations in these aerosols and the ratio of benzo[a]pyrene to benzo[e]pyrene (Bap/Bep) ranges from 0.4 to 0.8, which is high. The Bap/Bep values for samples from urban areas of China ranged from 0.1 to 1.0 [114, 197]. Benzo[a]pyrene degrades more rapidly than benzo[e]pyrene [136]. Thus, the high concentrations of benzo[a]pyrene indicate that these PAHs reside in tar particulate matter of coal smoke emissions (small fire type burning) and are thus protected from secondary reactions during transport [173].

Urban aerosols of China also contained concentrations of oxy-PAHs equivalent to those of the PAHs, both primary emissions (e.g., benzo[a]fluoren-11-one), as well as secondary oxidation products from the PAHs (e.g., 9,10-anthraquinone) [114]. Oxy-PAHs are not significant components of the samples collected on Gosan, probably reflecting their greater reactivity toward further oxidation upon emission to, for example, the benzene dicarboxylic acids.

2.2.9
Humic and Fulvic Acids

Preliminary yields and source assessments of humic and fulvic acids isolated from marine aerosol particles were reported [100]. The aerosol particles were collected off the coast of west Africa and the fulvic and humic acids were separated, after solvent extraction, with NaOH solution by the standard method [198]. Of the two, the humic acids were the major carbonaceous fraction. The assessment of a terrigenous origin for the humic acids from soils and dried lakebeds was based on the H/C values (1.0–1.4), typical of terrestrial humates, and the $\delta^{13}C$ values of -21 to -23 ‰, characteristic of terrestrial and lacustrine humic organic matter [100, 198]. Soil is a major component of aerosols derived from continents and transported to the ocean. Thus, the saccharide and the humic acid contents may prove to be the characteristic tracers for soil in the atmosphere [40, 185].

2.3
Elemental Carbon

The elemental or black carbon (or soot) and the organic carbon contents are determined routinely for aerosols [16, 199, 200]. Elemental carbon and organic carbon as particulate matter components have global importance because they affect direct radiative properties by absorption and scattering of solar radiation, which is an active research topic [15, 16, 201–204]. Elemental carbon and organic carbon have multiple origins from both biological and geological carbon sources. Biological contemporary carbon has also been quantified in total aerosol carbon mixtures by ^{14}C dating [205–208]. Typical amounts of contemporary carbon in the urban atmosphere of Los Angeles vary from 19 to 36% downtown and from 25 to 81% in suburban areas, and for Denver, CO, USA 23% (mean winter) and 97% (mean summer) of the total carbon (elemental carbon + organic carbon). The elemental carbon from vehicle emissions (especially diesel engines), power plants, and numerous other combustion sources has a small average particle size and is therefore readily transported over global distances [209].

2.4
Source Signatures

The emission profiles for sources of atmospheric particles have been determined mainly for urban Los Angeles in terms of elemental carbon and organic tracer compounds. Source testing with emission rates or particle loadings have been reported for ten major urban sources. These data were then used in conjunction with inventory modeling to apportion the emissions of the airborne particulate matter in the Los Angeles air basin [82, 91, 97, 141] and elsewhere [142, 210, 211]. Additional contributing sources are numerous and minor, but may have to be tested for other global areas where they are more important. Also, some source tests for Los Angeles (e.g., vehicles) may not have the same emission factors elsewhere owing to different attributes (e.g., different fleet and fuel compositions). Additional organic compound compositions of smoke from biomass burning and coal burning have been reported (reviewed by Simoneit [145]).

2.4.1
Ambient Vegetation

Globally, the emissions from vegetation-covered areas are major contributors of organic matter to the atmosphere. The organic matter of terrestrial aerosol particles in rural regions consists primarily of vascular plant waxes. The major components are n-alkanes, n-alkanols, and n-alkanoic acids, with minor amounts of n-alkan-2-ones, n-alkanedioic acids, wax esters, and other lipid compounds. An example of such an end-member aerosol composition is illustrated in Fig. 4. These are the total ion current traces from the GC-MS analyses of separated lipid fractions from an aerosol sample of a rural area near Corvallis, OR, USA (number 13 in Ref. [107]). The hydrocarbon fraction consists of higher-plant-wax n-alkanes ranging from C_{23} to C_{35}, with a strong odd-carbon-number predominance. The total fatty acid (analyzed as methyl esters) and ketone fraction from the same sample consists of n-alkanoic acids derived from vascular plant wax (C_{20}–C_{30}, strong even predominance) and microbial lipid residues (part of compounds shorter than C_{20}). The major ketone is 6,10,14-trimethylpentadecan-2-one (a in Fig. 4b) with a minor amount of 6,10-dimethylundecan-2-one (f in Fig. 4b). The minor n-alkan-2-ones generally have an odd-carbon-number predominance (b–e). The total fatty alcohol (analyzed as trimethylsilyl ethers) fraction consists mainly of n-alkanols ranging from C_{10} to C_{32} with a strong even-carbon-number predominance and C_{max} at 28. The homologs longer than C_{20} are typical of higher-plant wax and those shorter than C_{20} have multiple origins. Minor sterols are also present (a–c in Fig. 4c).

2.4.2
Fossil Fuel Utilization

Fossil fuels (mainly petroleum and coal) are used in vast quantities, especially in urban regions. The major emissions are from vehicular engine exhaust and thus typical end-member organic signatures are illustrated here [83, 101–103]. Vehicular traffic with the associated fuels and lubricants emits petroleum residues comprising n-alkanes with no carbon number predominance, UCMs of branched/cyclic hydrocarbon components, and biomarker tracers to the ambient atmosphere as shown for examples of auto exhaust

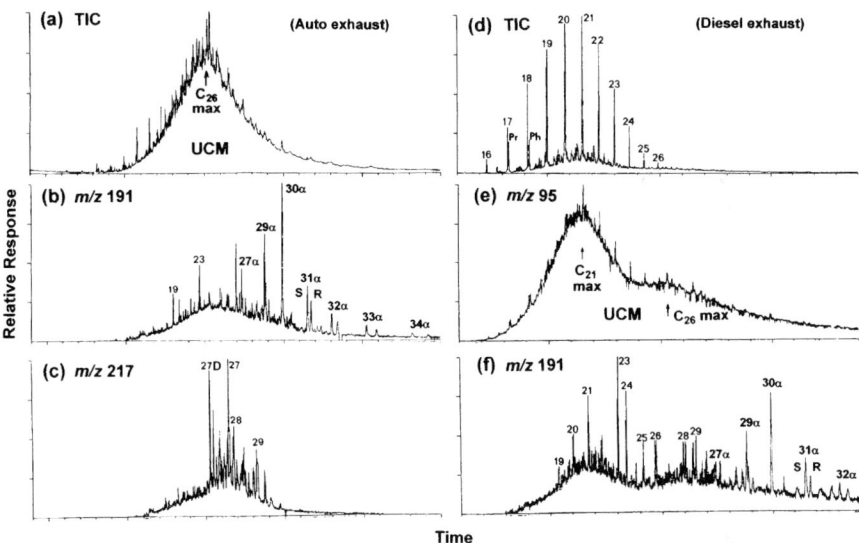

Fig. 6 Salient features of the GC-MS data for samples of automobile and diesel exhaust (total hydrocarbons). Automobile **a** TIC trace, **b** m/z 191 mass fragmentogram (key ion for hopanes and extended tricyclic terpanes), **c** m/z 217 mass fragmentogram (key ion for steranes). Diesel **d** TIC trace, **e** m/z 95 mass fragmentogram (key ion for the UCM hump), **f** m/z 191 mass fragmentogram (key ion for hopanes and extended tricyclic terpanes). *Numbers* refer to carbon-chain length or carbon number of biomarkers, *Pr* pristane, *Ph* phytane, *iα* $17\alpha(H), 21\beta(H)$-hopane configuration, *S* and *R* extended hopane epimer at C-22, *D* diasterane

and diesel exhaust in Fig. 6. Auto exhaust contains relatively low concentrations of *n*-alkanes longer than C_{15}. However, diesel exhaust does contain some uncombusted fuel components (as *n*-alkanes with UCM) and reformed alkanes with a second higher weight UCM of branched and cyclic hydrocarbons.

The associated terpenoid markers are present in all engine emissions and originate from the lubricants. These compounds include the terpenoid hydrocarbons characteristic of petroleum and are the geologically mature and environmentally stable isomers of the $17\alpha(H), 21\beta(H)$-hopane (**IX**) series (e.g., Fig. 6b, f). The identifications of these compounds are based primarily on their mass spectra and gas chromatographic retention times [101, 102]. They occur in aerosols usually at low concentrations, but their overall distribution signatures within samples can be easily determined by GC-MS and utilized for comparison purposes. This is based on the m/z 191 key ion intensity in the GC-MS data, which is the base peak of most of the hopanes as well as the extended tricyclic terpanes (**XIX**). The predominant analog is usually $17\alpha(H)$-hopane (**IX**, R is CH$_3$), with subordinate amounts of $17\alpha(H)$-22,29,30-trisnorhopane, $17\alpha(H)$-29-norhopane (**IX**, R is H), and the extended homohopanes from C_{31} to C_{35} (**IX**, R is $C_2H_5-C_6H_{13}$) with the C-22 diastereomers in a 22R to 22S ratio of about 0.6.

The tricyclic terpanes (**XIX**) generally range from C_{19} to C_{29} (no C_{22} or C_{27}), but the lower homologs (shorter than C_{25}) are volatile and thus not adequately collected with the aerosol particulate matter (Fig. 6b vs. f [101, 102]). These terpanes are common in diesel fuel and in part in lubricating oils and therefore are useful tracers for the volatile emissions from traffic.

Steranes (**XX**) and diasteranes (**XXI**) are additional biomarker tracers commonly found in petroleum that can be used for aerosol analyses [101-103]. These hydrocarbons are not present in contemporary biogenic materials. The distribution signature in GC-MS data of, for example, the m/z 217 fragmentogram (Fig. 6c) for steranes and diasteranes is useful supporting evidence for a petroleum component. These compounds are also introduced to the atmosphere from lubricants of vehicular engines, but at lower concentrations than the hopanes. They are not found in gasoline or diesel fuel.

The UCM from lubricants is the major organic input to the atmosphere from vehicular emissions in mainly urban regions. However, fossil fuel tracers are now recognizable in aerosols transported over longer distances [39, 40, 189]. Thus, the biomarker tracers in the lubricants are used to confirm and quantify the source strengths of these emissions.

2.4.3
Biomass Burning

Biomass burning emits numerous compounds into smoke and the compound classes include the following: homologous series of *n*-alkanes, *n*-alkenes, *n*-alkanoic acids, and *n*-alkanols; methoxyphenols from lignin; monosaccharide derivatives from cellulose; and steroid and terpenoid biomarkers [145]. The distributions and abundances of the biomass smoke constituents are strongly dependent on combustion temperature (smoldering versus flaming conditions), aeration, and burn duration. Only a few compounds are of utility as tracers owing to oxidation during transport (Table 3). A typical example of compounds in smoke from biomass burning is discussed next.

The GC-MS data of the total extract and separated fractions (by thin-layer chromatography, TLC) from smoke of a grass fire are shown in Fig. 7a–d. The total extract consists mainly of levoglucosan, *n*-alkanoic acids, and various methoxyphenolic compounds [e.g., catechol, dimethoxyphenol, syringic acid (**XXII**), vanillic acid (**VIII**)]. Grasses contain all three phenolic moieties from the lignin precursor alcohols; thus, the thermal breakdown products are not indicative tracers for this source, although they can be used as ratios with other compounds [125]. The dominant levoglucosan is derived from the thermal decomposition of cellulose, a confirming tracer for biomass burning. There are lesser contributions of the C_{16} and C_{18} *n*-alkanoic acids, C_{27}–C_{33} *n*-alkanes, and the C_{26} *n*-alkanol. The biomarkers identified in the

Table 3 Major compound groups identified in smoke particles from biomass burning

Compound group [a]	Plant source	Product
n-Alkanes ($> C_{23}$)	Epicuticular waxes	Natural
n-Alkenes ($> C_{24}$)	Epicuticular waxes/lipids	Altered
n-Alkanoic acids ($> C_{22}$)	Internal lipid substances	Natural
n-Alkanols ($> C_{22}$)	Epicuticular waxes	Natural
n-Alkan-2-ones ($> C_{23}$)	Epicuticular waxes/lipids	Altered
Alkyl amides	Lipids	Altered
Alkyl nitriles	Lipids	Altered
n-Acyl glycerides	Lipids	Altered
Diterpenoids	Conifer resin, wax	Natural/altered
Triterpenoids	Angiosperm gum, wax	Natural/altered
Monosaccharide anhydrides	Cellulose biopolymer	Altered
Methoxyphenols	Lignin biopolymer	Altered
Steroids	Internal lipid substances	Natural/altered
Wax esters	Lipid membrane, wax	Natural
Triterpenoid esters	Internal lipid substances	Natural
PAHs	Multiple sources	Altered

[a] Includes natural product precursors and/or thermally altered derivatives

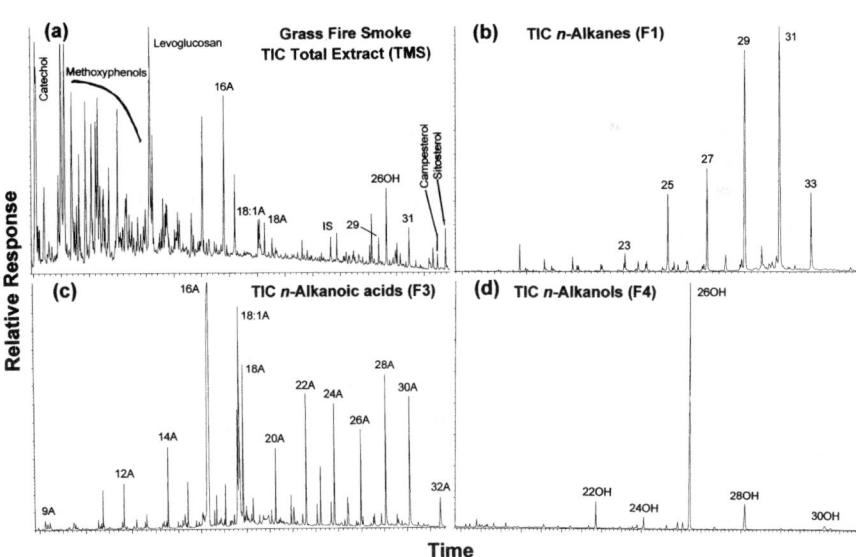

Fig. 7 Salient features of the GC-MS data (TIC traces) of grass smoke particulate matter: **a** total extract showing major compounds, **b** F1 fraction showing n-alkanes, **c** F3 fraction showing n-alkanoic acids (as methyl esters), and **d** F4 fraction showing n-alkanols (as TMS) (*numbers* refer to carbon-chain length of n-alkanes, $i:1$ monounsaturated, *A* alkanoic acid, *OH* alkanol)

total extract include the triterpenoids oleana-2,12-diene (**XXIII**, R is CH_3), ursa-2,12-diene (**XXIV**, R is CH_3), and lupa-2,22-diene (**XXV**), and the phytosterols (sitosterol, campesterol, and stigmasterol).

The relatively high concentration of *n*-alkanes (F1), with an odd-carbon-number to even-carbon-number predominance (CPI=12), range from C_{23} to C_{33}, and C_{max} at 31, is derived from the grass wax. Alkenes are not significant components. The major PAHs (F2) are phenanthrene, fluoranthene, and pyrene, with minor contributions of anthracene, C_1- and C_2-phenanthrenes, 11*H*-benzo[*b*]fluorene, C_1-pyrenes, cyclopenta[*c*,*d*]pyrene, benz[*a*]anthracene, and chrysene. The *n*-alkanoic acids (F3) have an even-carbon-number to odd-carbon-number predominance (CPI=4) and range from C_9 to C_{30}, with C_{max} at 16. The *n*-alkanols (F4) are even-carbon-numbered (CPI=20) and range from C_{22} to C_{30}, with C_{max} at 26. All three of these homologous lipid series (full homolog ranges) are ubiquitous in biomass burning smoke and are thus are not source-specific. However, the major biomarker compound groups, including the longer-chain lipids (longer than C_{22}), identified in smoke particulate matter from burning of vegetation (e.g., wood, litter; Table 3) are source-specific. This compound list will expand as more vegetation species are tested for smoke emission profiles [153].

2.4.4
Soil/Sand Resuspension

Transport of continental dust to the ocean has been documented extensively in the Atlantic (Sect. 1), and the Asian continent has also been inferred as such a source [30, 68]. Continental dust is advected into the atmosphere from sands of desert areas and from exposed soils during tilling in agricultural regions. The organic compositions of the fine particulate matter of both types of sources are currently being characterized and their effects on atmospheric processes are under extensive study. Results from these studies should appear in the near future.

The organic compositions from some examples of sand and soil particulate matter are illustrated here and compared with an aerosol sample (also compare Fig. 3a). The dominant organic tracers of desert sands are trace amounts of plant wax components comprising primarily *n*-alkanes and *n*-alkanols, with minor amounts of nonacosan-10-ol (**XXVII**), methyl alkanoates, sterols (**I**), and triterpenols (Fig. 8a [212]). The *n*-alkanes, *n*-alkanols, and *n*-alkanoic acid esters have the typical distributions and carbon number predominances of plant waxes. The organic tracers of desert sands are low in concentration and must be used in conjunction with mineralogical and trace metal compositions for source assessments. The lipid tracer compositions of different deserts are distinguishable and reflect primarily the regional climatic conditions affecting biomass fixation, with inputs of detritus mainly from grass waxes and secondarily from algae and woody plants.

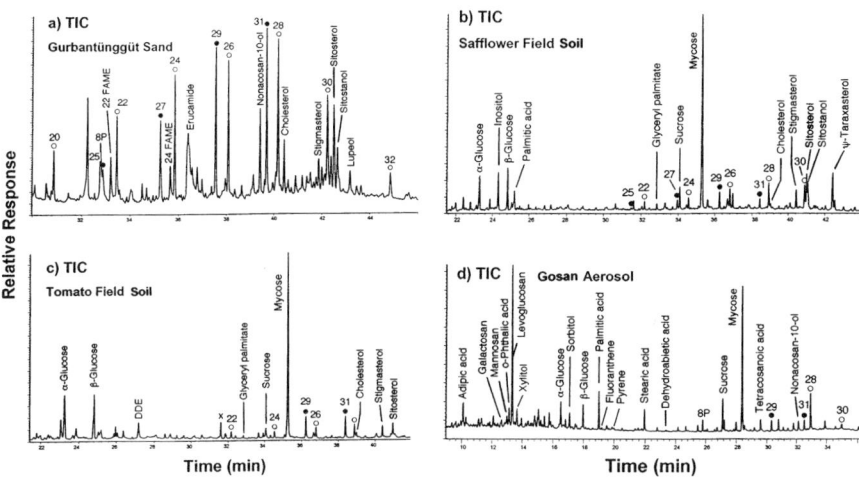

Fig. 8 Representative TIC traces from GC-MS analyses of total extracts (silylated): **a** sand (fines, Gurbantünggüt desert, Xinjiang, China), **b** soil from safflower field (San Joaquin Valley, CA, USA), **c** soil from tomato field (San Joaquin Valley, CA, USA), and **d** aerosol particulate matter (Gosan, Jeju Island, Korea, 11–12 April 2001) (compounds are labeled, *numbers* refer to carbon-chain length, see key for lipids)

Soils contain much higher concentrations of organic compounds than sands and those also comprise mainly lipids from higher vegetation. Soils contain the plant wax and biomarker compounds from the surficial or surrounding vegetation cover (e.g., resin acids, *XXVIII*–**XXXI**, from conifers in forest soil) [213]. Agricultural field soils contain the lipids from the crops, as well as the pesticide/herbicide residues remaining after use (e.g., DDE [212, 213]). The *n*-alkanols and *n*-alkanes are useful secondary tracers for fingerprinting soils, analogous to desert sands. However, these lipids have very high source concentrations in soils compared with trace levels in sands.

A novel and unique aspect is the presence of primary sugars (saccharides) in soils. The total extracts of agricultural soils from a safflower field and a tomato field are shown in Fig. 8b and c, respectively. The dominant compounds are sugars with only minor lipids and biomarkers. The most commonly encountered sugars are α- and β-glucose, inositols (**XXXII**, several isomers), sucrose (disaccharide), and mycose (fungal disaccharide metabolite) [185]. Fructose, mannose, xylose, and other monosaccharides can also occur. Another aspect of soils is their enhanced contents of sterols and triterpenoids with depleted contents of the aliphatic plant wax lipids (i.e., alkanes and alkanols). The phytosterols (sitosterol, campesterol, and stigmasterol) are dominant, but cholesterol is also present, indicating a component from algae and fauna. Some soils can contain species-specific tracers, as, for example, ψ-taraxasterol (**XXXIII**) in the safflower field (Fig. 8b). The presence of polar and water-soluble compounds in soil extracts indicates that both living and extracellular biomass was extracted. The enhanced sterol and triterpenol concentrations are probably due to their resistance toward biodegradation compared with the aliphatic plant wax lipids.

The impact of Asian dust on the northwestern Pacific region is currently under investigation and preliminary results have been reported [167]. A typical total ion current trace from the GC-MS analysis of a total extract (silylated) of aerosol particulate matter collected in Gosan (formerly Kosan) on Jeju (formerly Cheju) Island, Korea, is shown in Fig. 8d. A similar aerosol sample analysis is also shown in Fig. 3 for Sapporo. The high amounts of levoglucosan with mannosan and galactosan are tracers for smoke from biomass burning [126]. The primary sugars are α- and β-glucose, xylitol, sorbitol, sucrose, and mycose. Xylitol and sorbitol are probably microbial/fungal alteration products of primary saccharides in soil particles introduced into the atmosphere.

A minor homologous series of n-alkanoic acids, ranging from C_{12}–C_{30} with a strong even-to-odd carbon number predominance and a maximum at C_{16} (palmitic acid) was present, suggesting an input from terrestrial and marine biota, as well as from cooking. A trace of dehydroabietic acid further supports biomass burning. Significant amounts of dicarboxylic acids (e.g., glyceric, maleic, adipic, and benzene dicarboxylic acids) indicate secondary oxidation products [41]. The vascular plant wax components are minor and consist of the n-alkanes, n-alkanols, and nonacosan-10-ol, with traces of phytosterols. Thus, the primary inputs of organic compounds to these aerosols transported from Asia are (1) natural emissions from continental vegetation and marine lipids, (2) smoke from biomass burning, (3) soil resuspension due to agricultural activity, and (4) urban/industrial emissions from fossil fuel utilization, especially coal [39, 40, 167].

2.5
Secondary Reactions

All organic compounds in aerosols are susceptible to degradation by reaction with OH, ozone, NO_x, or ultraviolet radiation, provided that they are exposed to and accessible to the reactants [17]. Major secondary products are the short-chained dicarboxylic and keto acids, and aldehydes (carbonyls) [128, 129]. The biogenic origin can be discerned in secondary products after initial oxidation or nitration, but not in subsequent reactions where the chemical structure is lost. Then, the only possible correlatable parameter may be the stable isotope composition of the breakdown products and their ^{14}C content. Secondary products from volatile organic compounds, especially terpenes from vegetation, have an extensive literature base and constitute a separate topic on biogenic atmospheric components. Because this is not covered here, a few key references are cited [50, 214–218].

The formation of secondary reaction products in the atmosphere from source precursor compounds has been documented for the Los Angeles,

CA, USA, air basin [65]. A number of nitrated and oxygenated aromatic compounds were measured during a severe photochemical smog episode and indicated secondary formation via atmospheric chemical reactions. These included 1-nitronaphthalene, 2-nitronaphthalene, 3-nitrobiphenyl, and dinitrophenol, as well as certain oxygenated aromatic compounds such as cyclopenta[c, d]phenanthrone [65, 219]. The ambient concentrations of these compounds increase progressively with transport distance inland from the coastline. Thus, oxidative alteration of organic compounds occurs during transport. Source tracers must be encapsulated in particles to survive long-range transport.

2.6
Fallout and Washout

The lower troposphere represents one compartment in the global cycling of organic matter and thus it is important to understand the transfer of organic materials to ultimate sinks. Precipitation scavenging (washout) and dry fallout are important mechanisms for the removal of atmospheric particles, including organics, with diameters between 0.1 and 10 µm [17, 220]. Few detailed reports on organic matter in fallout and washout particulate matter have appeared [68, 108, 221–223]. Those studies identified contemporary biogenic and anthropogenic organic compounds in urban, rural, and remote oceanic areas. Some made comparisons of specific organic species in the particle versus aqueous phase (e.g., PAHs, halocarbons, lipids). Washout is efficient for removing water-soluble organic compounds and coarse particles, whereas fallout removes mainly coarse and dense particles with adsorbed/included organic matter. More work should be encouraged to assess particle removal processes from the atmosphere; organic tracers can be of utility here.

2.7
Modeling

Sources and their contributions to the ambient atmosphere need to be defined so that modelers can upgrade their global modeling tasks. Such data should also aid regulatory agencies in urban areas to make decisions on how to proceed for improvement or maintenance of air quality. Additional concerns are the suspected adverse health effects of low levels of airborne fine particulate matter. The extensive source testing and ambient aerosol characterization program for the Los Angeles air basin, concentrating on the fine particle size distribution, was carried out and then summarized by modeling [82, 91, 141]. An updated study for the Los Angeles area was carried out for 1993 with comparable results [97]. This model, using the same emission sources developed for Los Angeles, was applied with a good fit in the San

Joaquin Valley during winter [142]. The major discrepancies were the vegetation and soil emission uncertainties due to the different geographic area. The apportionment of fossil fuel combustion and other sources in $PM_{2.5}$ and PM_{10} aerosols by the chemical mass balance receptor model has been reviewed as applied to urban and regional areas [224].

Tracking the origins of major global dust events and air parcels is carried out by back-trajectory modeling and/or satellite observations [166, 225]. Major dust outbreaks are easily tracked by satellites for days [36, 226]. Air mass back-trajectory analysis is well established and provides evidence for geographic source regions of aerosols or the changes and mixing of air masses during transport [65, 70, 225].

2.8
Water-Soluble Organic Matter

The water solubility of the carbonaceous organic fraction of aerosols is a major open question in climate models [227, 228]. The water-soluble saccharides comprise from 13 to 26% of the total identified compound mass (TCM) in continental aerosols with isolated, higher values over the ocean (up to 63% [40]). These saccharides are interpreted to represent viable biomass and extracellular organic matter mainly in soils, but also in lesser amounts in road dust, dried lake sediments, etc., and possibly in marine particulate matter. Saccharides have been characterized in urban aerosols that contain entrained soil dust [191], and in remote and rural areas [39, 40, 121]. They were proposed as tracers for soil resuspension with associated microbiota from agricultural tilling and harvesting, wind erosion, or traffic [185]. The secondary oxidation products of organic compounds in aerosols, especially over the oceans, are oxalic acid and the other short-chain dicarboxylic acids. They have been documented as the dominant water-soluble components of organic aerosol matter [34, 41]. The sum of all identified water-soluble compounds ranges from 14 to 89% of the TCM for typical samples and is about 40–70% of the TCM during an Asian dust episode at the ground stations [40].

3
Sampling and Analytical Methods

3.1
Sampling

Aerosol samples are typically acquired by high-volume (Hi-Vol) filtration on quartz fiber filters with or without a particle size preseparator. Hi-Vol sampling is adequate for marine studies because the coarser particles have been winnowed out during transport. However, precautions should be taken to minimize sampling the resuspended sea slick (salt haze). Sampling is carried out from 1-h to up to 24-h periods and the extraction, separation, and analysis procedures are the same as used for urban aerosols and source tests. The quartz fiber filters are annealed prior to use at 550 °C for 3 h to reduce background contamination. Filters should be stored in a freezer or sterilized by addition of some extraction solvent prior to analysis to prevent microbial and mold activity.

3.2
Extraction and Fractionation

A schematic of a suggested sample treatment, extraction, and separation procedure is given in Fig. 9, as first used by Simoneit and Mazurek [107] with some modifications. The solvent mixture of dichloromethane and methanol is excellent for extracting both lipids and water-soluble organic compounds

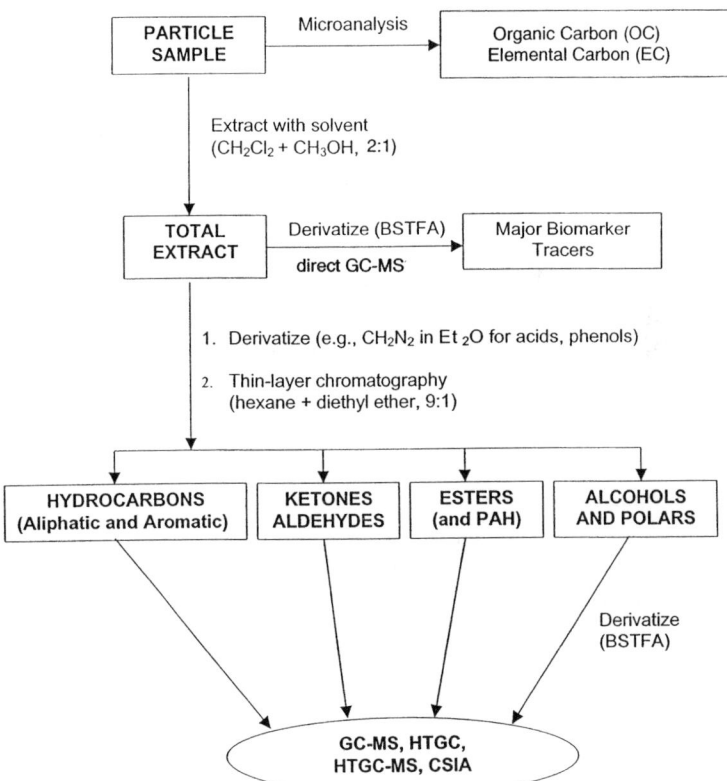

Fig. 9 Schematic of the extraction, separation, and analysis procedures for aerosol particulate matter on filters

such as sugars, levoglucosan, dicarboxylic acids, and phenols. Each filter or aliquot is extracted using ultrasonic agitation for three 20-min periods using a dichloromethane/methanol mixture (CH_2Cl_2/CH_3OH, 2 : 1, v/v). The solvent extract should be filtered to remove insoluble particles and filter fibers (Gelman Swinney filtration unit with an annealed glass-fiber filter) [107]. The filtrate is first concentrated by use of a rotary evaporator and then by a stream of filtered nitrogen gas. The final volume is adjusted exactly (e.g., 1.0 ml) by addition of CH_2Cl_2. Aliquots are then taken for derivatization or direct GC-MS analysis. Alternatively, carboxlic acid and phenolic moieties in the extracts can be methylated using diazomethane in diethyl ether prepared from the precursor N-methyl-N'-nitro-N-nitrosoguanidine. With current MS instrumentation, direct total extract analysis is recommended, with and without derivatization by N,O-bis(trimethylsilyl)trifluoracetamide (BSTFA) plus 1% trimethylchlorosilane, usually for 3 h at 70 °C [157, 190, 229, 230].

Methylated extracts can be separated by preparative TLC on silica gel plates with a mobile-phase eluent mixture of hexane and diethyl ether (9 : 1) or by column liquid chromatography. These procedures allow for the determination of chemical information on single molecular groups or homologous series, which may not be detected owing to the coelution in the total extract mixture. They also provide chemical information on molecular polarity or functional group constituents which further aids in structure elucidation and identification. Typically four fractions are removed from the TLC plates and contain the following classes of compounds (Fig. 9): (1) n-alkanes, n-alkenes, and saturated and unsaturated cyclic diterpenoid and triterpenoid hydrocarbons, (2) n-alkanones and PAHs, (3) n-alkanoic acids/alkenoic acids and resin acids (as methyl esters) and saturated and unsaturated diterpenoid and triterpenoid ketones, and (4) n-alkanols, sterols, terpenols, and polar organics. The fourth fraction is also converted to trimethylsilyl derivatives prior to analysis by reaction with BSTFA plus 1% trimethylchlorosilane.

3.3
Instrumental Analyses

The detection limits for organic compounds by GC-MS have decreased greatly since the initial studies and are currently in the 1 ng or less per compound range. The extracts or fractions can be analyzed by temperature-programmed capillary GC. However, with current instrument capabilities all samples, total extracts or fractions, can be analyzed directly by capillary GC-MS using benchtop quadrupole or ion trap mass spectrometers.

Samples that contain wax esters (plant wax) or triglycerides (e.g., beef fat from grilling) can also be analyzed by high-temperature GC and high-temperature GC-MS using custom-made or metal-covered capillary columns coated with OV-170-OH and GC oven temperatures to 400 °C [169]. All compound identifications should be based on comparisons with authentic standards whenever possible, GC retention time, literature mass spectra and interpretation of mass spectrometric fragmentation patterns. Homologous compound series should be defined with regard to carbon chain length by key ion searches on the GC-MS data and confirmed by mass spectrum and retention time comparison with standards.

The development of compound-specific stable carbon (also hydrogen and nitrogen) analysis (compound-specific isotope analysis) by GC combustion isotope ratio MS provides a secondary parameter in molecular characterization besides the chemical structure or elemental compositions [98, 179, 180, 231]. Any organic fraction that is amenable to GC analysis can be analyzed by this method. Compounds can also be measured as derivatives (e.g., methylated or trimethylsilyated), and a correction made for the isotope ratio of the derivatizing group. This is done by determining the isotope composition of suitable underivatized standards by both combustion analysis and then analyzing the standard mixture suitably derivatized versus the Pee Dee belemnite (PDB) standard for carbon, standard mean ocean water (SMOW) for hydrogen, and N_2 in air for nitrogen.

3.4
Data Modeling

A chemical mass balance approach, with specific organic compounds as tracers, is commonly used for source/receptor reconciliation [82, 91, 97, 141, 142]. In such models, the total ambient concentration of each of the organic compounds used in the mass balance model is reconstructed from the best linear combination of chemical profiles of emission sources that reproduce the composition of the ambient sample as a whole. The chemical mass balance model used in the references just cited can be expressed by the following set of linear equations:

$$c_{ik} = \sum_{j=1}^{m} a_{ij} s_{jk}, \qquad (1)$$

where c_{ik}, the concentration of compound i in particles at receptor site k, equals the sum over m source types of the product of a_{ij}, the relative concentration of compound i in the particle emissions from source j, multiplied by s_{jk}, the increment to the total particulate mass concentration at receptor site k originating from source j.

The system of Eq. 1 states that the ambient concentration of each mass balance species must result only from the m sources included in the model and that no selective loss or gain of compound i occurs in transport from the source to the receptor (sampling) site. Therefore, the selection of mass balance compounds must be limited to (1) species for which all major sources are included in the model, (2) species that do not undergo selective removal by chemical reaction or other mechanisms over the time scale for transport between the source and the receptor site, and (3) compounds which do not increase in concentration owing to chemical reactions in the atmosphere. The studies summarized by Schauer et al. [91, 97] and Schauer and Cass [142] used the CMB7 receptor modeling computer program (http://www.epa.gov/scram001/tt23.htm) [232] to solve Eq. 1. This program seeks an effective variance-weighted least-squares solution for the overdetermined set of mass balance equations and takes into account the known uncertainties in the atmospheric measurements and source emission data. This model can be applied to marine aerosols to define the source inputs of anthropogenic and terrestrial components [165].

4
Overview

This section presents a brief overview of the major, persistent organic compounds and organic carbon fractions that are transported from land to sea.

4.1
Major Terrestrial Organic Tracers and Sources

The major terrestrial organic tracer compounds recognizable over the oceans are derived from vegetation (plant waxes), with varying amounts of tracers from biomass burning, soil and sediment erosion, and anthropogenic (urban) emissions [39, 40, 113, 166]. The key organic compound tracers and their sources are summarized in Table 4. Once over the oceans, the continental aerosols are admixed with marine aerosol emissions which need to be distinguished, especially if sampling is carried out on ships or islands. The input of marine-derived compounds consists mainly of lipids and sterols [40, 166]. The common marine tracers are also summarized in Table 4. The source strengths of the tracers and their ambient concentration ranges are listed if known. A distinction is also made between the water-soluble and the hydrophobic compounds.

4.2
Major Secondary Products

The dominant secondary products from organic matter oxidation in the atmosphere over the ocean are mainly oxalic acid with lesser amounts of other short-chain dicarboxylic, hydroxycarboxylic, and oxocarboxylic acids. These are all water-soluble and are thus washed out by precipitation scavenging. Secondary products from other oxidative processes such as burning

Table 4 Major tracer compounds, their sources and concentrations in aerosols over the oceans

Compound or compound group	Tracer of source	Examples of reported concentration (ng m^{-3})	
		Near source	Marine areas
Terrestrial			
n-Alkanes ($> C_{23}$)	Higher plants	30 – 1340	0.5 – 6
n-Alkanols ($> C_{22}$)	Higher plants	15 – 1505	0.1 – 5
n-Alkanoic acids ($> C_{22}$)	Higher plants	3 – 7800	0.1 – 21
n-Alkan-2-ones	Lipids/Soils	0.4 – 124	NR
n-Alkyl amides[a]	Biomass burning	0.1 – 1070	NR
Diterpenoids[a]	Biomass burning	0.1 – 260	0.001 – 3.6
Triterpenoids[a]	Biomass burning	0.2 – 1.4	0.04 – 0.3
Monosaccharide anhydrides[a]	Biomass burning	7 – 33 400	0.2 – 27
Methoxyphenols[a]	Biomass burning	0.7 – 9	0.5 – 3.4
Phytosterols[a]	Higher plants/ biomass burning	0.5 – 10.3	0.1 – 3.2
Cholesterol	Urban/dry lakes	0.2 – 7.4	0.1 – 4.2
Wax esters	Biomass burning/ higher plants	0.2 – 15	0 – 1.2
PAH (also oxy-PAH)	Combustion processes	1 – 22	0.03 – 7
UCM ($C_{max} = 26$)	Urban (petroleum lubricants)	40 – 415	0 – 13
UCM ($C_{max} = 22$)	Dry lake beds	5 – 34	0 – 2
Saccharides[b]	Soils	14 – 574	0 – 12
Petroleum biomarkers	Urban (petroleum use)	0.3 – 4	0 – 0.1
Coal biomarkers	Coal burning	0.03 – 2	0 – 0.01
Plasticizers/antioxidants/ triphenylbenzene	Refuse burning	24 – 126	2 – 23
Marine			
n-Alkanols ($< C_{20}$)	Microbiota		0.04
n-Alkanoic acids ($< C_{20}$, with high $C_{16:1}$)	Microbiota		0.2 – 0.7
Sterols (mainly C_{27} and C_{28})	Microbiota		0.03 – 1.2
Lipids	Microbiota blooms		0.02 – 6
Squalene	Microbiota		0.1 – 2
Secondary Products			
Dicarboxylic acids (C_2-C_6)			1 – 170
Benzoic and benzenedicarboxylic acids			0.2 – 12
Oxy-PAH (mainly quinones)			0.01 – 0.03

NR not reported
[a] High values occur during wildfire seasons.
[b] High values occur during spring/summer seasons.

of biomass and fossil fuel combustion survive long-range transport, because they are in part entrapped within particles and thus protected from further oxidation. Therefore, levoglucosan in smoke from fires, and PAH in soot from urban emissions, as well as some secondary products (benzene dicarboxylic acids, phenolic acids) are detectable downwind over the ocean. The key indicator compounds for the oxidative secondary reactions in the atmosphere are also summarized in Table 4.

5
Conclusions

Organic matter of aerosol particles of both marine and continental areas is derived from two major sources and is admixed depending on environmental conditions. These particle sources are natural biogenic detritus (e.g., plant wax, terpenoids, and marine lipids) and anthropogenic emissions (e.g., soot, smoke, and oils). Combustion, both natural and anthropogenic, yields particle emissions from fossil fuel (e.g., coal and petroleum products) and biomass burning, which is superimposed on the other-source emissions. Wind erosion and resuspension of soil (agricultural activities) and desert sand (dust) advects mainly mineral components with respective major and minor organic tracers into continental atmospheres. In general, the continental aerosol mixtures are transported over the oceans by dominant weather and wind systems. Continental aerosols are diluted, oxidized, and mixed with marine aerosols during longer-range transport. These processes have been illustrated here with examples of ambient aerosols collected in urban, rural, and remote marine areas and discussed in terms of source emission profiles.

In summary, the organic compounds of atmospheric particles from marine and continental regions can be characterized as specific tracers for sources and alteration during transport from (1) natural emissions (vegetation waxes, terpenes), (2) fossil fuels' use (vehicle emissions, coal burning), (3) biomass burning (taxon-specific, wildfires, heating, etc.), (4) anthropogenic emissions (industry, cooking, etc.), and (5) soil and desert dust resuspension (agriculture, wind erosion). The precursor–product chemistry can be used to assess the secondary reactions (thermal or atmospheric) and the fate of aerosol organic matter.

Acknowledgements I thank my numerous collaborators over the years and especially one of my mentors/collaborators, Prof. Glen R. Cass, in memoriam, for advancing the topic of organics in the atmosphere.

References

1. Weyl PK (1970) Oceanography—an introduction to the marine environment. Wiley, New York
2. Prospero JM (1996) The atmospheric transport of particles to the ocean. In: Ittekot V, Schäfer P, Honjo S, Depetris PJ (eds) Particle flux in the ocean. Wiley, New York, p 19
3. Griffin DW, Kellogg CA, Garrison VH, Shinn EA (2002) Am Sci 90:228
4. Chester R (1972) Geological, geochemical and environmental implications of the marine dust veil. In: Dyrssen D, Jagner D (eds) The changing chemistry of the oceans. Nobel symposium 20. Wiley, New York, p 291
5. Darwin C (1846) Q J Geol Soc Lond 2:26
6. Ehrenberg C (1847) Abh König Akad Wiss Berlin 269
7. Radczewski OE (1937) Wissenschaftliche Ergebnisse der Deutschen Atlantik Expedition "Meteor", 1925–1929, B.3, T1-3, 262
8. Radczewski OE (1939) Eolian deposits in marine sediments. In: Trask PD (ed) Recent marine sediments. American Association of Petroleum Geologists, Tulsa, OK, p 496
9. Delany AC, Delany AC, Parkin DW, Griffin JJ, Goldberg ED, Reimann BEF (1967) Geochim Cosmochim Acta 31:885
10. Kolbe RW (1957) Science 126:1053
11. Rex RW, Goldberg ED (1962) Insolubles. In: Hill MN (ed) The sea, vol I. Interscience, New York, p 295
12. Nicolet M (1960) The properties and constitution of the upper atmosphere. In: Ratcliffe JA (ed) Physics of the upper atmosphere. Academic, New York, p 17
13. Griffin JJ, Windom HC, Goldberg ED (1968) Deep-Sea Res 15:433
14. Charlson RJ, Schwartz SE, Hales JM, Cess RD, Coakley JA Jr, Hansen JE, Hofmann DJ (1992) Science 255:423
15. Chameides WL, Bergin M (2002) Science 297:2214
16. Menon S, Hansen J, Nazarenko L, Luo Y (2002) Science 297:2250
17. Seinfeld JH, Pandis SN (1998) Atmospheric chemistry and physics. Wiley, New York, p 700
18. Abelson PH (1998) Science 281:1609
19. Hannigan MP, Cass GR, Penman BW, Crespi CL, Lafleur AL, Busby WF Jr, Thilly WG, Simoneit BRT (1998) Environ Sci Technol 32:3502
20. Chester R, Johnson CR (1971) Nature 229:105
21. Chester R, Stoner JH (1973) Nature 246:138
22. Chester R, Stoner JH (1974) Mar Chem 2:158
23. Chester R, Elderfield H, Griffin JJ, Johnson LR, Padgham RC (1972) Mar Geol 13:91
24. Aston SR, Chester R, Johnson LR, Padgham RC (1973) Mar Geol 14:15
25. Folger DW (1970) Deep-Sea Res 17:337
26. Folger DW, Burckle L, Heezen FC (1967) Science 155:1243
27. Prospero JM, Bonatti E, Schubert C, Carlson TN (1970) Earth Planet Sci Lett 9:287
28. Windom HL (1969) Geol Soc Am Bull 80:761
29. Chester R, Elderfield H (1970) New Sci 47:432
30. Kawamura K (1995) Land-derived lipid class compounds in the deep-sea sediments and marine aerosols from North Pacific. In: Sakai H, Nozaki Y (eds) Biogeochemical processes and ocean flux in the western Pacific. Terra, Tokyo p 31
31. Poynter JG, Farrimond P, Robinson N, Eglinton G (1989) Aeolian-derived higher plant lipids in the marine sedimentary record: links with palaeoclimate. In:

Leinen M, Sarnthein M (eds) Paleoclimatology and paleometeorology: modern and past patterns of global atmospheric transport. Kluwer, Dordrecht, p 435
32. Buat-Ménard P, Cachier H, Chesselet R (1989) Sources of particulate carbon in the marine atmosphere. In: Riley JP, Chester R, Duce RA (eds) Chemical oceanography, vol 10. Academic, New York, p 252
33. Kawamura K, Gagosian RB (1990) Naturwissenschaften 77:25
34. Kawamura K, Usukura K (1993) J Oceanogr 49:271
35. Prospero JM (1996) Saharan dust transport over the North Atlantic Ocean and Mediterranean: an overview. In: Guerzoni S, Chester R (eds) The impact of desert dust across the Mediterranean. Kluwer, Dordrecht, p 133
36. Prospero JM (2001) Geotimes Nov 24
37. Simoneit BRT (1977) Mar Chem 5:443
38. Simoneit BRT, Cox RE, Standley LJ (1988) Atmos Environ 22:983
39. Simoneit BRT, Kobayashi M, Mochida M, Kawamura K, Huebert BJ (2004) J Geophys Res Atmos 109:D19S09. DOI 10.1029/2004JD004565
40. Simoneit BRT, Kobayashi M, Mochida M, Kawamura K, Lee M, Lim HJ, Turpin BJ, Komazaki Y (2004) J Geophys Res Atmos 109:D19S10. DOI 10.1029/2004JD004598
41. Kawamura K, Sakaguchi F (1999) J Geophys Res 104:3501
42. Kawamura K, Gagosian RB (1987) Nature 325:330
43. Mayol-Bracero OL, Rosario O, Corrigan CE, Morales R, Torres I, Pérez V (2001) Atmos Environ 35:1735
44. Simoneit BRT (2002) Atmos Environ 36:4563
45. Gagosian RB, Peltzer ET (1986) Org Geochem 10:661
46. Kranck K (1973) Nature 246:348
47. Gordon DC Jr (1970) Deep-Sea Res 17:175
48. Gimalt JO, Simoneit BRT, Gómez-Belinchón JI, Fischer K, Dymond J (1990) Nature 345:147
49. Simoneit BRT, Grimalt JO, Fischer K, Dymond J (1986) Naturwissenschaften 73:322
50. Claeys M, Graham B, Vas G, Wang W, Vermeylen R, Pashynska V, Cafmeyer J, Guyon P, Andreae MO, Artaxo P, Maenhaut W (2004) Science 303:1173
51. Alves C, Pio C, Duarte A (2000) Environ Sci Technol 34:4287
52. Alves C, Pio C, Durate A (2001) Atmos Environ 35:5485
53. Arpino P, van Dorsselaer A, Sevier KD, Ourisson G (1972) C R Acad Sci Paris 275D:2837
54. Azevedo DA, Moreira LS, Siqueira DS (1999) Atmos Environ 33:4987
55. Barbier M, Tuseau D, Marty JC, Saliot A (1981) Oceanol Acta 4:77
56. Broddin G, Cautreels W, van Cauwenberghe D (1980) Atmos Environ 14:895
57. Conte MH, Weber JC (2002) Glob Biogeochem Cycles 16:1142. DOI 10.1029/2002GB001922
58. Cox RE, Mazurek MA, Simoneit BRT (1982) Nature 296:848
59. Duce RA, Mohnen VA, Zimmerman PR, Grosjean D, Cautreels W, Chatfield R, Jaenicke R, Ogren JA, Pellizzari ED, Wallace GT (1983) Rev Geophys Space Phys 21:921
60. Eichmann R, Neuling P, Ketseridis G, Hahn J, Jaenicke R, Junge C (1979) Atmos Environ 13:587
61. Eichmann R, Ketseridis G, Schebeske G, Jaenicke R, Hahn J, Warneck P, Junge C (1980) Atmos Environ 14:695
62. Fang M, Zheng M, Wang F, To KL, Jaafar AB, Tong SL (1999) Atmos Environ 33:783
63. Fang M, Zheng M, Wang F, Chim KS, Kot SC (1999) Atmos Environ 33:1803

64. Fraser MP, Cass GR, Simoneit BRT, Rasmussen RA (1997) Environ Sci Technol 31:2356
65. Fraser MP, Cass GR, Simoneit BRT, Rasmussen RA (1998) Environ Sci Technol 32:1760
66. Fraser MP, Cass GR, Simoneit BRT (1998) Environ Sci Technol 32:2051
67. Fraser MP, Cass GR, Simoneit BRT (1999) Atmos Environ 33:2715
68. Gagosian RB, Peltzer ET, Zafiriou OC (1981) Nature 291:312
69. Gagosian RB, Zafiriou OC, Peltzer ET, Alford JB (1982) J Geophys Res 87:11133
70. Gagosian RB, Peltzer ET, Merrill JT (1987) Nature 291:321
71. Gogou A, Stephanou EG, Stratigakis N, Grimalt JO, Simo R, Aceves M, Albaiges J (1994) Atmos Environ 28:1301
72. Gogou A, Stratigakis N, Kanakidou M, Stephanou E (1996) Org Geochem 25:79
73. Hildemann LM, Rogge WF, Cass GR, Mazurek MA, Simoneit BRT (1996) J Geophys Res 101(D14):19541
74. Limbeck A, Puxbaum H (1999) Atmos Environ 33:1847
75. Marty J-C, Saliot A (1982) Nature 298:144
76. Matsumoto G, Hanya T (1980) Atmos Environ 14:1409
77. Mazurek MA, Simoneit BRT (1984) Characterization of biogenic and petroleum-derived organic matter in aerosols over remote, rural and urban areas. In: Keith LH (ed) Identification and analysis of organic pollutants in air. ACS symposium. Ann Arbor/Butterworth, Woburn, MA, p 353
78. Mazurek MA, Simoneit BRT (1997) High molecular weight terpenoids as indicators of organic emissions from terrestrial vegetation. In: Eganhouse R (ed) Molecular markers in environmental geochemistry. American Chemical Society symposium series 671. American Chemical Society, Washington, DC, p 92
79. Pio C, Alves C, Duarte A (2001) Atmos Environ 35:389
80. Pio C, Alves C, Duarte A (2001) Atmos Environ 35:1365
81. Rogge WF, Hildemann LM, Mazurek MA, Cass GR, Simoneit BRT (1991) Environ Sci Technol 25:1112
82. Rogge WF, Mazurek MA, Hildemann LM, Cass GR, Simoneit BRT (1993) Atmos Environ 27A:1309
83. Rogge WF, Mazurek MA, Hildemann LM, Cass GR, Simoneit BRT (1993) Environ Sci Technol 27:636
84. Rogge WF, Mazurek MA, Hildemann LM, Cass GR, Simoneit BRT (1993) Environ Sci Technol 27:1892
85. Rogge WF, Mazurek MA, Hildemann LM, Cass GR, Simoneit BRT (1993) Environ Sci Technol 27:2700
86. Rogge WF, Mazurek MA, Hildemann LM, Cass GR, Simoneit BRT (1993) Environ Sci Technol 27:2736
87. Rogge WF, Mazurek MA, Hildemann LM, Cass GR, Simoneit BRT (1994) Environ Sci Technol 27:1375
88. Rogge WF, Mazurek MA, Hildemann LM, Cass GR, Simoneit BRT (1997) Environ Sci Technol 27:2726
89. Rogge WF, Mazurek MA, Hildemann LM, Cass GR, Simoneit BRT (1997) Environ Sci Technol 27:2731
90. Rogge WF, Mazurek MA, Hildemann LM, Cass GR, Simoneit BRT (1998) Environ Sci Technol 27:13
91. Schauer JJ, Rogge WF, Hildemann LM, Mazurek MA, Cass GR, Simoneit BRT (1996) Atmos Environ 30:3837
92. Schauer JJ, Kleeman MJ, Cass GR, Simoneit BRT (1999) Environ Sci Technol 33:1566

93. Schauer JJ, Kleeman MJ, Cass GR, Simoneit BRT (1999) Environ Sci Technol 33:1578
94. Schauer JJ, Kleeman MJ, Cass GR, Simoneit BRT (2001) Environ Sci Technol 33:1716
95. Schauer JJ, Kleeman MJ, Cass GR, Simoneit BRT (2002) Environ Sci Technol 36:567
96. Schauer JJ, Kleeman MJ, Cass GR, Simoneit BRT (2002) Environ Sci Technol 36:1169
97. Schauer JJ, Fraser MP, Cass GR, Simoneit BRT (2002) Environ Sci Technol 36:806
98. Schefuss E, Ratmeyer V, Stuut J-BW, Fred Jansen JH, Sinninghe Damsté JS (2003) Geochim Cosmochim Acta 67:1757
99. Simoneit BRT (1979) Biogenic lipids in eolian particulates collected over the ocean. In: Novakov T (ed) Proceedings of carbonaceous particles in the atmosphere. NSF-LBL, p 233
100. Simoneit BRT (1980) Eolian particulates from oceanic and rural areas—their lipids, fulvic and humic acids and residual carbon. In: Douglas AG, Maxwell JR (eds) Advances in organic geochemistry 1979. Pergamon, Oxford, p 343
101. Simoneit BRT (1984) Atmos Environ 18:51
102. Simoneit BRT (1984) Sci Tot Environ 36:61
103. Simoneit BRT (1985) Int J Environ Anal Chem 22:203
104. Simoneit BRT (1986) Int J Environ Anal Chem 23:207
105. Simoneit BRT (1989) J Atmos Chem 8:251
106. Simoneit BRT (1999) Environ Sci Pollut Res 6:153
107. Simoneit BRT, Mazurek MA (1982) Atmos Environ 16:2139
108. Simoneit BRT, Mazurek MA (1989) Aerosol Sci Technol 10:267
109. Simoneit BRT, Chester R, Eglinton G (1977) Nature 267:682
110. Simoneit BRT, Mazurek MA, Cahill TA (1980) J Air Pollut Control Assoc 30:387
111. Simoneit BRT, Mazurek MA, Reed WE (1983) Characterization of organic matter in aerosols over rural sites: phytosterols. In: Bjorøy M et al. (eds) Advances in organic geochemistry 1981. Wiley, Chichester, p 355
112. Simoneit BRT, Cardoso JN, Robinson N (1990) Chemosphere 21:1285
113. Simoneit BRT, Cardoso JN, Robinson N (1991) Chemosphere 21:447
114. Simoneit BRT, Sheng G, Chen X, Fu J, Zhang J, Xu Y (1991) Atmos Environ 25A:2111
115. Simoneit BRT, Crisp PT, Mazurek MA, Standley LJ (1991) Environ Internat 17:405
116. Simoneit BRT, Radzi bin Abas M, Cass GR, Rogge WF, Mazurek MA, Standley LJ, Hildemann LM (1996) Natural compounds as tracers for biomass combustion in aerosols. In: Levine JS (ed) Biomass burning and global change, vol 1. MIT Press, Cambridge, MA, p 504
117. Simoneit BRT, Rogge WF, Lang Q, Jaffé R (2000) Chemosphere Glob Change Sci 2:107
118. Simoneit BRT, Oros DR, Elias VO (2000) Chemosphere Glob Change Sci 2:101
119. Zheng M, Wan TSM, Fang M, Wang F (1997) Atmos Environ 31:227
120. Zheng M, Fang M, Wang F, To KL (2000) Atmos Environ 34:2691
121. Graham B, Mayol-Bracero OL, Guyon P, Roberts GC, Decesari S, Facchini MC, Artaxo P, Maenhaut W, Köll P, Andreae MO (2002) J Geophys Res 107(D20):8047. DOI 10.1029/2001JD000336
122. Mazurek MA, Simoneit BRT, Cass GR (1989) Aerosol Sci Technol 10:408
123. Schneider JK, Gagosian RB, Cochran JK, Trull TW (1983) Nature 304:429
124. Sicre MA, Marty JC, Saliot A, Aparicio X, Grimalt J, Albaiges J (1987) Atmos Environ 21:2247
125. Simoneit BRT, Rogge WF, Mazurek MA, Standley LJ, Hildemann LM, Cass GR (1993) Environ Sci Technol 27:2533
126. Simoneit BRT, Schauer JJ, Nolte CG, Oros DR, Elias VO, Fraser MP, Rogge WF, Cass GR (1999) Atmos Environ 33:173

127. Kawamura K, Sempéré R, Imai Y, Fujii Y, Hayashi M (1996) J Geophys Res 101:18721
128. Kawamura K, Kasukabe H, Barrie LA (1996) Atmos Environ 30:1709
129. Kawamura K, Steinberg S, Kaplan IR (2000) Atmos Environ 34:4175
130. Graedel TE, Hawkins DT, Claxton LD (1986) Atmospheric chemical compounds, sources, occurrence, and bioassay. Academic, New York
131. Lamb S, Petrowski C, Kaplan IR, Simoneit BRT (1980) J Air Pollut Control Assoc 30:1098
132. Pankow JF (1987) Atmos Environ 21:2275
133. Simoneit BRT, Mazurek MA (1981) CRC Crit Rev Environ Control 11:219
134. Clar E (1964) Polycyclic hydrocarbons, vol 1 and 2. Academic, London
135. Harvey RG (1991) Polycyclic aromatic hydrocarbons: chemistry and carcinogenicity. Cambridge University Press, Cambridge
136. Lane DA (1989) The fate of polycyclic aromatic compounds in the atmosphere during sampling. In: Vo-Dinh T (ed) Chemical analysis of polycyclic aromatic compounds. Chemical analysis, vol 101. Wiley, New York, p 31
137. Lee ML, Novotny MV, Bartle KD (1981) Analytical chemistry of polycyclic aromatic compounds. Academic, London
138. Neff JM (1979) Polycyclic aromatic hydrocarbons in the aquatic environment. Applied Science, London.
139. National Research Council (1983) Polycyclic aromatic hydrocarbons: evaluation of sources and effects. National Academy Press, Washington, DC
140. Simoneit BRT (1998) Biomarker PAHs in the environment. In: Neilson AH (ed) The handbook of environmental chemistry. PAHs and related compounds, vol 3, part I. Springer, Berlin Heidelberg New York p 176
141. Rogge WF, Hildemann LM, Mazurek MA, Cass GR, Simoneit BRT (1996) J Geophys Res 101:19379
142. Schauer JJ, Cass GR (2000) Environ Sci Technol 34:1821
143. Lunde G, Bjørseth A (1977) Nature 268:518
144. Bjørseth A, Lunde G, Lindskog A (1979) Atmos Environ 13:45
145. Simoneit BRT (2002) Appl Geochem 17:129
146. Crutzen PJ, Andreae MO (1990) Science 250:1669
147. Levine JS (ed) (1991) Global biomass burning: atmospheric, climatic, and biospheric implications. MIT Press, Cambridge, MA.
148. Levine JS (ed) (1996) Biomass burning and global change, vols 1 and 2. MIT Press, Cambridge, MA
149. Abas MR, Simoneit BRT, Elias VO, Cabral JA, Cardoso JA (1995) Chemosphere 30:995
150. Fine PM, Cass GR, Simoneit BRT (2001) Environ Sci Technol 35:2665
151. Fine PM, Cass GR, Simoneit BRT (2002) Environ Sci Technol 35:1442
152. Fine PM, Cass GR, Simoneit BRT (2002) J Geophys Res Atmos 107(D21):8349. DOI 10.1029/2001JD00061
153. Hawthorne SB, Miller DJ, Barkley RM, Krieger MS (1988) Environ Sci Technol 22:1191
154. Hawthorne SB, Krieger MS, Miller DJ, Mathiason MB (1989) Environ Sci Technol 23:460
155. Hays MD, Geron CD, Linna KJ, Smith MD, Schauer JJ (2002) Environ Sci Technol 36:2281
156. Moreira dos Santos CY, Azevedo DA, Aquino Neto FR (2002) Atmos Environ 36:3009
157. Nolte CG, Schauer JJ, Cass GR, Simoneit BRT (2001) Environ Sci Technol 35:1912
158. Ramdahl T (1983) Nature 306:580

159. Standley LJ, Simoneit BRT (1987) Environ Sci Technol 21:163
160. Standley LJ, Simoneit BRT (1990) Atmos Environ 24B:67
161. Standley LJ, Simoneit BRT (1994) J Atmos Chem 18:1
162. Echalar F, Gaudichet A, Cachier H, Artaxo P (1995) Geophys Res Lett 22:3034
163. Olmez I, Sheffield AE, Gordon GE, Houch JE, Pritchett LC, Cooper JA, Dzuby TG, Bennett RL (1988) J Air Pollut Control Assoc 38:1392
164. Sheffield AE, Gordon GE, Currie LA, Riederer GE (1994) Atmos Environ 28:1371
165. Mochida M, Umemoto N, Kobayashi M, Matsunage S, Kawamura K, Bates TS, Simoneit BRT (2003) J Geophys Res Atmos 108(D23):8638. DOI 10.1029/2002JD003249
166. Peltzer ET, Gagosian RB (1989) In: Duce RA, Riley JP, Chester R (eds) Chemical oceanography, vol 10. Academic, London, p 281
167. Kobayashi M, Simoneit BRT, Kawamura K, Mochida M, Lee M, Lee G (2002) Levoglucosan, other saccharides and tracer compounds in the Asian dust and marine aerosols collected during the ACE-Asia campaign. In: Extended abstracts of the 6th international aerosol conference, Taipei, Taiwan, September 8–12
168. Simoneit BRT (1978) The organic chemistry of marine sediments. In: Riley JP, Chester R (eds) Chemical oceanography, vol 7, 2nd edn. Academic, New York, p 233
169. Elias VO, Simoneit BRT, Pereira AS, Cabral JA, Cardoso JN (1999) Environ Sci Technol 33:2369
170. Leif RN, Simoneit BRT, Kvenvolden KA (1992) Hydrous pyrolysis of n-$C_{32}H_{66}$ in the presence and absence of inorganic components. In: American Chemical Society, Division of Fuel Chemistry, 204th National Meeting Preprints 37:1748
171. Simoneit BRT (1986) Cyclic terpenoids of the geosphere. In: Johns RB (ed) Biological markers in the sedimentary record. Methods in geochemistry and geophysics, vol 24. Elsevier, Amsterdam, p 43
172. Tulloch AP (1976) Chemistry of waxes of higher plants. In: Kolattukudy PE (ed) Chemistry and biochemistry of natural waxes. Elsevier, Amsterdam, p 235
173. Oros DR, Simoneit BRT (2000) Fuel 79:515
174. Abas MR, Simoneit BRT (1996) Atmos Environ 30:2779
175. Ballentine DC, Macko SA, Turekian VC, Gilhooly WP, Martincigh B (1996) Chemical and isotropic characterization of aerosols collected during sugar cane burning in South Africa. In: Levine JS (ed) Biomass burning and global change, vol 1. MIT Press, Cambridge, MA, p 460
176. Conte MH, Weber JC (2002) Nature 417:639
177. Fang J, Kawamura K, Ishimura Y, Matsumoto K (2002) Environ Sci Technol 36:2598
178. Norman AL, Hopper JF, Blanchard P, Ernst D, Brice N, Alexandrov N, Klouda G (1999) Atmos Environ 33:2807
179. Rudolph J, Czuba E, Norman AL, Huang L, Ernst D (2002) Atmos Environ 36:1173
180. Simoneit BRT (1997) Atmos Environ 31:2225
181. Swap R, Garstang M, Greco S, Talbot R, Kållberg P (1992) Tellus Ser B 44:133
182. Swap R, Garstang M, Macko SA, Kållberg P (1996) Comparison of biomass burning emissions and biogenic emissions to the tropical South Atlantic. In: Levine JS (ed) Biomass burning and global change, vol 1. MIT Press, Cambridge, MA, p 396
183. Turekian VC, Macko SA, Gilhooly WP, Ballentine DC, Swap RJ, Garstang M (1996) Bulk and compound-specific isotope characterization of the products of biomass burning. In: Levine JS (ed) Biomass burning and global change, vol 1. MIT Press, Cambridge, MA, p 422
184. Narukawa M, Kawamura K, Takeuchi N, Nakajima T (1999) Geophys Res Lett 26:3101
185. Simoneit BRT, Elias VO, Kobayashi M, Kawamura K, Rushdi AI, Medeiros PM, Rogge WF, Didyk BM (2004) Environ Sci Technol 38:5939

186. Sempéré R, Kawamura K (1994) Atmos Environ 28:449
187. Fraser MP, Lakshmanan K (2000) Environ Sci Technol 34:4560
188. Sheesley RJ, Schauer JJ, Chowdhury Z, Cass GR, Simoneit BRT (2003) J Geophys Res Atmos 108(D9):4285. DOI 10.1029/2002JD002981
189. Simoneit BRT, Elias VO (2000) Mar Chem 69:231
190. Zdráhal Z, Oliveira J, Vermeylen R, Claeys M, Maenhaut W (2002) Environ Sci Technol 36:747
191. Didyk BM, Simoneit BRT, Pezoa AL, Riveros LM, Flores AA (2000) Atmos Environ 34:1167
192. Simoneit BRT, Rushdi AI, Otto A (2001) Source correlations for carbonaceous matter of aerosol particles in the atmosphere of the Santiago metropolitan and the Concon areas of Chile. Oregon State University report
193. Grimalt JO, Fernández P, Berdié L, Vilanova RM, Catalan J, Psenner R, Hofer R, Appleby PG, Rosseland BO, Lien L, Massabuau JC, Battarbee RW (2001) Environ Sci Technol 35:2690
194. Fernández P, Vilanova RM, Grimalt JO (1999) Environ Sci Technol 33:3716
195. Fernández P, Carrera G, Grimalt JO, Ventura M, Camarero L, Catalan J, Nickus U, Thies H, Psenner R (2003) Environ Sci Technol 37:3261
196. Fu J-M, Sheng G-Y, Chen Y, Wang X-M, Min YS, Peng PA, Lee SC, Chan LY, Wang ZS (1997) Preliminary study of organic pollutants in air of Guangzhou, Hong Kong, and Macao. In: Eganhouse RP (ed) Molecular markers in environmental geochemistry. American Chemical Society, Washington, DC, p 164–176
197. Okuda T, Kumata H, Naraoka H, Takada H (2002) Org Geochem 33:1737
198. Stuermer DH, Peters KE, Kaplan IR (1978) Geochim Cosmochim Acta 42:989
199. Penner JE (1995) Carbonaceous aerosols influencing atmospheric radiation: Black and organic carbon. In: Charlson R, Heintzenberg J (eds) Aerosol forcing of climate. Environmental sciences research report 17. Wiley, Chichester, p 91
200. Penner JE, Charlson RJ, Hales JM, Laulainen N, Leifer R, Novakov T, Ogren J, Radke LF, Schwartz SE, Travis L (1994) Bull Am Meteorol Soc 75:375
201. Penner JE, Novakov T (1996) J Geophys Res 101:19373
202. Turpin BJ, Huntzicker JJ (1991) Atmos Environ 25A:207
203. Turpin BJ, Saxena P, Andrews E (2000) Atmos Environ 34:2983
204. White WH, Macias ES (1989) Aerosol Sci Technol 10:111
205. Hildemann LM, Klinedinst DB, Klouda GA, Currie LA, Cass GR (1994) Environ Sci Technol 28:1565
206. Kaplan IR, Gordon RJ (1994) Aerosol Sci Technol 21:343
207. Klinedinst DB, Currie LA (1999) Environ Sci Technol 33:4146
208. Masiello CA, Druffel ERM, Currie LA (2002) Geochim Cosmochim Acta 66:1025
209. Huebert B, Bertram T, Kline J, Howell S, Eataugh D, Blomquist B (2004) J Geophys Res Atmos 109:DS19S11. DOI 10.1029/2004JD004700
210. Aceves M, Grimalt JO (1993) Environ Sci Technol 27:2896
211. Kavouras IG, Lawrence J, Koutrakis P, Stephanou EG, Oyola P (1999) Atmos Environ 33:4977
212. Al-Mutlaq K, Rushdi AI, Simoneit BRT (2002) Arab J Sci Res 20:141
213. Oros DR, Mazurek MA, Baham JE, Simoneit BRT (2002) Air Water Soil Pollut 137:203
214. Arey J, Winer AM, Atkinson R, Aschmann SM, Long WD, Morrison CL (1991) Atmos Environ 25A:1063
215. Aschmann S, Reissell A, Atkinson R, Arey J (1998) J Geophys Res 103:25553
216. Fruekilde P, Hjorth J, Jensen NR, Kotzias D, Larsen B (1998) Atmos Environ 32:1893

217. Paulson SE, Pandis SN, Baltensperger H, Seinfeld JH, Flagan RC, Palen EJ, Allen DT, Schaffner C, Giger W, Portmann M (1990) J Aerosol Sci 21:S245
218. Yu J, Griffin RJ, Cocker DR, Flagan RC, Seinfeld JH (1999) Geophys Res Lett 26:1145
219. Arey J, Atkinson R, Zielinska B, McElroy PA (1989) Environ Sci Technol 23:321
220. Slinn WGN (1983) Precipitation scavenging. In: Atmospheric sciences and power production—1979. Division of Biomedical Environmental Research, US Department of Energy, Washington, DC, p 1–60
221. Atlas E, Giam CS (1981) Science 211:163
222. Zafiriou OC, Gagosian RB, Peltzer ET, Alford JB, Yoder T (1985) J Geophys Res 90:2409
223. Aboul-Kassim TAT, Simoneit BRT (1995) Environ Sci Technol 29:2473
224. Chow JC, Watson JG (2002) Energy Fuels 16:222
225. Merrill JT (1989) Atmospheric long-range transport to the Pacific Ocean. In: Duce RA, Riley JP, Chester R (eds) Chemical oceanography, vol 10. Academic, London, p 15
226. Wilkening KE, Barrie LA, Engle M (2000) Science 290:65
227. Decesari S, Facchini MC, Matta E, Lettini F, Mircea M, Fuzzi S, Tagliavini E, Putaud J-P (2001) Atmos Environ 35:3691
228. Facchini MC, Mircea M, Fuzzi S, Charlson RJ (1999) Nature 400:257
229. Nolte CG, Schauer JJ, Cass GR, Simoneit BRT (1999) Environ Sci Technol 33:3313
230. Nolte CG, Schauer JJ, Cass GR, Simoneit BRT (2002) Environ Sci Technol 36:4273
231. Freeman KH, Hayes JM, Trendel JM, Albrecht P (1989) Nature 353:254
232. Watson JG, Robinson NF, Chow JC, Henry RC, Kim BM, Pace TG, Meyer EI, Nguyen Q (1990) Environ Software 5:38
233. Arrhenius G (1959) Sedimentation on the ocean floor. In: Abelson PH (ed) Researches in geochemistry. Wiley, New York, p 1

Appendix—Chemical Structures Cited in the Text

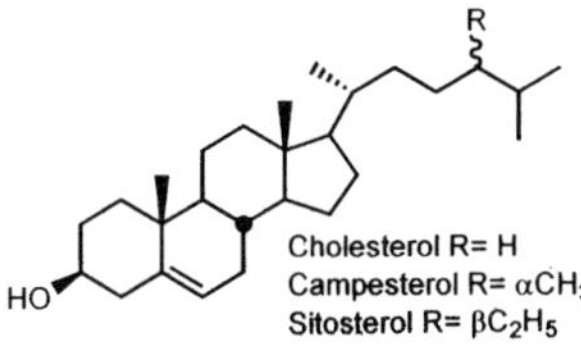

I. Sterols
- Cholesterol R= H
- Campesterol R= αCH₃
- Sitosterol R= βC₂H₅

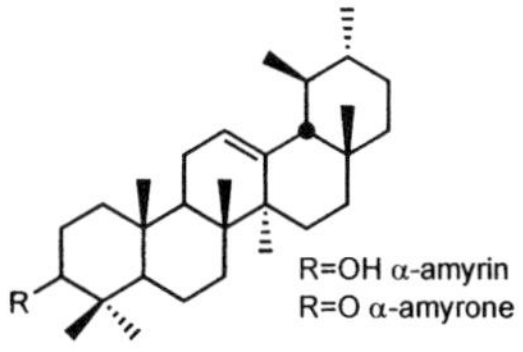

II. Ursenoids
- R=OH α-amyrin
- R=O α-amyrone

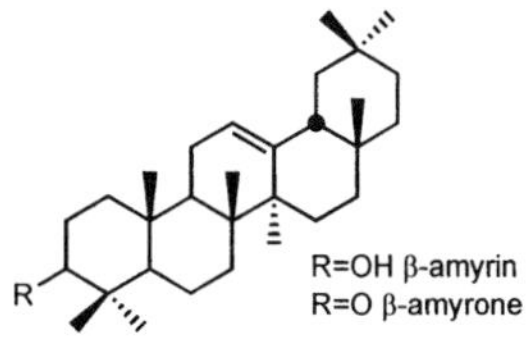

III. Oleanoids
- R=OH β-amyrin
- R=O β-amyrone

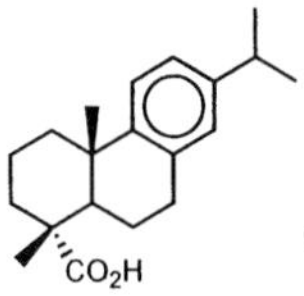

IV. Dehydroabietic acid

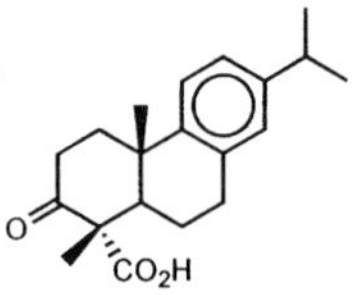

V. 3-Oxodehydroabietic acid

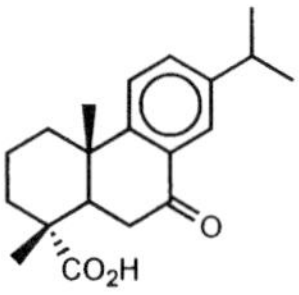

VI. 7-Oxodehydroabietic acid

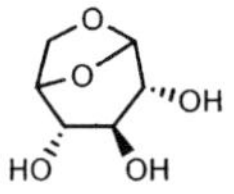

VII. Levoglucosan

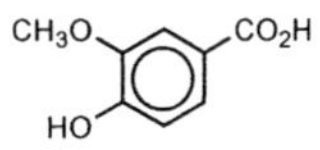

VIII. Vanillic acid

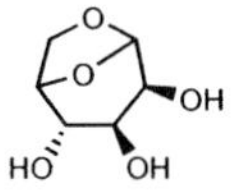

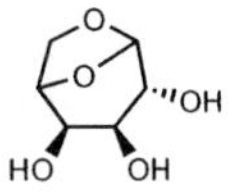

IX. 17α(H),21β(H)-Hopanes R= H, CH₃-C₆H₁₃

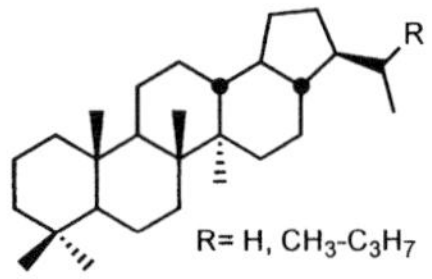

X. 17β(H),21α(H)-Hopanes R= H, CH₃-C₃H₇

XI. Mannosan

XII. Galactosan

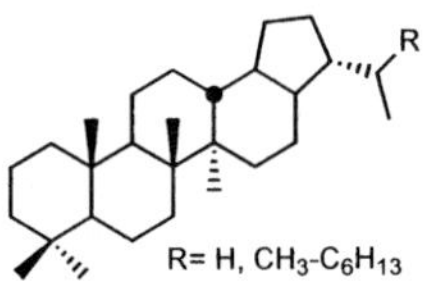

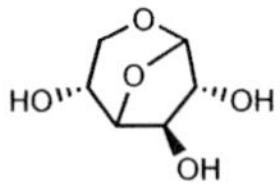

XIII. 1,6-Anhydro-β-glucofuranose

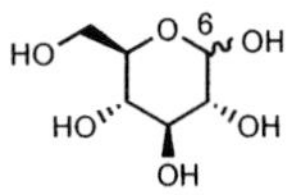

XIV. Glucose (α=6αOH, β=6βOH)

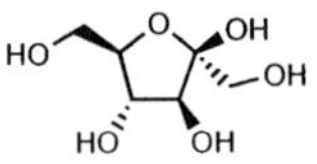

XV. Fructose

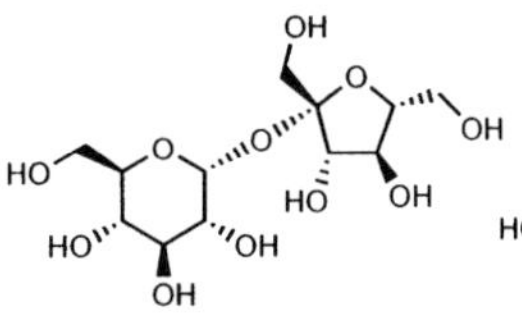

XVI. Sucrose

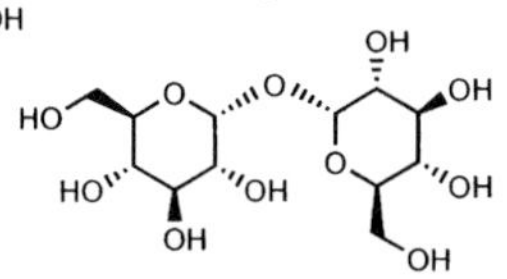

XVII. Mycose (trehalose)

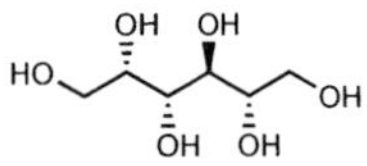

XVIII. Sorbitol (glucitol)

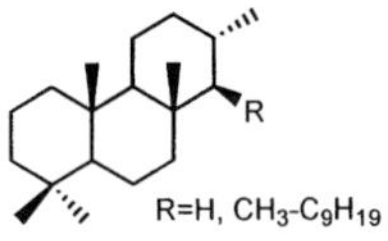

XIX. Tricyclic terpanes R=H, CH₃-C₉H₁₉

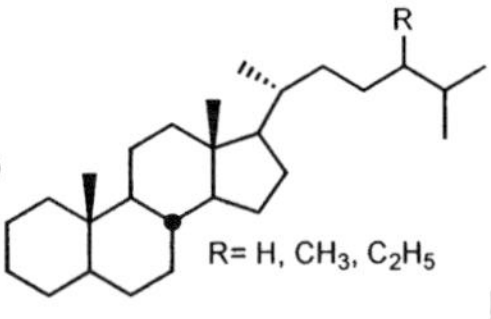

XX. 5α,14α,17α-Steranes R= H, CH₃, C₂H₅

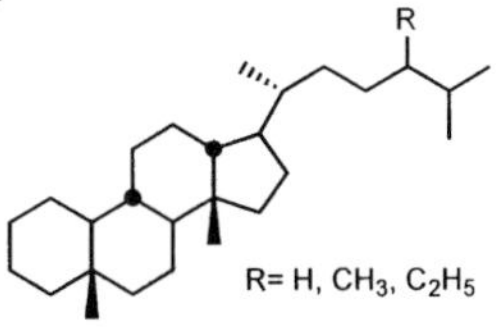

XXI. 10α,13β,17α-Diasteranes R= H, CH₃, C₂H₅

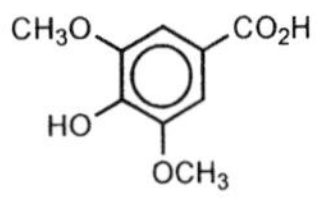

XXII. Syringic acid

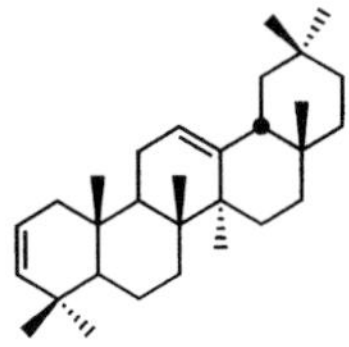

XXIII. Oleana-2,12-diene

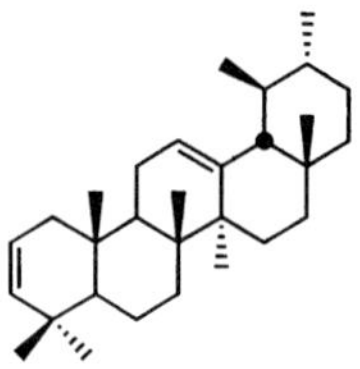

XXIV. Ursana-2,12-diene

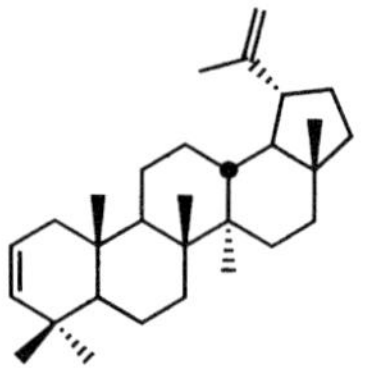

XXV. Lupa-2,22(29)-diene

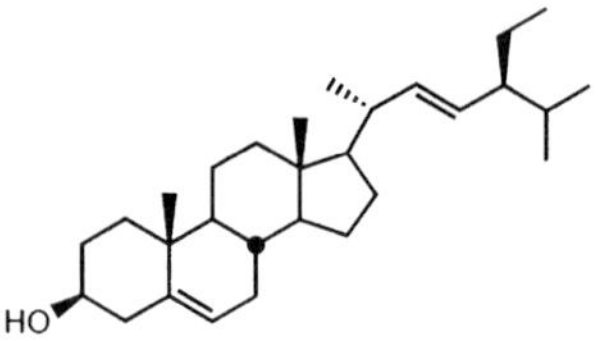

XXVI. Stigmasterol

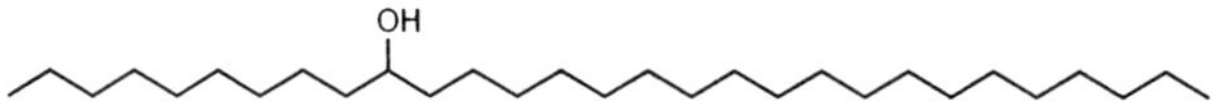

XXVII. Nonacosan-10-ol

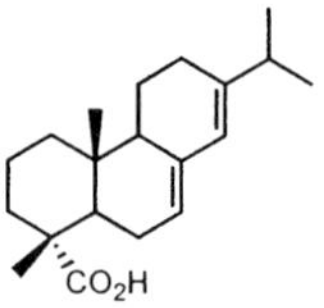

XXVIII. Abietic acid

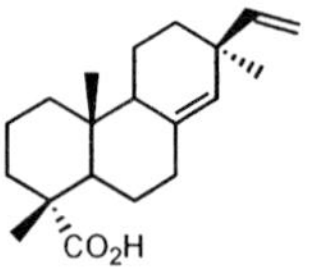

XXIX. Pimaric acid

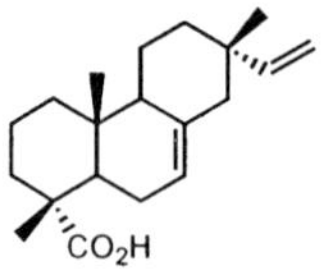

XXX. *Iso*-pimaric acid

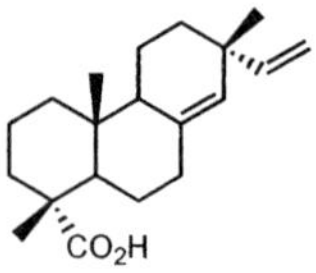

XXXI. Sandaracopimaric acid

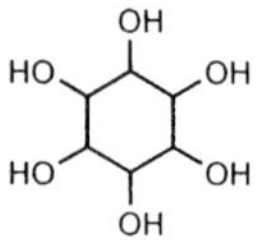

XXXII. Inositols

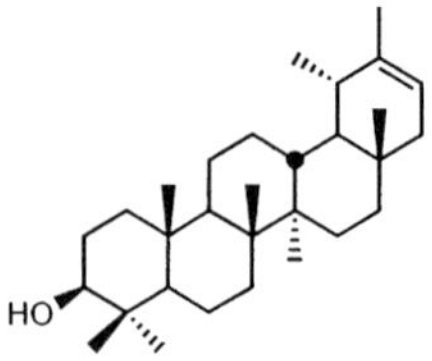

XXXIII. ψ-Taraxasterol

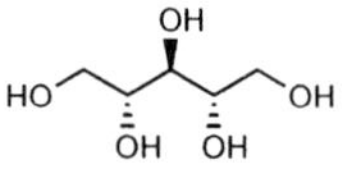

XXXIV. Xylitol

Controls on the Carbon Isotopic Compositions of Lipids in Marine Environments

Richard D. Pancost[1] (✉) · Mark Pagani[2]

[1]Organic Geochemistry Unit, Biogeochemistry Research Centre, School of Chemistry, University of Bristol, Cantock's Close, Bristol BS8 1TS, UK
R.D.Pancost@bristol.ac.uk

[2]Department of Geology and Geophysics, Yale University, PO Box 208109, New Haven, CT 06520, USA

1	Introduction	210
1.1	Background	211
1.2	Isotope Analysis	213
2	Common Prokaryotic and Eukaryotic Biomarkers in Marine Sediments	214
2.1	Algal Lipids	214
2.2	Lipids from Higher Plants	215
2.3	Bacterial and Archaeal Lipids	216
3	Controls on Algal $\delta^{13}C$ Values	217
3.1	The Carbon Isotopic Composition of Dissolved Inorganic Carbon (DIC)	218
3.2	Carbon Isotope Fractionation During Carbon Assimilation	220
3.2.1	Calvin Cycle	221
3.2.2	Fractionation in the Absence of Active Uptake	221
3.2.3	The Impact of Active Uptake of Dissolved Inorganic Species on Algal $\delta^{13}C$ Values	223
4	Controls on Higher Plant $\delta^{13}C$ Values	226
4.1	The Carbon Isotopic Composition of Atmospheric Carbon Dioxide	226
4.2	The Carbon Isotopic Compositions of C_3 Plants	226
4.3	C_3 versus C_4 Plants	227
5	Controls on Prokaryote δ^3C Values	229
5.1	The Carbon Isotopic Composition of Microbial Substrate Carbon	229
5.2	Isotope Fractionation During Carbon Assimilation by Bacteria and Archaea	230
6	Biosynthetic Controls on Lipid $\delta^{13}C$ Values	231
6.1	Microalgae	232
6.2	Higher Plants	233
6.3	Prokaryotes	234
7	Applications	235
7.1	Alkenone $\delta^{13}C$ Values as a Paleo-pCO_2 Proxy	235
7.2	Identifying Methane Cycling in Modern and Ancient Settings	239
8	Conclusions and Directions for Future Research	242
	References	243

Abstract Organic carbon isotopes have long been used to study modern and ancient biogeochemical processes. Controls on the isotopic composition of organic matter in marine sediments range from the sources of the organic matter (e.g. allochthonous vs. autochthonous inputs) to the physiology of the predominant photoautotrophs, the nutrient status of the photoautotrophs' growth environment and the degree of organic matter reworking by heterotrophic eukaryotes and bacteria. The diversity and antagonistic effects of these variables make it difficult to interpret the bulk organic carbon isotope record; consequently, the use of gas chromatography–combustion–isotope ratio monitoring mass spectrometry (GC–IRMS) to determine the carbon isotopic compositions of specific compounds (compound-specific carbon isotope analysis; CSIA) has become a widely used tool in palaeoclimate and ecological investigations. Here, we summarize the main controls on the $\delta^{13}C$ values of lipids found in marine environments and illustrate their utility with several case studies.

1
Introduction

The organic matter occurring in modern or ancient settings represents an archive of potential insight into the organisms and environmental conditions of those settings. Commonly, this information is extracted via the chemical composition of the organic matter—with potentially useful compounds ranging from diagenetically and catagenetically altered lipids that can be traced to specific biological precursors (biomarkers) to pigment distributions that can be used to reconstruct modern phytoplankton assemblages. In addition to the chemical structure of organic matter, much additional information is preserved in the isotopic composition of that organic matter. In particular, the carbon isotopic composition of marine organic matter has commonly been used in a variety of ecological and palaeobiological investigations. These initial investigations utilized the fact that organic matter produced by organisms employing the Calvin cycle (also referred to as the Calvin–Benson cycle) is characteristically depleted in ^{13}C relative to inorganic carbon substrates. Hence, Schidlowski [1] invoked the ^{13}C-depleted character of ancient organic matter as evidence for photosynthesis on the ancient earth, while others recognized it as further evidence for the biological origin of petroleums. Scientists also realized that a variety of processes, including diagenesis [2], catagenesis [3] and heterotrophic reworking of organic matter [4–6], imposed secondary controls on its carbon isotopic composition. This latter observation, combined with analogous observations for organic matter $\delta^{15}N$ values, served as the basis for trophic studies of marine food webs [4].

More recently, recognizing that photoautotrophs have widely varying carbon isotopic compositions, scientists have attempted to constrain the controls on carbon isotope fractionation during photosynthesis; this has served as the basis for additional applications in palaeoenvironmental research.

Calder and Parker [7], Morris [8] and Degens et al. [9] all observed that microalgae grown in cultures became more depleted in ^{13}C as the amount of supplied CO_2 increased. These observations led workers to interpret anomalously light carbon isotopic values in geological samples as indicators of high pCO_2 [10, 11]. Subsequent developments have refined the approach of these early workers, but the principles remain the same and organic matter δ^{13}C values remain a widely used palaeo-pCO_2 proxy [12], although further work indicates that other variables also govern algal δ^{13}C values and must be considered.

While the diversity of controls on organic matter δ^{13}C values presents the opportunity for their application to diverse investigations, they also make it difficult to interpret the bulk organic carbon isotope record. Consequently, the use of gas chromatography–combustion–isotope ratio monitoring mass spectrometry (GC–IRMS) to determine the carbon isotopic compositions of specific compounds (compound-specific carbon isotope analysis; CSIA) has become widespread. By using this approach, primary versus secondary and allochthonous versus autochthonous organic materials can be isotopically distinguished, and pCO_2 reconstruction can be based on compounds derived from a narrow range of organisms rather than the physiologically diverse sources of organic matter present in most marine settings. Moreover, the information recovered from a single sample is much more diverse: insight into algal photosynthesis, higher plant community structure on adjacent land masses and bacterial recycling of organic matter can be elucidated from a few analyses [13]. However, this approach also introduces additional complications related to carbon isotope fractionation during the synthesis of specific compound classes [14, 15]. Here, we discuss the main controls on the δ^{13}C values of lipids found in marine environments and illustrate their utility with two case studies.

1.1
Background

The relative abundances of stable isotopes are usually expressed as ratios, with the lighter isotope in the denominator, e.g. $^{13}C/^{12}C$ ($^hX/^1X$). Because variations in isotope ratios are small and measured against standards of known isotopic composition, the isotopic compositions of natural materials are reported as $\delta^h X$ values (parts per mil ‰).

$$\delta X(\%) = \left(\frac{R_{sample} - R_{standard}}{R_{standard}} \right) \times 10^3 \quad \text{where } R = {^h X}/{^1 X} \tag{1}$$

Thus, a $\delta^h X$ value of 0‰ indicates that the isotopic composition of the sample is identical to that of the standard, positive $\delta^h X$ values indicate that the sample is enriched in the heavy isotope relative to the standard, and negative $\delta^h X$ values indicate that the sample is depleted in the heavy isotope relative to the standard. Carbon isotopic compositions are expressed relative to the primary standard Vienna Pee Dee Belemnite (VPDB) as described above:

$$\delta^{13}C = (^{13}R_{SA}/^{13}R_{VPDB} - 1) \times 10^3 \qquad (2)$$

where $^{13}R_{SA}$ and $^{13}R_{VPDB}$ represent the $^{13}C/^{12}C$ abundance ratios for the sample and VPDB, respectively.

Variations in relative isotopic abundance are a consequence of the preferential reaction of one isotopic species over another. This arises as a direct result of differences in the energies of molecules comprised of different isotopes of the same element, with molecules containing the light isotope generally existing at a higher energy level. This means, for example, that a bond between the light isotope and another atom will break more rapidly than a comparable bond involving the heavy isotope. This leads to isotopic fractionation, without which natural isotopic variation would not exist. Fractionation can occur during a variety of physiochemical processes, for example chemical reactions, phase changes and molecular diffusion, but is always defined as:

$$\alpha = \frac{R_{products}}{R_{reactants}} \qquad (3)$$

The fractionation factor represents the degree of depletion in the instantaneous product of a reaction. In a closed system, characterized by a normal isotope effect (where the light isotope reacts more quickly), the accumulated product (initially the instantaneous product) gradually becomes less depleted in the heavy isotope. When the reaction goes to completion, the isotopic composition of the product is the same as that of the initial reactant, i.e. no enrichment or depletion of the product is observed. This is known as Rayleigh fractionation and is represented by:

$$\frac{R}{R_0} = f^{(\alpha-1)} \qquad (4)$$

where R_0 is the isotopic ratio of the reactant at time zero and f is the fraction of the reactant remaining.

Changes in isotopic composition associated with a reaction are also commonly reported as isotope effects or enrichment factors (ε) [16]; if the fractionation factor differs from 1 by less than 5%, then:

$$\varepsilon = (\alpha - 1) \times 1000 \qquad (5)$$

The latter relationship is similar to the commonly used isotope discrimination term Δ:

$$\Delta = (\alpha - 1) \times 1000 \sim \delta_{products} - \delta_{reactants} \qquad (6)$$

Again, this relationship is only valid if $\alpha \sim 1$.

1.2
Isotope Analysis

All mass spectrometers can measure isotope ratios, but in practice only Nier-type isotope ratio mass spectrometers (IRMS) have sufficient precision to measure variations at natural abundances (4–6 significant figures). These mass spectrometers are fitted with double or triple ion collectors, and traditionally use dual-inlet systems allowing rapid switching between standard and samples. One of the most important advances in analytical chemistry in the past 20 years has been the development and widespread application of continuous-flow isotope ratio mass spectrometry (CF-IRMS) [17–19]. Continuous-flow mass spectrometers can operate at the higher pressures associated with the continual influx of helium, allowing online introduction of small quantities of analyte. Consequently, CF-IRMS allows the isotopic composition of chemically diverse materials to be determined much more rapidly and at significantly lower quantitites (< 100 ng of the element of interest) than traditional offline approaches. Moreover, the development of sophisticated interfaces for IRMS now allows the isotopic characterization of diverse organic and inorganic materials. Coupling of a gas chromatograph (GC) to the IRMS via a combustion interface, which converts eluting compounds into CO_2 and H_2O, enables carbon isotopic compositions to be determined for specific compounds in complex mixtures [5].

Determination of compound-specific $\delta^{13}C$ values is obviously contingent not only on the mass spectrometer but also on the analytical work-up prior to analysis and the resolvability of compounds by gas chromatography. The methods traditionally used in the organic geochemistry field are well established, involving a combination of extraction, compound class fractionation and chemical degradation approaches, and have been discussed in detail elsewhere. Upon appropriate derivatization of polar compounds (e.g. methylation of carboxylic acids, conversion of alcohols into their corresponding trimethylsilyl ethers), compound fractions can be analysed by GC–IRMS. Although specific approaches vary depending on the analyte and instrument used, the general operating principles are the same for all GC–IRMS systems. Sample aliquots are injected into the GC and the compounds are partitioned on a capillary column as in normal GC or GC–MS operation; however, eluting compounds are passed through an oxidizing furnace, which converts organic compounds into CO_2 and H_2O. The H_2O is subsequently removed and CO_2 enters the mass spectrometer through a continuous-flow interface.

2
Common Prokaryotic and Eukaryotic Biomarkers in Marine Sediments

The utility of compound-specific isotope analysis is directly related to the specificity of the compounds being analysed. Compounds with very specific sources, such as alkenones derived from certain species of haptophyte algae, provide more precise data, allowing the most constrained interpretation, but less diagnostic compounds can still provide useful information in the proper context. It is not the role of this chapter to provide a complete discussion of compounds typically examined during isotopic investigations nor the caveats associated with their source assignments; however, it is useful to describe some of the most commonly analysed compound classes.

2.1
Algal Lipids

The lipids of marine and lacustrine algae, their degradation pathways and representative biomarkers in the geologic record have been studied for decades [20–23], and this literature is only briefly summarized here, with particular emphasis on compounds used in carbon isotope studies. Alkenones, long-chain ($C_{37} - C_{39}$) unsaturated ethyl and methyl ketones produced by only a few species of Haptophyte algae in the modern ocean [24–26], are the most commonly used algal biomarkers in environmental investigations due to their relative ease of preparation and isotopic analysis, source specificity, diagenetic robustness and the use of their distributions as a sea surface temperature proxy [26, 27]. Controls on their carbon isotopic composition and application to environmental studies are described in Sect. 7.1. After alkenones, steroids are the algal lipids most commonly investigated using isotopic approaches. Because of their structural diversity [22, 28], certain sterols are relatively diagnostic for specific taxa. For example, 24-methylcholesta-5,22E-dien-3β-ol and especially 24-methylcholesta-5,24(28)-dien-3β-ol have both been invoked as diatom biomarkers, although these sterols are also present in other algae [28]. More diagnostic are the 4-methylsterols, especially 4α,23,24-trimethyl-5α-cholest-22E-en-3β-ol (dinosterol), as biomarkers for dinoflagellates [29–31]. Several workers have determined sterol $\delta^{13}C$ values in modern surface waters [32–35] and shallow marine sediments. Sterol $\delta^{13}C$ values have been used to either clarify the sterol sources (e.g. that 24-ethylcholesterols in Peru surface waters derive from diatoms and not higher plants [33]), or evaluate controls on algal growth rates. In ancient sediments, steranes are among the most abundant preserved hydrocarbons; however, thermal isomerization typically results in a complex distribution of steranes and determination of the $\delta^{13}C$ values of specific compounds is difficult [36].

Other diagnostic compounds that could serve as useful algal biomarkers in isotopic studies are C_{20}, C_{25} and C_{30} highly branched isoprenoids (derived from diatoms [37, 38]), long-chain alkyl diols (derived from eustigmatophytes and other microalgae [28, 39]) and highly unsaturated fatty acids [40, 41]. However, isotopic studies of such compounds are relatively uncommon and future research on cultured organisms and modern marine environments is critical.

Chlorophylls and their degradation products, while not as diagnostic as the aforementioned compounds, can be used as tracers for the isotopic composition of the entire algal community. Because they are not amenable to GC, it is difficult to directly determine chlorophyll (and porphyrin) $\delta^{13}C$ values, and most efforts have focussed on their degradation products. Phytol, the esterified side chain of most chlorophylls, and inferred hydrocarbon degradation products (pristane, phytane) have been the subject of compound-specific carbon isotopic analysis since the advent of the technique [5, 13]. However, sedimentary pristane and phytane can have higher plant sources [42] and some workers have focussed on examining S-bound phytane as a more specific photoautotroph biomarker [43]. Maleimides, 1-H-pyrrole-2,5-diones, are direct degradation products of the chlorophyll and bacteriochlorophyll tetrapyrrole structure, are common in extracts of ancient sediments and are amenable to GC [44]. Thus, maleimide $\delta^{13}C$ values are relatively easy to determine and have been used to gain insight into carbon cycling in ancient settings (e.g. Permian Kupferschiefer [44]).

2.2
Lipids from Higher Plants

Long-chain n-alkyl compounds are major components of epicuticular waxes from the leaves of vascular plants [45]. These compounds are relatively resistant to degradation, which makes them suitable for use as higher plant biomarkers [46], and include n-alkanes, n-alkanols, n-alkanoic acids and wax esters. n-Alkanes occur in vascular plant leaf extracts with carbon chain-lengths ranging from C_{25} to C_{35} [45, 47] and with a strong predominance of odd-carbon-number homologues over even-numbered ones. The n-alkanols and n-alkanoic acids typically occur in higher plants as the $C_{26} - C_{34}$ homologues with a strong even-over-odd predominance, reflecting their biosynthesis from acetyl moieties [47, 48]. It is relatively easy to determine carbon isotopic compositions of higher plant n-alkyl compounds using GC–IRMS, because they commonly occur in high abundances and relatively simple adduction procedures can be used to obtain pure fractions. Thus, the carbon isotopic compositions of these compounds in modern plants, soils and lacustrine and marine sediments have been extensively published. A variety of pentacyclic triterpenoids, typically with structures based on the ursene, oleanene or lupene hydrocarbon skeletons, are common in higher plants. Due to their ubiquity, such compounds are typically only used as general tracers of higher plant input. However, oleanoids and lupanoids, deriving only from angiosperms, and taraxeroids, thought to derive predominantly from mangrove leaves [49], can be more specific higher plant tracers. Although triterpenoids are often abundant in marine and terrestrial sediments, their δ^{13}C values have been rarely published. This is mainly due to co-elution with other compounds, including steroids and bacterial hopanoids.

Higher plant macromolecular components in sediments include resin- and lignin-derived compounds. Lignin, a relatively stable and microbially resistant heteropolymeric structure comprised of phenylpropanoid subunits [50], is a significant component of wood and occurs in the cell walls of all vascular plant tissue. Moreover, lignin monomers have different sources [51, 52] and the isotopic compositions of syringyl, vanillyl and cinnamyl phenols can be used to distinguish isotopic signals of different plant types [53, 54]. Although preparation of lignin monomers for isotopic analysis requires careful chemical work-up, typically involving CuO oxidation of the lignin macromolecule, generated fractions are readily analysable by GC–IRMS.

2.3
Bacterial and Archaeal Lipids

The most common bacterial biomarkers in marine sediments are free and bound (phospholipid) fatty acids, of which the latter comprise the membranes of bacteria; however, eukaryotes also contain membranes comprised of phospholipid fatty acids and these compounds are not diagnostic as a class. The most common, such as saturated C_{16} and C_{18} fatty acids, are particularly widespread and appear to have little utility as tracers of explicit prokaryotic processes. However, some fatty acids, characterized by site-specific methyl groups, double bonds or cyclic moieties, are less common. Other bacterial membrane lipids are the hopanoids, pentacyclic triterpenoids common in cyanobacteria, methanotrophs and aerobic heterotrophic bacteria [55–57]. Until recently, hopanoid structures had not been found in cultures of anaerobic bacteria and were thought to be largely diagnostic for aerobic organisms [55]. However, the occurrence of ^{13}C-depleted hopanoids in Black Sea sediments is evidence for at least one anaerobic source [58]. Moreover, hopanoids have recently been recovered in anaerobic cultures containing predominantly anaerobic ammonia-oxidizing bacteria (Annamox) [59]. The most commonly observed hopanoids are diplopterol and bacteriohopanpolyol derivatives. During diagenesis, vicinal cleavage of the bacteriohopanpolyol side chain results in the formation of, typically, bishomohopanol and eventually bishomohopanoic acid. Both of these compounds are typically well resolved during GC analysis of marine sediments and their δ^{13}C values can be readily determined. In contrast to the widely occurring hopanoids, the methylhopanoids appear to be more source-diagnostic, with 3β-methylbacteriohopanoids being relatively specific for some methanotrophs [60] and 2β-methylbacteriohopanoids being common in and largely restricted to cyanobacteria [57]. However, due to co-elution with the more abundant hopanoids, precise determination of methylhopanoid δ^{13}C values is difficult.

In addition to the above widespread bacterial biomarkers, a variety of compounds are biomarkers for photosynthetic bacteria. These include biomarkers for cyanobacteria, including monomethyl alkanes [61, 62], and diverse pigments, including isorenieratene [63], chlorobactene and bacteriochlorophylls *d* and *e*, which are diagnostic for green sulphur bacteria. Of these, it is relatively easy to determine $\delta^{13}C$ values for isorenieratene derivatives (especially isorenieratane) due to their high molecular weight and long retention time eliminating most co-elution problems. Even in samples where co-elution does occur, clean fractions can be readily obtained by thin-layer chromatography [64].

Archaeal membrane lipids are distinct from those of the bacteria and eukarya because they contain isoprenoidal alkyl units ether-bound to a glycerol backbone rather than ester-linked alkyl components [65]. Archaeol is the most common of the archaeal diethers [66], while hydroxyarchaeol has the same core structure as archaeol but contains an additional hydroxyl group on the third carbon of the phytanyl moiety ether-linked to either the third (*sn*-3-hydroxyarchaeol) or second (*sn*-2-hydroxyarchaeol) glycerol carbon. Like diethers, glycerol tetraethers (glycerol dialkyl glycerol tetraether; GDGT) are diagnostic for, and common in, archaea. In archaeal tetraethers, the alkyl groups are biphytane units with zero to four cyclopentane rings. The GDGT composed of two acyclic biphytanyl units is apparently widespread in archaea [67], whilst GDGTs containing biphytanyls with cyclopentane units were previously thought to be present only in thermophilic archaea [65]. However, recent work has revealed that GDGTs containing biphytanyl groups bearing one or two cyclopentane rings occur in mesophilic settings [66–72]. In general, archaeol and hydroxyarchaeol elute during gas chromatography after other common biomarkers such as steroids and alkenones, such that it is relatively straightforward to determine their $\delta^{13}C$ values in samples where they are abundant. In contrast, GDGTs are not amenable to GC using standard approaches and determination of their $\delta^{13}C$ values requires chemical degradation, typically with HI acid followed by $LiAlH_4$ reduction, to release biphytanes [66]. Other archaeal biomarkers include the irregular isoprenoids 2,6,10,15,19-pentamethylicosane (PMI) and crocetane and their unsaturated derivatives; these compounds have only been identified in association with methane cycling and appear to derive from either methanogens or anaerobic methane-oxidizing archaea [73–78].

3
Controls on Algal $\delta^{13}C$ Values

Although biomass and lipid $\delta^{13}C$ values can differ significantly among phytoplankton, bacteria, archaea and higher plants, the underlying controls on the carbon isotopic compositions in all organisms are intrinsically the same and depend on the isotopic composition of the carbon source and the mechanisms by which that carbon is assimilated.

3.1
The Carbon Isotopic Composition of Dissolved Inorganic Carbon (DIC)

Marine photoautotrophs utilize various forms of dissolved inorganic carbon that derive from interactions of seawater with carbon dioxide. The solubility of gaseous carbon dioxide in seawater, and its subsequent reaction with H_2O, results in several dissolved forms of inorganic carbon including aqueous carbon dioxide (CO_{2aq}), carbonic acid (H_2CO_3), bicarbonate (HCO_3^-) and carbonate ion (CO_3^{2-}). The hydration of CO_{2aq} produces H_2CO_3, whereas HCO_3^- and CO_3^{2-} result from the dissociation of H_2CO_3. However, due to the low concentration of H_2CO_3 relative to CO_{2aq} (the other neutral carbon species), and the fact that H_2CO_3 and CO_{2aq} are chemically inseparable, it is conventional to treat the concentrations of H_2CO_3 and CO_{2aq} as singular (i.e. $[CO_{2aq}]$). At equilibrium, the activity of all dissolved carbon species (or less precisely, the concentration of the carbon species), can be determined from Henry's law, and application of the first and second dissociation constants of carbonic acid (Fig. 1):

$$[CO_{2aq}] = [pCO_2]K_H \tag{7}$$
$$[HCO_3^-] = K_1^*[CO_{2aq}]/[H^+] \tag{8}$$
$$[CO_3^{2-}] = K_2^*[HCO_3^-]/[H^+] \tag{9}$$

where K_H is the temperature-, salinity-, and pressure-dependent equilibrium constant of CO_2 solubility [79], and K_1^*, K_2^* are the first and second apparent dissociation constants of carbonic acid.

Associated with the chemical equilibria of the carbonate system are temperature-dependent isotopic fractionations of carbon and oxygen atoms that lead to distinct isotopic compositions for each carbonate species. Carbon isotopic fractionations of the carbonate system maintain quantifiable relationships between aqueous species, with CO_{2aq} depleted in ^{13}C and CO_3^{2-} and HCO_3^- enriched in ^{13}C relative to gaseous CO_2 (CO_{2g}). For example, the isotopic composition of HCO_3^- is $\sim$ 10 and 9‰ greater than that of CO_{2aq} at 15 and 25 °C, respectively. The results of Mook et al. [80] are represented in Fig. 2 as the equilibrium fractionation of each carbon species relative to bicarbonate. Also displayed are the conclusions of Romanek et al. [81], which indicate that calcite and aragonite precipitated in equilibrium are enriched in ^{13}C relative to HCO_3^- by $\sim 1.0 \pm 0.2$ and 2.7 ± 0.6‰, respectively. Furthermore, carbon isotopic fractionations associated with carbonate precipitation are insensitive to temperature between 10–40 °C. In general, recent experimental determinations of carbon fractionations for the carbonate system are in broad agreement. One notable exception is provided by the work of Mook

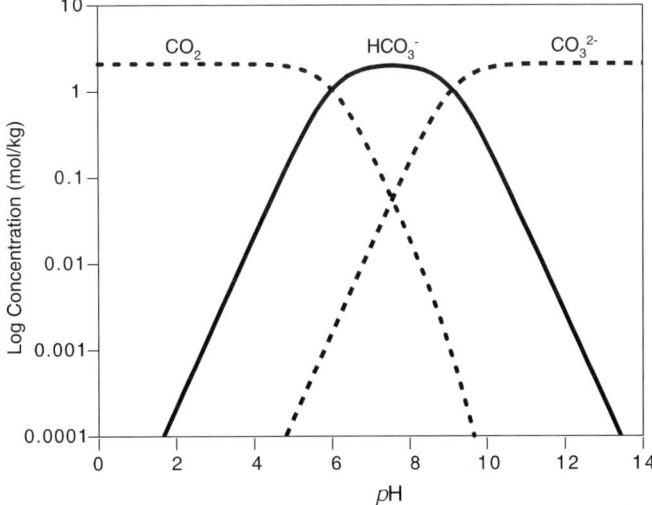

Fig. 1 Concentrations of the dissolved carbon species as a function of pH. $T = 25\,°C$, $S = 35$, $\sum CO_2 = 2.1$ mmol

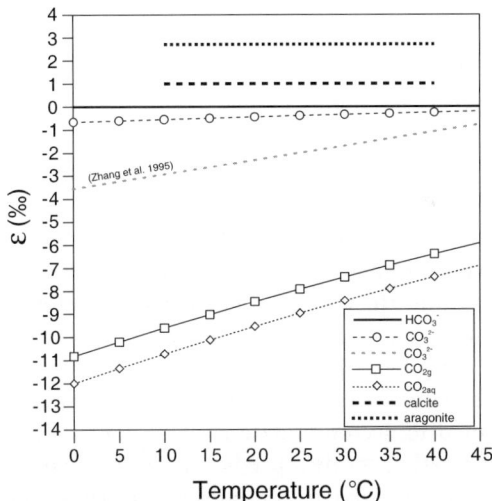

Fig. 2 Carbon isotope fractionations between carbon species of the carbonate system relative to HCO_3^-. Fractionations between dissolved species are from the results of Mook et al. [80] and Zhang et al. [82]. Fractionations between solid phases are from the results of Romanek et al. [81]

et al. [80] and Zhang et al. [82] for the $CO_3^{2-} - HCO_3^-$ equilibrium, which present values differing by ∼ 2.5‰ at 0 °C (Fig. 2). The true nature of the carbon isotopic fractionation that occurs between CO_3^{2-} and other aqueous carbon species remains unresolved. However, the recent conclusions of Zhang et al. [82], which predict a larger fractionation for CO_3^{2-} – HCO_3^- relative to that found by Mook et al. [80], are supported by the experimental results of Halas et al. [83] and recent theoretical calculations [84]. A more detailed description of the topic can be found in the article by Zeebe and Wolf-Gladrow [84].

All of these variables must be considered when examining the isotopic composition of ancient organic matter. Past changes in the isotopic composition of the entire ocean–atmosphere reservoir of carbon dioxide are significant [85] and arise from changes in the relative magnitudes of carbon sources (e.g. volcanism, metamorphism, methane clathrate dissolution) and sinks (organic carbon and carbonate burial). Thus, during times of high organic burial, such as the Mesozoic oceanic anoxic events, DIC $\delta^{13}C$ values were about 2 to 3‰ greater [86], while during inferred clathrate release events, ocean DIC $\delta^{13}C$ values were as much as 4‰ lower [87]. Past pH variations in open ocean settings probably exert only a secondary control on substrate and algal $\delta^{13}C$ values (although its influence on carbon availability as discussed below is important), but pH does need to be considered when interpreting isotopic data from alkaline lakes, hydrothermal springs or similar settings.

3.2
Carbon Isotope Fractionation During Carbon Assimilation

Of the dissolved carbon species in seawater, CO_{2aq} and HCO_3^- are the primary forms available for marine algae during photosynthesis. However, the precise pathway of inorganic carbon uptake and utilization for various marine algae is actively debated in the literature. In general, two models for marine carbon fixation are probable: (1) a passive model requiring diffusive transport of CO_{2aq} across the cell membrane, and (2) an active process where HCO_3^- and/or CO_{2aq} is enzymatically transported across the cell membrane or, alternatively, the concentration of extracellular CO_{2aq} is enzymatically amplified to accommodate low ambient $[CO_{2aq}]$. Both diffusive and carbon-concentrating models predict a large carbon isotope fractionation during photosynthesis due to the influence of ribulose-1,5-bisphosphate carboxylase/oxygenase (rubisco) during carbon fixation via the Calvin cycle. Rubisco displays a fractionation of ca. 29‰ with respect to aqueous CO_{2aq} during in vitro experiments [88, 89]. However, this maximum value is rarely expressed in marine photoautotrophs from natural settings. Smaller in situ fractionations have been attributed to differences in growth rate and cell geometry [90, 91], the influence of β-carboxylase PEPC (phosphoenolpyruvate carboxylase) [92] during carboxylation and/or the active transport of inorganic carbon to the site of fixation [89, 93, 94].

3.2.1
Calvin Cycle

The first step of the Calvin cycle involves the reaction of a molecule of CO_2 with ribulose-1,5-bisphosphate to form an unstable six-carbon intermediate molecule that degrades into two molecules of 3-phosphoglycerate (PGA). PGA, the first component synthesized during the carboxylase reaction, is a three-carbon molecule. Hence, the Calvin cycle is also known as the C_3 pathway. Three molecules of CO_2 are necessary to produce six molecules of PGA. Addition of phosphate groups to six molecules of PGA (derived from ATP) to synthesize six molecules of 1,3-biphosphoglycerate, and subsequent reduction of a carboxyl group and hydrolysis of the phosphate group, forms one molecule of glyceraldehyde 3-phosphate. Glyceraldehyde 3-phosphate is the primary molecule transferred from the chloroplast for the synthesis of carbohydrates.

3.2.2
Fractionation in the Absence of Active Uptake

If CO_{2aq} is supplied to the cell by simple diffusion or if diffusion dominates the transfer of CO_{2aq} from the ambient environment to the site of carboxylation, then the magnitude of total carbon isotope discrimination during photosynthesis (ε_p) is a function of the isotope fractionations associated with carbon transport and fixation, and the concentrations of extra- and intercellular CO_{2aq}:

$$\frac{\delta_a - \delta_{org}}{1 + \delta_{org}/1000} = \varepsilon_t + (\varepsilon_f - \varepsilon_t)\frac{C_i}{C_e} \tag{10}$$

where ε_p is equivalent to the left side of Eq. 10, and δ_a and δ_{org} are the carbon isotopic compositions of the substrate inorganic carbon and algal organic carbon, respectively. The terms ε_f and ε_t are carbon fractionations associated with carbon fixation and diffusive transport, and C_e and C_i are extra- and intercellular [CO_{2aq}], respectively. This model derives from, and is equivalent to, the model of C_3 photosynthesis in higher plants [95].

Equation 10 indicates that the intercellular pool of carbon dioxide (C_i) exerts considerable influence on the magnitude of ε_p values for marine algae. However, the concentration of intercellular CO_{2aq} is rarely known. Thus, C_i can be described in terms of C_e and the specific growth rate (μ) of the cell (i.e. the net flux of CO_{2aq} divided by the carbon content per cell), allowing Eq. 10 to be recast [96, 97]:

$$\varepsilon_p = \varepsilon_t + (\varepsilon_f - \varepsilon_t)\left(1 - \frac{\mu C}{kC_e}\right) \tag{11}$$

where C is the carbon content of the cell and k is the rate constant for the diffusion of CO_{2aq} into and out of the cell, which is equivalent to the permeability of the cell membrane to CO_{2aq} [96].

This formulation reasonably assumes that resistance to diffusion of CO_{2aq} into and out of the cell are the same. Furthermore, the permeability of the cell membrane is likely to be proportional to the surface area of the cell [91, 96, 98], whereas the carbon content of the cell is proportional to the cell volume [91, 99].

Equation 11 predicts a linear relationship between ε_p versus the ratio $\mu/[CO_{2aq}]$ with maximum fractionation occurring under high $[CO_{2aq}]$ and/or as μ approaches zero. Support for the model described in Eq. 11 is found in chemostat incubations performed under nitrate-limited conditions for some algae including alkenone-producing haptophyte algae *Emiliania huxleyi* [90], and the diatoms *Phaeodactylum tricornutum* [96] and *Porosira glacialis* [91]. In these three experiments, ε_p varied linearly with respect to $\mu/[CO_{2aq}]$, each with different slopes and identical y-intercepts (Fig. 3a). The y-intercept (~ 25‰) reflects the value of ε_f as $\mu/[CO_{2aq}]$ approaches zero. A similar value for both coccolithophorids and diatoms suggests that 25‰ is the maximum value of ε_f for marine algae utilizing a combination of rubisco and β-carboxylases [100]. A wider range of values for ε_f (~ 22.5–26.6‰) is statistically plausible from evaluation of the same data set [12]. However, a value of 25‰ is consistent with calculated estimates

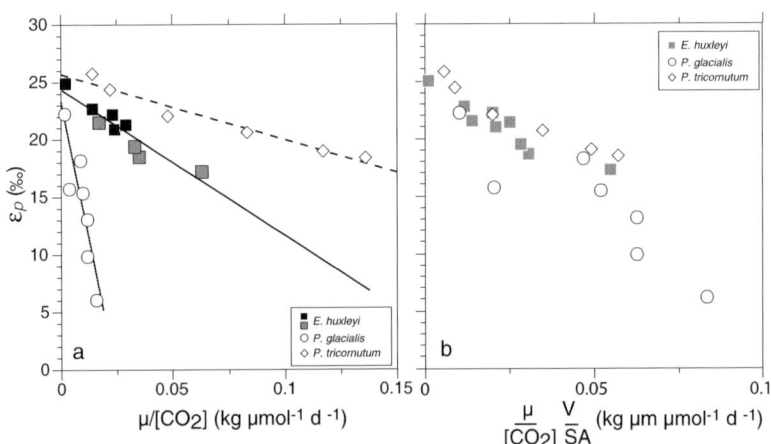

Fig. 3 Comparison of ε_p versus $\mu/[CO_{2aq}]$. **a** Chemostat incubations (from [91]). All experiments were conducted under nitrate-limited, continuous light conditions, except for *P. tricornutum* which includes both continuous light and 12 : 12-hour light/dark cycles. *P. tricornutum, open diamonds* (from [211]); *E. huxleyi, squares* (calcifying: *filled squares*, noncalcifying: *shaded squares*) (from [90]); *P. glacialis, open circles* (from [91]). **b** Comparison of ε_p versus ($\mu/[CO_{2aq}]$) (volume/surface area) (from [91])

of 25–28‰ for algae with C_3-type metabolisms, assuming a 2–10% contribution of β-carboxylation to the total carboxylation [101]. Importantly, differences in slopes for ε_p versus $\mu/[CO_{2aq}]$ can be normalized by accounting for differences in cell geometries, specifically the ratio of volume to surface area (Fig. 3b). This observation supports assumptions regarding the relationships between permeability and cell surface area, carbon content and biovolume [91]. However, a linear relationship for ε_p versus $\mu/[CO_{2aq}]$ can also be explained by models incorporating the effects of active carbon uptake and, thus, is not proof of a diffusive carbon uptake model [91].

3.2.3
The Impact of Active Uptake of Dissolved Inorganic Species on Algal $\delta^{13}C$ Values

Various algae have been shown to have an inducible carbon concentrating mechanism (CCM) when $[CO_{2aq}]$ is lowered such that it becomes limiting to growth [93, 102, 103]. Burns and Beardall [102] demonstrated on the basis of intercellular pH that the marine microalgae *Phaeodactylum tricornutum, Dunaliella tertiolecta, Isochrysis galbana,* and *Porphyridium purpureum* grown in batch cultures under continuous light had the capacity to increase internal $[CO_{2aq}]$ by three to seven times relative to ambient concentrations ($\sim 12\,\mu M$). Further, these authors argue [102] that the presence of extracellular carbonic anhydrase, an enzyme that catalyses the dehydration of HCO_3^-, does not support the active transport of HCO_3^-. More likely, CO_{2aq} is actively or diffusively transported across the plasmalemma, with the potential for active transport of HCO_3^- at the chloroplast envelope.

Sharkey and Berry [93] assessed the carbon source utilized during the growth of *Chlamydomonas reinhardtii* by tracking changes in the $\delta^{13}C$ value of residual air. In these experiments, microalgae grown under high $[CO_2]$ ($3300\,\mu L\,L^{-1}$) were transferred to a medium with low $[CO_2]$ ($200\,\mu L\,L^{-1}$). Initially, carbon fractionation under high $[CO_2]$ was high (20–29‰) and remained high after the transfer to low $[CO_2]$ conditions even as the rate of photosynthesis fell to 25% of the original rate. Carbon isotope discrimination decreased over the following 3 h to ~ 4‰, with respect to the carbon isotopic composition of CO_{2g}, as the rate of photosynthesis increased. Importantly, when the activity of carbonic anhydrase was chemically suppressed, the rate of photosynthesis decreased while ^{13}C discrimination increased. These results clearly indicate that *C. reinhardtii* has the capacity to enhance carbon availability when $[CO_{2aq}]$ is limiting and, by doing so, alter the isotopic character of the organic carbon produced.

In general, diatoms have demonstrated a capacity to increase the concentration of intercellular dissolved inorganic carbon ([DIC]) relative to ambient concentrations. The carbon uptake pathway of diatom assemblages, characterized by large-diameter diatoms including *Asterionella, Nitzchia* and *Rhizoslenia* from the Delaware Bay, were studied using ^{14}C-labelled HCO_3^- [103]. These uptake experiments demonstrate that diatoms utilize newly introduced HCO_3^- within 10 s, increasing intercellular [DIC] up to ten times that of ambient concentrations and supporting high photosynthetic rates at low ambient [DIC]. The specific mechanism of carbon transport could not be clearly determined. Inhibition of carbonic anhydrase activity had little effect on intercellular [DIC], but led to a decrease in carbon fixation. These results allow the possibility that HCO_3^- is actively transported across the plasmalemma, while carbonic anhydrase is used to catalyse the dehydration of HCO_3^- within the cell [103].

Laws et al. [96] provided organic isotope evidence for a carbon concentrating mechanism in the diatom P. tricornutum. Under very low $[CO_{2aq}]$ ($< 2.4\,\mu mol\,kg^{-1}$), the relationship between ε_p and $\mu/[CO_{2aq}]$ for P. tricornutum deviates from linearity (Fig. 4) as a result of high growth rates and low carbon isotope fractionation. This non-linearity is consistent with either active uptake of HCO_3^- or CO_{2aq} or the catalysed addition of extracellular CO_2 leading to an increased diffusive flux [96].

Other culture experiments suggest that different growth and environmental conditions trigger different carbon uptake pathways, as well as carbon isotopic responses [104]. For example, strains of E. huxleyi, P. tricornutum and P. glacialis, grown in dilute batch cultures under nitrate-replete conditions and varying light/dark cycles, demonstrate that although ε_p varies with respect to $[CO_{2aq}]$ and $\mu/[CO_{2aq}]$, these relationships differ from those established in nutrient-limited, predominantly continuous light, chemostat incubations [105]. Relative to chemostat cultures, dilute batch cultures result in substantially lower absolute ε_p values, and different, as well as non-linear, species-specific slopes for ε_p versus $\mu/[CO_{2aq}]$ (Fig. 5).

Low ε_p values associated with carbon-concentrating mechanisms are likely a consequence of reservoir effects. Francois et al. [106] argued that ε_p is

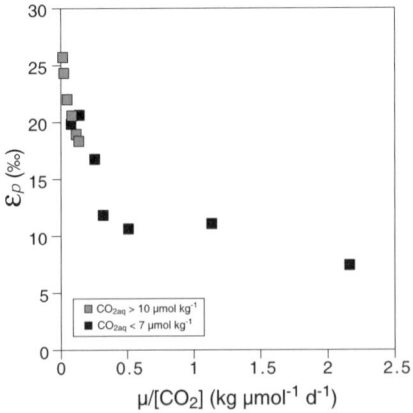

Fig. 4 The effect of low $[CO_{2aq}]$ on P. tricornutum determined from chemostat incubations. $[CO_{2aq}] > 10\,\mu mol\,kg^{-1}$: *filled squares*; $[CO_{2aq}] < 7\,\mu mol\,kg^{-1}$: *shaded squares* [211]

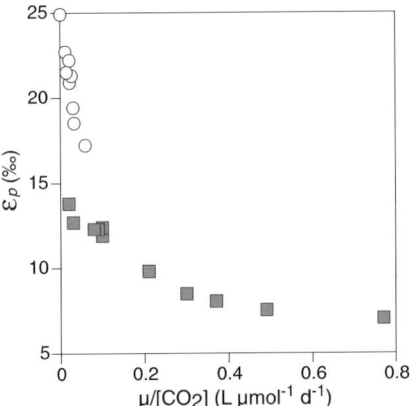

Fig. 5 Cultures of E. huxleyi: ε_p versus $\mu/[CO_{2aq}]$ from chemostat incubations and dilute batch cultures. Chemostat incubations: *open circles* [90], conducted under nitrate-limited, continuous light conditions. Dilute batch cultures: *shaded squares* [105], performed under nitrate-saturated, 16 : 8-h light/dark regime

a function of the fraction of intercellular inorganic carbon that diffuses back into the environment:

$$\varepsilon_p = \varepsilon_t + f(\varepsilon_f - \varepsilon_t) \tag{12}$$

where f is the fraction of the total intercellular carbon that back-diffuses from the cell. Low ε_p values suggest that the internal reservoir of CO_2 is finite with a low back-diffusive flux. Depletion of the internal carbon reservoir leads to a Rayleigh-type fractionation that drives the product (i.e. organic carbon) to approach the isotopic composition of the reactant (i.e. CO_{2aq}). Other considerations include the form of inorganic carbon transported. If HCO_3^- is actively transported and completely catalysed intercellularly, the $\delta^{13}C$ value of the source CO_{2aq} would be approximately 8‰ enriched in ^{13}C with respect to ambient CO_{2aq} and consequently lead to lower apparent ε_p values. Although the culture experiments for *P. tricornutum* yielded low ε_p values associated with low $[CO_{2aq}]$ and high growth rates, Laws et al. [96] found that Eq. 11 (based on Eq. 12) did not fit their experimental results. Considering the additional energetic expense associated with active uptake, the authors constructed a model that allows for both passive and active transport of inorganic carbon while providing a minimal energetic cost to the organism:

$$\frac{\mu}{C_e - C_o} = \frac{k_{-1}}{C(1+\beta)} \left(\frac{\varepsilon'_2 - \varepsilon_p}{\varepsilon_p - \varepsilon'_1} \right) \tag{13}$$

where $\varepsilon'_2 = \varepsilon_2 - \varepsilon_{-1} + \varepsilon_1$, and $\varepsilon'_1 = \varepsilon_1 + (\beta/(1+\beta))(\varepsilon_2 - \varepsilon_{-1})$, and β and C_o are constants. ε_2, ε_1 and ε_{-1} represent the fractionations associated with carbon fixation, and transport into and out of the cell, respectively.

4
Controls on Higher Plant $\delta^{13}C$ Values

Higher plant biomass, including specific biomarkers, is common in marine sediments. Like those of algal biomass, higher plant $\delta^{13}C$ values are governed by the isotopic composition of substrate carbon, fractionation during carbon assimilation (ε_p) and environmental conditions that influence ε_p values. In addition, higher plant physiology, specifically the differences between C_3 and C_4 plants, exerts an important control on their carbon isotopic compositions.

4.1
The Carbon Isotopic Composition of Atmospheric Carbon Dioxide

The majority of higher plant lipids found in marine or lacustrine sediments are thought to derive from sub-aerial land plants rather than submerged plants. Thus, atmospheric carbon dioxide is their sole carbon source and controls on the $\delta^{13}C$ values of atmospheric CO_2 represent a major control on the carbon isotopic compositions of higher plants. In the recent past, the primary control on atmospheric CO_2 $\delta^{13}C$ values has been the burning of fossil fuels—the Suess effect—which has caused CO_2 to become ca. 1.5‰ more depleted in ^{13}C since the industrial revolution [107]. Prior to that, atmospheric CO_2 $\delta^{13}C$ values appear to have varied little in the recent past; since the last glacial period maximum values have varied from -6.4 to -6.9‰ despite significant changes in CO_2 concentrations. On longer timescales, atmospheric CO_2 $\delta^{13}C$ values have changed along with the entire ocean–atmosphere reservoir, as discussed above. In addition to these global changes, the carbon isotopic composition of air in a plant's immediate growth environment can differ. An example of this is the canopy effect [108], wherein ^{13}C-depleted CO_2, generated by soil respiration, accumulates and is assimilated into plants growing below the forest canopy. Similarly, ^{13}C-depleted CO_2 derived from methane oxidation has been invoked to explain variations in peat bog vegetation $\delta^{13}C$ values [109].

4.2
The Carbon Isotopic Compositions of C_3 Plants

The majority of extant and fossil vascular plants are characterized by C_3 physiology, and various studies have established that photosynthetic fixation of atmospheric CO_2 by plants is accompanied by significant discrimination against ^{13}C [110]. As with aquatic photoautotrophs, the primary cause of photosynthetic carbon isotope fractionation during C_3 plant photosynthesis is the discrimination against ^{13}C by rubisco during carboxylation of ribulose-1,5-bisphosphate. Also in common with aquatic photoautotrophs, numerous studies have revealed that $\delta^{13}C$ values of C_3 plants vary significantly.

The carbon isotopic composition of C_3 plants is related to that of atmospheric CO_2 by Eq. 10, described above for aquatic photoautotrophs, and is contingent upon the same assumptions as described above. Distinct from aquatic photoautotrophs, however, the carbon isotope fractionation associated with diffusive transport (ε_t) is estimated to be 4.4‰; the carbon isotope fractionation during carboxylation mediated by rubisco (ε_f) is generally thought to be ca. 30‰. As with microalgae, Eq. 10 indicates that if the $\delta^{13}C_{air}$ value increases or the C_i/C_a ratio decreases, the $\delta^{13}C$ value of plant tissues increases. On geological timescales, the most important control on C_i/C_a ratios in higher plants is atmospheric pCO_2 [11]; lower pCO_2 levels result in low C_i/C_a ratios and lower plant $\delta^{13}C$ values. However, higher plant $\delta^{13}C$ values are not as useful a palaeo-pCO_2 barometer as those of algal-derived organic matter, because the latter organisms are effectively more carbon-limited by slow CO_2 diffusion in water [11]. C_i/C_a ratios also decrease if stomatal conductance in the plant leaf is reduced, which can occur at low humidity to minimize loss of water via evapotranspiration [95, 111]. Indeed, reduced water availability associated with low precipitation or hypoxic soil correlates with higher $\delta^{13}C$ values in different Hawaiian populations of a single C_3 species [112].

The rate of carbon assimilation, reflecting the plant's growth rate but also specific rates at individual leaves, will also affect C_i/C_a ratios and plant $\delta^{13}C$ values [113, 114], although this has not been as well-studied as for algae. Controls on carbon assimilation rates include nutrient status and light intensity; while the former probably influences the entire plant, it is possible that $\delta^{13}C$ values could vary among leaves on a single plant depending on their orientation with respect to the sun. Indeed, Lockheart et al. [115] observed that leaves on the southern side of (northern hemisphere) trees were enriched in ^{13}C relative to leaves on the northern side of the same tree and attributed this difference to enhanced rates of photosynthesis in the south-facing leaves. Temperature is expected to exert a minor control on isotopic fractionation by the rubisco enzyme [116, 117]; this has been demonstrated by in vitro studies [118] and some field investigations [119–122], while others have failed to find any effect [110]. However, the relationship between temperature and fractionation varies widely in both magnitude and sign, probably due to the difficulty in resolving multiple variables. Imposed on these environmental controls, there appear to be significant ecosystem and interspecies variations in $\delta^{13}C$ values among C_3 plants. For example, Chikaraishi and Naraoka [123] observed that C_3 angiosperm species are depleted in ^{13}C relative to co-occurring C_3 gymnosperms (– 32.8 ± 2.5‰ and – 26.9 ± 1.1‰, respectively).

4.3
C_3 vs. C_4 Plants

In contrast to C_3 plants, C_4 plants initially fix carbon as bicarbonate via carboxylation of phosphoenolpyruvate [110], forming oxaloacetate (Fig. 6).

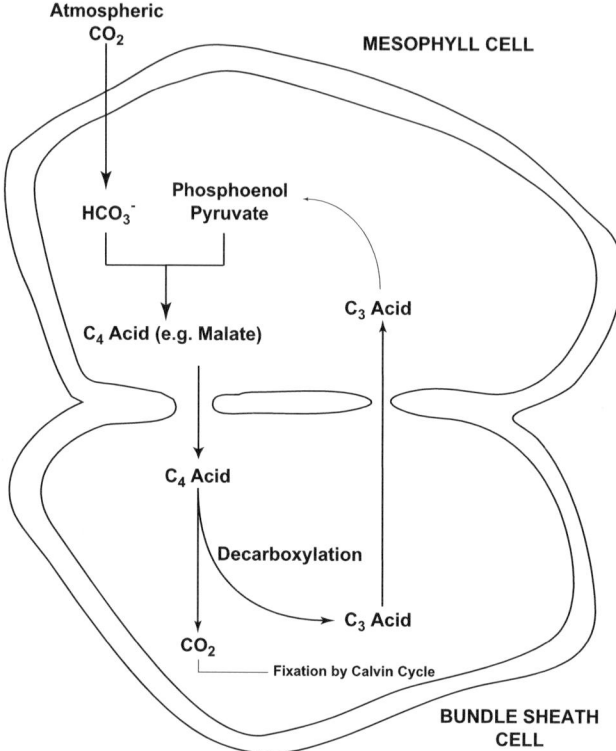

Fig. 6 Schematic of a C$_4$ plant mesophyll and bundle sheath cells and how carbon is shuttled between them

This occurs in the outer layer of photosynthetic cells (mesophyll), and the fixed carbon, as either malate or aspartate, is transported to the inner bundle sheath cells where decarboxylation occurs, ultimately regenerating phosphoenolpyruvate and releasing CO_2. Critically, it is thought that all of this released CO_2 is then fixed by rubisco. Because fractionation by phosphoenolpyruvate carboxylase is small (2.2‰), and the complete utilization of CO_2 in the bundle sheath results in no expression of the isotope effect associated with rubisco-mediated carboxylation, the net fractionation during C$_4$ photosynthesis is very small (– 1.3‰) [124]. Numerous studies have confirmed that C$_4$ plant biomass (– 10 to – 16‰) is significantly enriched relative to that of C$_3$ plants (– 25 to – 30‰). Moreover, $\delta^{13}C$ values for C$_4$ plants are much less dependent on the C_i/C_a ratio and environmental variables than those for C$_3$ plants [124]. The fact that C$_3$ and C$_4$ plants can be distinguished through their $\delta^{13}C$ values is of great use in palaeoclimatic investigations. C$_3$- and C$_4$-type plants have different optimum conditions for growth, and it is generally thought that high temperatures and low pCO_2 favour C$_4$-dominated ecosystems [125].

5
Controls on Prokaryote $\delta^{13}C$ Values

Prokaryotes inhabit a vast range of environments and have diverse metabolisms, ranging from methanogenesis to photoautotrophy to heterotrophy; consequently, their $\delta^{13}C$ values are much more variable than those of eukaryotes.

5.1
The Carbon Isotopic Composition of Microbial Substrate Carbon

Carbon substrates for bacterial and archaeal growth include CO_{2aq}, HCO_3^-, CH_4 and a variety of low molecular weight organic compounds. For autotrophic organisms utilizing either CO_{2aq} or HCO_3^-, the same controls on DIC $\delta^{13}C$ values discussed for aquatic photoautotrophs are applicable. However, bacterial and archaeal autotrophs include chemoautotrophs, which are not dependent on sunlight, and they occupy a far greater range of environmental settings—including shallow and deep marine sediments and the water column—than photoautotrophs. In such settings, DIC $\delta^{13}C$ values can vary significantly; near cold seeps or in organic-rich sediments, where a significant quantity of DIC is ultimately derived from oxidation of methane or organic matter, DIC is more ^{13}C-depleted [126]. In contrast, in closed or semi-closed settings, autotrophic, and particularly methanogenic, assimilation of DIC can cause the residual pool to become enriched in ^{13}C [126, 127].

Methane $\delta^{13}C$ values are generally strongly depleted in ^{13}C [126] and methanotroph lipids and biomass have lower $\delta^{13}C$ values—often lower than −100‰ [128]—than any other organic material. However, methane $\delta^{13}C$ values are also quite variable, depending on (1) whether the methane is thermogenic or biogenic and (2) if biogenic, whether methane was generated autotrophically or heterotrophically and correspondingly, the carbon isotopic compositions of DIC or acetate [126]. However, extensive methane oxidation and resultant Rayleigh distillation also exert an important control on methane $\delta^{13}C$ values in the environment.

Microbial heterotrophs obviously make use of substrates with widely divergent $\delta^{13}C$ values depending on the ecology of the source organism(s). For example, lipids of inferred bacterial heterotrophs living in methane-oxidizing environments have low $\delta^{13}C$ values [129]. In addition, different substrates (glucose, acetate) derive from different biosynthetic pathways in the source organism and this can impart different carbon isotopic compositions to the consumer. For example, carbohydrates are typically enriched in ^{13}C relative to lipids [130–133]. The $\delta^{13}C$ values of substrates generated from the degradation of these different compound classes vary widely, and thus heterotrophic microorganisms will reflect those differences. Indeed, preferential utilization of carbohydrate-derived carbon has been invoked as an explanation for ^{13}C-enriched bacterial lipids in peats [129, 134] and soils [135].

5.2
Isotope Fractionation During Carbon Assimilation by Bacteria and Archaea

Unlike eukaryotic photoautotrophs, all of which utilize the Calvin cycle during carbon assimilation, bacteria and archaea employ a wide range of carbon assimilation mechanisms. Many chemoautotrophs utilize the Calvin cycle and, thus, rubisco as the carboxylating enzyme, although studies of this are limited [136, 137]. Rubisco-catalysed carboxylation appears to have a similar fractionation (27‰) in bacteria and archaea as in eukaryotic photoautotrophs. Other pathways, apparently restricted to the bacteria and archaea, are the reverse tricarboxylic acid (TCA) cycle, the carbon assimilation pathway utilized by the green sulphur bacterium *Chlorobium* [138, 139], and the 3-hydroxypropionate pathway, utilized by the green non-sulphur bacterium *Chloroflexus* [140, 141] as well as some archaea [142].

The reverse TCA cycle involves a C_4 substrate, starting with oxaloacetate, to which two carbon atoms are sequentially added to form citrate; cleavage of citrate forms acetyl coenzyme A (acetyl-CoA), which is then carboxylated to pyruvate. Thus, in three discrete steps, two of which are catalysed by reduced ferredoxin, a total of three carbon atoms are assimilated. Critically, none of the carboxylations is apparently characterized by a large carbon isotopic fractionation and the net fractionation is much lower than for organisms using the Calvin cycle. Sirevåg et al. [138] reported $\Delta\delta^{13}C$ values ($\delta^{13}C_{biomass} - \delta^{13}C_{CO_2} \sim \varepsilon_p$) of 2.5 to 5.2‰, while Quandt et al. [139] reported $\Delta\delta^{13}C$ values of 12.2‰. Similar low values for the reverse TCA cycle have been observed for non-photoautotrophic organisms utilizing this pathway [137, 143].

The 3-hydroxypropionate pathway [140, 141] involves two carboxylations of acetyl-CoA to form methylmalonyl-CoA, which is then rearranged and cleaved to regenerate the acetyl-CoA and yield glyoxylate. The 3-hydroxypropionate pathway is also characterized by relatively low carbon isotopic fractionation, with Holo and Sirevåg [140] observing a 14‰ difference between biomass and dissolved inorganic carbon and van der Meer et al. [144] observing a 12.2‰ difference. It is thought that one reason for the low carbon isotope fractionation relative to DIC is that *Chloroflexus* might assimilate bicarbonate rather than dissolved CO_2 [144], consistent with the presence of the PEP carboxylase enzyme [142]. Organisms other than *Chloroflexus* that use the 3-hydroxypropionate pathway also exhibit low carbon isotope fractionation during carbon assimilation [137, 142] (0.6 to 3.2‰ relative to CO_{2aq}), and utilization of this pathway has been invoked to explain the ^{13}C-enriched composition of pelagic crenarchaeal lipids [145, 146].

There is relatively little known about the controls on the carbon isotopic composition of methanogenic archaea. Methanogens assimilate either CO_2 or other C_1 substrates (methylotrophs) or acetate, and it has generally been thought that such organisms will have a highly ^{13}C-depleted biomass, similar to the low δ^{13}C values of generated methane. This has been shown for several species of methanogens utilizing C_1 substrates grown in culture [147], and has been invoked as an explanation for ^{13}C-depleted isoprenoids in the Messel Shale [13]. However, methanogen biomarkers found in a variety of other settings, including Ace Lake, Antarctica [148] and peats [134], tend to have δ^{13}C values similar to or even enriched relative to those of co-occurring plant biomarkers. Potentially, this reflects reduced carbon isotope fractionation during acetate utilization [134] or complete utilization of the substrate pool in a closed system [147].

Methanotroph δ^{13}C values can be similarly variable. As with autotrophs, significant but variable kinetic carbon isotope effects are associated with methane assimilation [149]. Jahnke et al. [149] showed that methanotroph biomass δ^{13}C values can be as much as 24.8‰ depleted relative to substrate methane, but also observed 30‰ variability depending on whether the methanotrophs utilized the ribulose monophosphate or serine pathways and whether they used the membrane-bound particulate monooxygenase enzyme or the soluble isozyme. Combined with the low δ^{13}C values typical of methane, this results in the methanotroph biomass being highly depleted in ^{13}C. Methane is also the carbon substrate for anaerobic methane-oxidizing archaea [128, 150, 151]. These organisms have not yet been isolated and the mechanism that they employ for methane uptake is unknown. Environmental studies where biomarker and methane δ^{13}C values have both been determined are uncommon and should be interpreted with caution due to the spatial and temporal variability of the latter, but these studies do suggest that lipids of methane-oxidizing archaea can be 10 to 55‰ depleted relative to the methane substrate [150, 152].

In contrast to multicellular heterotrophs in which respiratory loss of carbon can be significant, prokaryotic heterotroph carbon isotopic compositions are generally thought to be similar to that of the food [14]. However, laboratory incubations of *Shewanella putrefaciens* by Teece et al. [153] revealed that small differences (< 2.5‰) between substrate and biomass δ^{13}C values can occur and are dependent on the growth conditions. More strikingly, recent work on the iron-reducing bacteria, *Geobacter metallireducens* and *Shewanella algae*, indicates that these heterotrophs' biomass can be as much as 7‰ depleted relative to substrate acetate and lactate, respectively [154]. Thus, assumptions that the bacterial heterotroph biomass will always be the same as that of substrate carbon need to be more thoroughly tested.

6
Biosynthetic Controls on Lipid δ^{13}C Values

After carbon assimilation, isotope effects during biosynthesis further affect the δ^{13}C values of specific lipids within an organism. This can cause differences in δ^{13}C values among compound classes (e.g. lipids vs. carbohydrates as described above) and also between δ^{13}C values of individual lipids within a structural class. The traditional paradigm regarding biosynthetic isotope effects is that fractionation occurs during formation of acetyl-CoA with the strongest depletion occurring at the carboxyl carbon [155]; presumably this fractionation is associated with either the oxidation of pyruvate to acetyl-CoA by pyruvate dehydrogenase [156] or conversion of acetyl phosphate to acetyl-CoA by phosphotransacetylase [157]. Monson and Hayes [155] determined that the kinetic isotope effect associated with pyruvate dehydrogenase in *E. coli* was 23‰ (ε_{PDH}), but this full effect is rarely expressed because of Rayleigh distillations (such that the residual pyruvate carbon pool becomes progressively enriched in ^{13}C as the reaction progresses). Instead, the fractionation expressed during decarboxylation is $(1-f)\varepsilon_{PDH}$ [155], where f is the fraction of pyruvate flowing to acetyl-CoA.

Because these reactions occur during both photoautotrophic and heterotrophic lipid synthesis, acetogenic lipids in most organisms are expected to be ca. 4‰ depleted relative to biomass [14]. Isoprenoids synthesized by the mevalonic acid pathway, which results in a 3:2 ratio of methyl to carboxyl atoms, are expected to be depleted relative to biomass by only ca. 2‰. However, it is now recognized that isoprenoids are not synthesized via the mevalonic acid pathway in many or most organisms; instead, the C_5 isoprene unit can be formed by condensation of a C_2 subunit derived from pyruvate decarboxylation and a C_3 subunit derived from triose phosphate [158–161]. This pathway actually appears to be predominant in bacteria and has also been identified in higher plants [162], unicellular algae, and cyanobacteria [163]. The effect of the glyceraldehyde phosphate/pyruvate pathway on the carbon isotopic composition of isoprenoids remains unclear. It is possible that isoprenoids derived from this pathway could be enriched in ^{13}C relative to mevalonate-derived isoprenoids, because the former contain only one ^{13}C-depleted carboxyl carbon.

Generally, biomarker lipid δ^{13}C values tend to match these general considerations, but significant exceptions are common. In the following sections, we discuss algal, higher plant and prokaryotic biomarker δ^{13}C values in the context of the above considerations.

6.1
Microalgae

The carbon isotopic compositions of microalgae have been studied by a number of workers. In general, they exhibit relationships with bulk organic matter similar to those described above, but there are a number of exceptions and the range of values can be considerable. Bidigare et al. [90] reported that alkenones were 3.8‰ depleted in ^{13}C relative to haptophyte biomass, and subsequent work has generally agreed with that finding. However, Schouten et al. [164] showed that the offsets between lipid and bulk organic matter δ^{13}C values in microalgal batch cultures are quite variable. In general, fatty acid δ^{13}C values varied by ca. 2‰ within a given organism; however, the fatty acids ranged from 0.8 to 8.7‰ depleted relative to bulk biomass. Alkenones in *Isochrysis galbana* generally had δ^{13}C values 3 to 3.5‰ depleted relative to biomass consistent with the work of Bidigare et al. [90], but were enriched by 4 to 4.5‰ relative to fatty acids, ostensibly derived from similar biosynthetic pathways. Isoprenoidal lipids, phytol and sterols had δ^{13}C values consistent with theoretical predictions, being enriched relative to fatty acids and depleted relative to bulk biomass, but their values ranged from being nearly the same as biomass to being > 8‰ depleted relative to it. Evaluating the isotopic offset between lipids and biomass is harder to do in field investigations due to the range of compounds and organisms that contribute to particulate organic matter. Consequently, a range of differences have been observed. In the Peru upwelling region [90, 165], alkenones, phytol and a variety of sterols inferred to derive from sterols are depleted relative to particulate organic carbon (POC) δ^{13}C values by typically 1 to 4‰, but no systematic offset is observed between alkenones and isoprenoids. In contrast, Popp et al. [166] observed differences of 5 to 12‰ between δ^{13}C values of POC and steroids thought to derive from diatoms. More recent work has reported biomass–lipid offsets falling between these two extremes, further illustrating that the controls on lipid δ^{13}C values are complex and their relationship to bulk organic carbon δ^{13}C values is still poorly understood.

6.2
Higher Plants

n-Alkanes are the terrestrial biomarkers that have been most thoroughly characterized isotopically. Collister et al. [167] observed that in C$_3$ and C$_4$ species, the *n*-alkanes were depleted relative to biomass by ca. 5.9 and 9.9‰, respectively. *n*-Aldehydes had isotopic compositions similar to those of the *n*-alkanes, and it is thought that other biosynthetically related *n*-alkyl compounds (e.g. *n*-acids and *n*-alkanols) have similar δ^{13}C values. Chikaraishi and Naraoka [123] observed similar differences between biomass and *n*-alkane δ^{13}C values in a range of C$_3$ angiosperms and gymnosperms and C$_4$ plants. Lignin-derived phenols are depleted in ^{13}C compared to bulk plant tissue, but not to the same extent as lipids. Goñi and Eglinton [168] measured the δ^{13}C values of lignin phenols (produced from the oxidation of lignin polymers), and found that those from C$_3$ plants had δ^{13}C values of -30.4 ± 3.9‰ and those from C$_4$ plants had δ^{13}C values of -16.9 ± 2.7‰.

6.3
Prokaryotes

In general, a 2 to 4‰ offset between fatty acids and biomass is observed for bacteria grown under aerobic conditions, including *E. coli* grown on glucose [155], *S. putrefaciens* [153], and Type I and Type X methanotrophs using the ribulose monophosphate pathway (RuMP) [149, 169]. However, the fatty acids of cultured methanotrophs utilizing the serine pathway can be 10 to 15‰ depleted relative to biomass [149]. The acetogenic lipids of heterotrophic bacteria are also more depleted than expected when grown under anaerobic conditions, with fatty acid δ^{13}C values being: 7.3 to 10.7‰ more depleted than biomass in *S. putrefaciens* grown on lactate and with NO_3 as the oxidant [153]; 4.5 to 8.6‰ more depleted than biomass in *G. metallireducens* grown on acetate [154]; and 10.4 to 14.7‰ more depleted than biomass in *S. algae* grown on acetate [154].

The isotopic relationships between lipids and biomass can also vary in autotrophic prokaryotes. Theoretical considerations and culture studies indicate that in organisms using the reverse TCA cycle, such as green sulphur bacteria, acetogenic lipids are ca. 4‰ enriched in ^{13}C relative to biomass [170]. In contrast studies of *Chloroflexus aurantiacus* grown photoautotrophically indicate that alkyl lipids of organisms using the 3-hydroxypropionate pathway are 1 to 2‰ depleted in ^{13}C relative to biomass [132]. This is to be expected because these organisms also use the TCA cycle during lipid synthesis, such that the isotopic difference between biomass and lipids should be much the same as in photoautotrophs using the Calvin cycle.

As mentioned earlier, isotope fractionations are also dependent on partitioning of carbon at branch points in carbon metabolic pathways [14]. This is particularly well illustrated by culture studies of the cyanobacterium *Synechocystis* [171]. This organism uses the TCA cycle and lipids are biosynthesized from acetyl-CoA generated by decarboxylation of pyruvate, so that acetogenic lipids are expected to be ca. 4‰ depleted relative to biomass. Instead, Sakata et al. [171] observed that fatty acids were depleted by 9.1‰. This additional depletion suggests that the fraction of pyruvate flowing to acetyl-CoA is small, which is consistent with the relatively low abundance of lipids in cyanobacteria.

Different carbon assimilation pathways and mechanisms of acetyl-CoA formation will also affect isoprenoid lipid δ^{13}C values. For example, in the study by Jahnke et al. [169], hopanoids in methanotrophs utilizing the serine pathway had low δ^{13}C values similar to those of the fatty acids. Thus, although paradigms for the biosynthetic controls on lipid δ^{13}C values can provide a first approximation of expected biomarker δ^{13}C values in some situations, unresolved issues regarding the magnitude of isotopic effects, the influence of growth conditions and the biological distribution of different metabolic pathways remain a limitation in interpreting sedimentary lipid δ^{13}C values.

7
Applications

7.1
Alkenone $\delta^{13}C$ Values as a Paleo-pCO_2 Proxy

The stable carbon isotopic compositions of C_{37} di-unsaturated alkenones collected from surface waters across a range of oceanic environments have been studied in relationship to CO_2, nutrients, and other variables [34, 90, 100, 104, 172]. The influence of growth rate and cell geometry on ε_p values (see Eq. 5) derived from C_{37} di-unsaturated alkenones ($\varepsilon_{p37:2}$) are, however, difficult to evaluate given that these data are rarely, if ever, collected for specific species in natural settings. Therefore, the model described in Eq. 11 is not applicable in the analysis of field-based $\varepsilon_{p37:2}$ values. Instead, these data are commonly assessed using a convention proposed by Rau et al. [173] and Jasper et al. [174] and further modified by Bidigare et al. [90]:

$$\varepsilon_p = \varepsilon_f - \frac{b}{CO_{2aq}} \quad (14)$$

where b represents an integration of the physiological variables, such as growth rate and cell geometry, affecting the total carbon isotope fractionation during photosynthesis [90, 174].

The available data provide evidence for a significant correlation between the physiological-dependent term b and the concentration of reactive soluble phosphate (Fig. 7). It is likely that this relationship stems from the influence

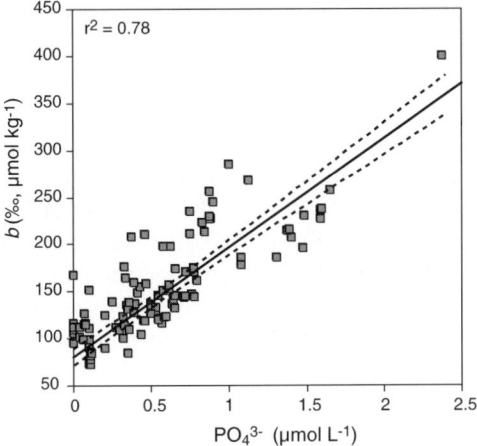

Fig. 7 Compilation of b versus soluble phosphate for natural haptophyte populations [34, 91, 100, 172, 212]. Values for b are calculated using a value of 25‰ for ε_f. *Solid line*: geometric mean regression; *dotted lines*: 95% confidence intervals

of growth rate on the expression of $\varepsilon_{p37:2}$ [90]. However, it is unlikely that [PO_4^{3-}] alone is responsible for the variability in growth rate inferred from variation in b. Instead, Bidigare et al. [90] and Laws et al. [104] propose that the availability of specific trace elements, such as Se, Co and Ni, is ultimately impacting on the growth characteristics of natural haptophyte populations. In this sense, [PO_4^{3-}] is acting as a proxy for these growth-limiting trace elements that also exhibit phosphate-like distributions in the modern ocean.

The robust relationship between [PO_4^{3-}] and the physiologically dependent term b determined from $\varepsilon_{p37:2}$ values provides the opportunity to apply alkenone $\delta^{13}C$ values in the reconstruction of ancient CO_2 levels. That is, b can be cast in terms of [PO_4^{3-}] and applied to Eq. 11 [12, 90, 104]. If phosphate concentrations can be constrained for a specific location and ancient values of $\varepsilon_{p37:2}$ reconstructed, CO_{2aq} concentrations can be estimated. In general, ancient $\varepsilon_{p37:2}$ records can be reconstructed by measuring the carbon isotopic composition of di-unsaturated alkenones and coeval near surface-dwelling planktonic foraminifera. The $\delta^{13}C$ values of foraminiferal carbonate are then used to estimate the $\delta^{13}C$ of CO_{2aq} in equilibrium with calcite. This approach was developed by Jasper and Hayes [175] and applied to evaluate palaeoceanographic dynamics and surface-water [CO_{2aq}] during the Pleistocene [174]. These earlier studies calibrated ε_p–CO_2 relationships from measurements of particulate organic carbon in the modern ocean, derived primarily from highly productive waters. Subsequent studies have applied $\varepsilon_{p37:2}$ values in conjunction with phosphate estimates to reconstruct Late Quaternary surface water pCO_2 in the South Atlantic [176] and global Miocene-age pCO_2 trends [12, 177].

The alkenone–pCO_2 proxy rests on the assumption that intercellular CO_{2aq} fixed during photosynthesis arrives by diffusive flux, or that the proportion of diffusive and actively transported carbon flux is constant under a variety of environmental conditions. Results from chemostat culture experiments lend support to, but do not prove, a photosynthetic diffusion model for *Emiliania huxleyi*, the dominant haptophyte in the modern ocean [91]. The presence of carbonic anhydrase (CA) was not found in low-calcifying strains of *E. huxleyi* [178], but CA was detected in the chloroplasts of high-calcifying cells [179]. Inhibition of CA activity led to only a 30% decrease in $^{14}CO_2$ fixation, suggesting that a CA-mediated carbon flux constitutes a minor contribution to the overall inorganic carbon flux [179]. Further, there appear to be differences between exponential versus stationary growth phases for *E. huxleyi*. During the stationary phase *E. huxleyi* appears to maintain substantially higher intercellular DIC concentrations relative to ambient levels, suggesting that stationary-phase cells have the capacity to transport inorganic carbon against a concentration gradient [180].

Many, if not most, marine microalgae demonstrate characteristics of a carbon concentrating mechanism. This includes haptophytes such as *Isochrysis galbana* [89, 102], as well as stationary-phase cells of high-calcifying strains of *E. huxleyi* [180]. Moreover, batch cultures suggest that carbon isotope fractionation may be influenced differently by light-limited [105, 181] versus nutrient-limited growth performed in chemostat experiments [90]. Whether or not haptophyte populations in their natural settings control intercellular concentrations of inorganic carbon by actively transporting bicarbonate or CO_{2aq} remains unresolved.

Recently, Pagani et al. [182] tested the accuracy of the alkenone ε_p–CO_2 technique by comparing reconstructed pre-industrial water-column [CO_{2aq}] from sedimentary alkenone $\delta^{13}C$ values from 20 sites across a North Pacific transect against both observed water-column [CO_{2aq}] values and estimated pre-industrial concentrations at the depth of alkenone production at each site. Sedimentary alkenone-based CO_{2aq} estimates ([CO_{2aq}]$_{alk}$) were established by estimating the depth of alkenone production at each site as defined by $U_{37}^{K'}$ temperature estimates [183] and seasonal temperature–depth relationships established by the northwest Pacific carbon cycle study [184]. Production depths inferred by these records (Fig. 8a) were then used to identify the appropriate $\delta^{13}C_{CO_{2aq}}$ values and phosphate concentrations (Fig. 8b) required in the calculation of ε_p and [CO_{2aq}]$_{alk}$.

The pattern of ε_p values across the Pacific transect (Fig. 9), that is, lower values in high-nutrient environments (high and low latitudes) and higher ε_p values in the oligotrophic subtropics, is consistent with a growth rate control on carbon isotope fractionation. Although [CO_{2aq}]$_{alk}$ values track measured water-column values, they are clearly lower than modern values across the subtropical region. This offset likely reflects the contributions of anthropogenic CO_2 in modern surface waters relative to pre-industrial concentrations at the time of alkenone production. The smallest offsets are observed at the higher latitudes, consistent with expectations, since in regions subjected to intense seasonal upwelling, a component of surface CO_{2aq} derives from deeper, pre-industrial waters. The offset is greater across well-stratified waters in the subtropical latitudes, which track the rise in anthropogenic CO_2 more closely. When a model-based estimate of anthropogenic CO_2 is removed from the modern observed values, a majority of the [CO_{2aq}]$_{alk}$ values fall within 20% of modelled pre-industrial values. Thus, both the magnitude and pattern of offset between pre-industrial [CO_{2aq}]$_{alk}$ and modern [CO_{2aq}] are consistent with a 30% anthropogenic increase in pCO_2 over the past 150 years. When the effects of anthropogenic [CO_{2aq}] are removed from the modern signal, alkenone CO_{2aq} estimates accurately reproduce pre-industrial water column concentrations (Fig. 10).

The results of Pagani et al. [182] support the use of alkenone-based ε_p values as a proxy for palaeo-pCO_2 reconstruction and suggest that the alkenone approach can resolve relatively small differences in water-column CO_2 when phosphate concentrations and temperatures are well constrained. Their

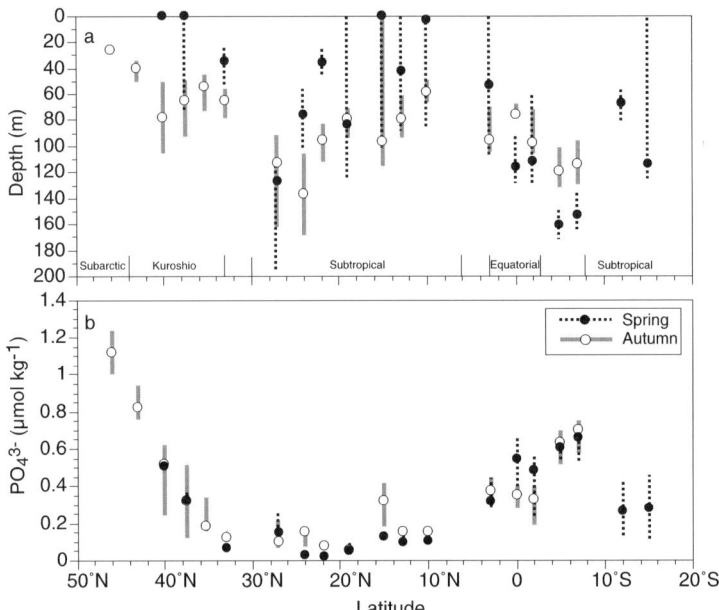

Fig. 8 Relationship between depth of haptophyte production and phosphate concentration across 175°E [182]. **a** Estimated seasonal depth of haptophyte production based on $U_{37}^{K'}$ temperature values and seasonal temperature–depth relationships established by NOPACCS. *Lines* represent the possible range of production based on $U_{37}^{K'}$ values ±1 °C; *squares* represent depths as defined by exact $U_{37}^{K'}$ values. Temperatures calculated using the results of Prahl et al. [213]. **b** Seasonal phosphate concentrations as defined by inferred haptophyte production depths and seasonal phosphate–depth relationships established by NOPACCS. *Open circles*: Autumn (August and September); *filled circles*: Spring (April, May and June)

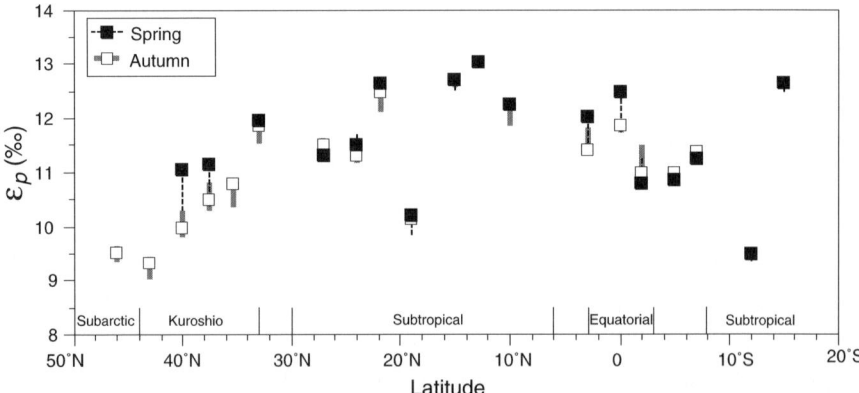

Fig. 9 $\varepsilon_{p37:2}$ values across 175°E reconstructed from sedimentary alkenones [182]. *Filled squares*: $\varepsilon_{p37:2}$ values assume Spring production. *Open squares*: $\varepsilon_{p37:2}$ values assume Autumn production

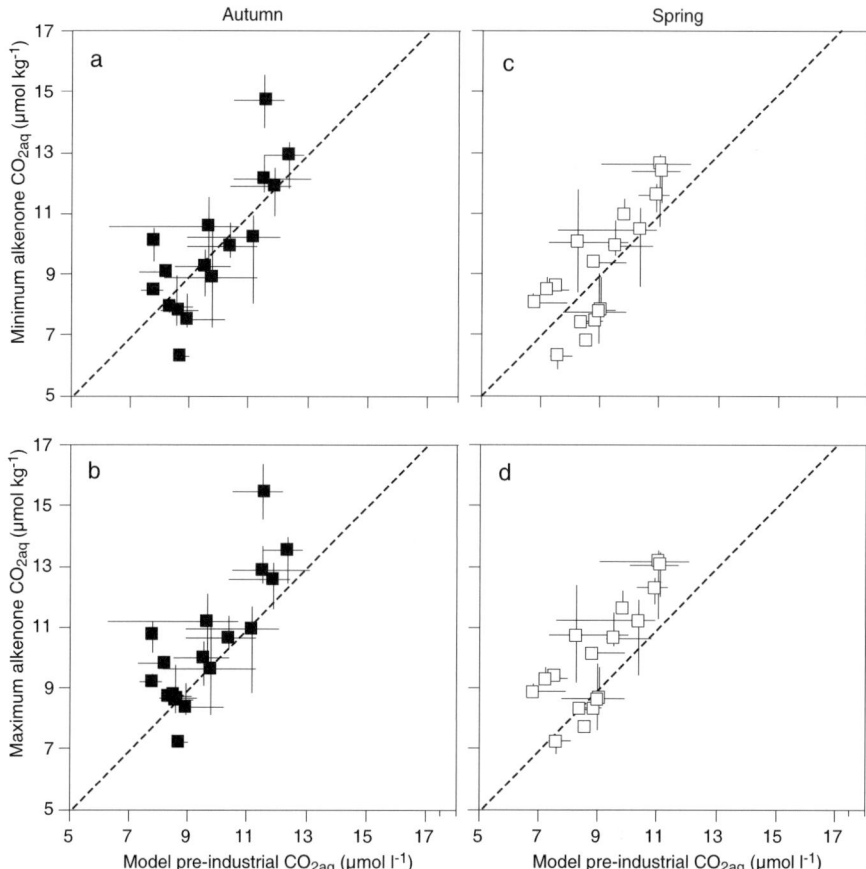

Fig. 10 Modelled pre-industrial $[CO_{2aq}]$ versus alkenone-based $[CO_{2aq}]$. Crossplot of modelled pre-industrial $[CO_{2aq}]$ versus minimum and maximum alkenone-based $[CO_{2aq}]$. **a, b** Autumn production assumed. **c, d** Spring production assumed. Maximum alkenone-based $[CO_{2aq}]$ calculated using $\varepsilon_f = 27‰$ and the equation: $[CO_{2aq}] = (4.14[PO_4^{3-}] + 125.48[PO_4^{3-}] + 107.85)/(27 - \varepsilon_p)$. Minimum alkenone-based $[CO_{2aq}]$ calculated using $\varepsilon_f = 25‰$ and the equation: $[CO_{2aq}] = (116.12[PO_4^{3-}] + 81.5)/(25 - \varepsilon_p)$. *Dashed line*: 1 : 1 correlation

results further suggest that light-limited growth effects and/or active carbon uptake are negligible, or that these processes have negligible effects on carbon isotopic compositions of haptophytes in the natural environment.

7.2
Identifying Methane Cycling in Modern and Ancient Settings

Due to the highly ^{13}C-depleted character of methane in most settings, biomass derived from organisms either directly or indirectly supported by methane can be readily identified. Although this applies to all methane-utilizing organisms, in marine sediments it is particularly useful in examining anaerobic oxidation of methane (AOM). In marine environments, methane oxidation occurs under anaerobic conditions and is coupled to sulphate reduction [185–189]. Consequently, only a small proportion of the methane generated in marine sediments is available for aerobic oxidation and the flux of methane from marine sediments to the atmosphere is small compared to other sources [190]. Multiple lines of evidence show that in such diverse settings as Hydrate Ridge [77, 191], the California Margin [150], Mediterranean mud volcanoes [151], Gulf of Mexico hydrocarbon seeps [192, 193], Black Sea sediments [152, 194] and water column [195, 196], and the Guyamas Basin [197], methane is consumed anaerobically in a process mediated by archaea. Whilst the nature of the AOM community has subsequently been elucidated by a variety of microbial techniques, including 16S rRNA analyses [150, 198] and fluorescence in situ hybridization [191], the identification of ^{13}C-depleted diagnostic archaeal biomarkers was critical in confirming the role of archaea in AOM (Fig. 11). Among the archaeal biomarkers found at cold seep sites are archaeol, sn-2- and sn-3-hydroxyarchaeol, glycerol dialkyl glycerol tetraethers, pentamethylicosane and crocetane (and unsaturated analogues). The diversity of ^{13}C-depleted archaeal biomarkers amongst different sites has been proposed as evidence that multiple species of archaea mediate AOM.

^{13}C-depleted bacterial biomarkers have also been detected at cold seeps, providing insight into the co-existing bacterial community [151, 191, 198–203]. Apparently, some or all AOM archaea are unable to reduce sulphate during methane oxidation, and sulphate-reducing bacteria (SRB) are required to syntrophically consume accumulating reduced substrates [204] (acetate, formate, H_2) if AOM is to remain thermodynamically viable [205]. Consistent with this, highly ^{13}C-depleted C_{15} and C_{17} iso and anteiso fatty acids, common in bacteria but particularly abundant in SRB, are abundant in cold seep sediments. Other ^{13}C-depleted fatty acids are also commonly found in cold seep sediments and likely derive from SRB [203]. Also present are ^{13}C-depleted hopanoids and non-isoprenoidal dialkyl glycerol diether lipids [199, 201], the latter also inferred to be SRB biomarkers. While the bacterial biomarkers are all strongly depleted in ^{13}C relative to typical algal biomarkers, indicating that a significant portion of the carbon must derive from methane, Pancost et al. [129, 201], Hinrichs et al. [199] and Thiel et al. [152] have observed that they are 10 to 40‰ enriched relative to the co-occurring archaeal biomarkers (Fig. 12). Such differences have led to speculation about the nature of carbon flow amongst these syntrophic organisms. For example, the enrichment can be explained if the SRB are autotrophic and assimilating a mixture of seawater and methane-derived DIC [129].

In addition, archaeal and bacterial biomarkers can be used to identify related processes in ancient cold seep deposits. Among the ancient cold seep

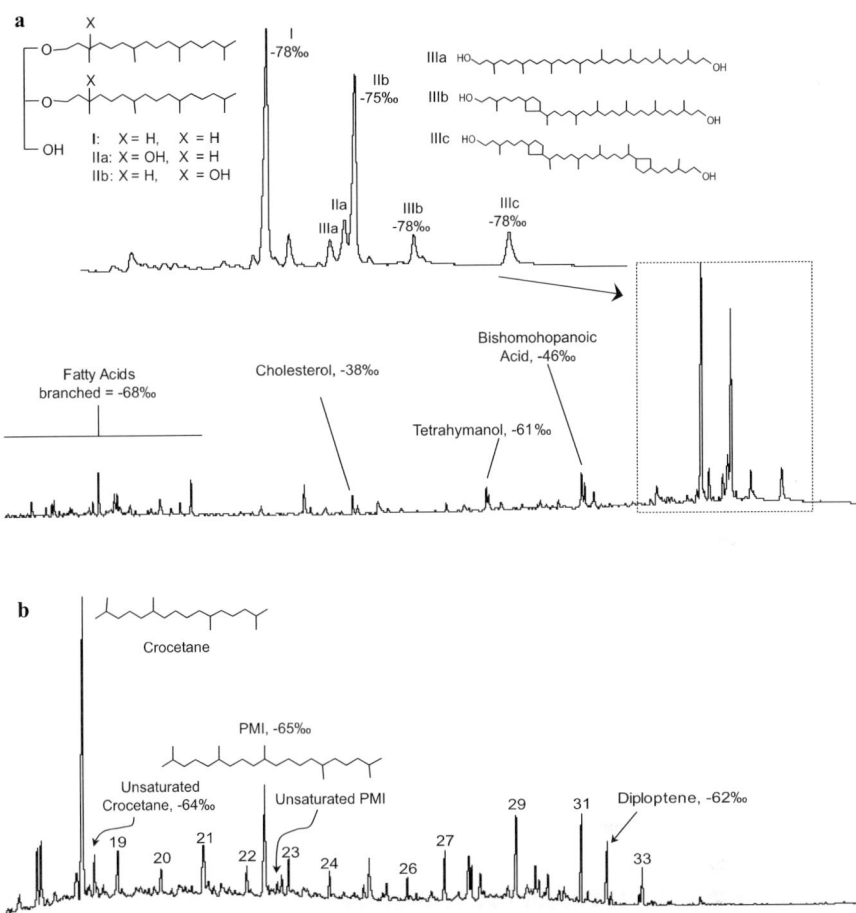

Fig. 11 Partial gas chromatograms showing the relative abundances of archaeal and bacterial biomarkers (and their carbon isotopic compositions) in the polar fraction (**a**) and apolar fraction (**b**) of the total lipid extract of a Napoli Mediterranean mud volcano cold seep. Adapted from [151] and [200]

sites that have been examined are those associated with the Oligocene Lincoln Creek Formation [152, 206], the Miocene Marmorito Limestone [194], the Eocene "Whiskey Creek" deposit [207] and the Jurassic Beauvoisin Formation [208]. These sites contain many of the archaeal (PMI, crocetane, biphytane) and bacterial (hopanes and methyl-branched alcohols, alkanes and alkanoic acids) biomarkers identified in modern cold seep sediments and crusts. Such compounds are typically depleted in ^{13}C (<– 100‰), indicating assimilation of methane and suggesting that the microbial consortia mediating methane oxidation in these ancient sites were similar to those active in modern cold seeps. Thus, determination of biomarker abundances and isotopic compositions in inferred ancient cold seeps can be a powerful additional tool in the identifica-

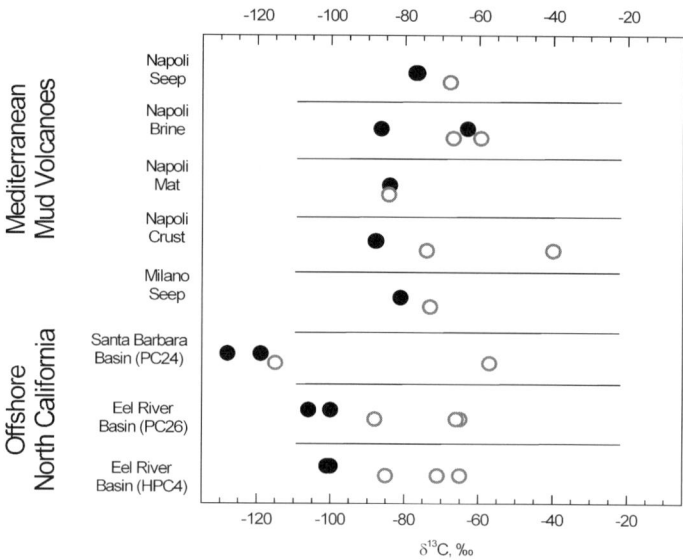

Fig. 12 Plot showing the isotopic offsets between archaeal and bacterial lipid $\delta^{13}C$ values in a range of samples from Mediterranean mud volcano and California margin cold seeps. Adapted from [129] and [199]

tion of AOM and provide a microbiological fingerprint, complementing faunal assemblages in the characterization of such sites.

8
Conclusions and Directions for Future Research

Compound-specific carbon isotopic compositions can help resolve pathways of carbon flow in diverse modern and ancient ecosystems. However, the application of such approaches requires an understanding of: (1) the carbon isotopic composition of likely substrates; (2) the mechanisms of carbon assimilation; (3) carbon isotope effects during assimilation and their sensitivity to environmental conditions; and (4) subsequent isotope effects during lipid biosynthesis. Given these constraints, it is particularly important that assignments of the sources of organic matter used in environmental studies are accurate. Although compound-specific isotope analysis is an important tool in constraining the sources of organic matter being examined, it is vital to continue to identify biomarkers diagnostic for specific organisms and processes. Here we have summarized how alkenone $\delta^{13}C$ values can be used to estimate past pCO_2 levels and how ^{13}C-depleted archaeal and bacterial lipids provide insight into the mechanisms of methane oxidation under anaerobic conditions.

Acknowledgements The authors would like to thank numerous past and present collaborators, including (but certainly not exclusively): Katherine Freeman, Stuart Wakeham, Bob Bidigare, Brian Popp, Jaap Sinninghe Damsté, Stefan Schouten, Giovanni Aloisi, Ioanna Bouloubassi and James Zachos.

References

1. Schidlowski M (1987) Ann Rev Earth Planet Sci 15:47
2. Spiker EC, Hatcher PG, Orem WH (1985) Estuaries 8(2B):A90
3. Hayes JM, Des Marais DJ, Lambert IB, Strauss H, Summons RE (1992) Proterozoic biogeochemistry. In: Schopf JW, Klein C (eds) The Proterozoic biosphere: a multidisciplinary study. Cambridge University Press, Cambridge, p 81–134
4. Ostrom PH, Fry B (1993) Sources and cycling of organic matter within modern and prehistoric food webs. In: Engle MH, Macko SA (eds) Organic geochemistry: principles and applications. Plenum, New York, p 785–798
5. Hayes JM, Freeman KH, Popp BN, Hoham C (1990) Compound-specific isotopic analyses: a novel tool for the reconstruction of ancient biogeochemical processes. In: Durand B, Behar F (eds) Advances in organic geochemistry. Pergamon, Oxford, p 1115–1128
6. Logan GA, Summons RE, Hayes JM (1997) Geochim Cosmochim Acta 61:5391
7. Calder JA, Parker PL (1973) Geochim Cosmochim Acta 37:133
8. Morris I (1980) Paths of carbon assimilation in marine phytoplankton. In: Falkowski P (ed) Primary productivity of the sea. Plenum, New York, p 139–159
9. Degens ET, Guillard RRL, Sackett WM, Hellebust JA (1968) Deep Sea Res 15:1
10. Dean WE, Arthur MA, Claypool GE (1986) Mar Geol 70:119
11. Popp BN, Takigiku R, Hayes JM, Louda JW, Baker EW (1989) Am J Sci 289:436
12. Pagani M, Arthur MA, Freeman KH (1999) Paleoceanography 14:273
13. Freeman KH, Hayes JM, Trendel JM, Albrecht P (1990) Nature 343:254
14. Hayes JM (1993) Mar Geol 1:111
15. Schouten S, Hoefs MJL, Koopmans MP, Bosch H-J, Sinninghe Damsté JS (1998) Org Geochem 29:1305
16. Mariotti A, Germon JC, Hubert P, Kaiser P, Letolle R, Tardieux A, Tardieux P (1981) Plant Soil 62:413
17. Preston T, Owens NJP (1985) Biomed Mass Spectrom 12:510
18. Merritt DA, Hayes JM (1994) Anal Chem 66:2336
19. Meier-Augenstein W (1999) J Chromatogr A 842:351
20. Boon JJ, Rijpstra WIC, de Leeuw JW, Schenck PA (1975) Nature 258:414
21. Grimalt J, Alsaad HT, Douabul AAZ, Albaiges J (1985) Naturwissenschaften 72:35
22. Volkman JK (1986) Org Geochem 9:83
23. Wakeham SG, Farrington JW, Gagosian RB, Lee C, de Baar H, Nigrelli GE, Tripp BW, Smith SO, Frew NM (1980) Nature 286:798
24. Marlowe IT, Brassell SC, Eglinton G, Green JC (1984) Org Geochem 6:135
25. Volkman JK, Eglinton G, Corner EDS, Forsberg TEV (1980) Phytochemistry 19:2619
26. Conte MH, Volkman JK, Eglinton G (1994) Lipid biomarkers of the Haptophyta. In: Green JC, Leadbeater BSC (eds) The Haptophyte algae. Clarendon, Oxford, p 351–377
27. Brassell SC, Eglinton G, Marlowe IT, Pflaumann U, Sarnthein M (1986) Nature 320:129

28. Volkman JK, Barrett SM, Blackburn SI, Mansour MP, Sikes EL, Gelin F (1998) Org Geochem 29:1163
29. Withers NW, Tuttle RC, Goad LJ, Goodwin TW (1979) Phytochem 18:71
30. Robinson N, Eglinton G, Brassell SC, Cranwell PA (1984) Nature 308:439
31. Piretti MV, Pagliuca G, Boni L, Pistocchi R, Diamante M, Gazzotti T (1997) J Phycol 33:61
32. Pancost RD, Freeman KH, Wakeham SG, Robinson CY (1997) Geochim Cosmochim Acta 61:4983
33. Pancost RD, Freeman KH, Wakeham SG (1999) Org Geochem 30:319
34. Eek ME, Whiticar MJ, Bishop JKB, Wong CS (1999) Deep-Sea Res II 46:2863
35. Popp BN, Trull T, Kenig F, Wakeham SG, Rust TM, Tilbrook B, Griffiths FB, Wright SW, Marchant HJ, Bidigare RR, Laws EA (1999) Global Biogeochem Cycles 13:827
36. Schouten S, Van Kaam-Peters HME, Rijpstra WIC, Schoell M, Sinninghe Damsté JS (2000) Am J Sci 300:1
37. Volkman JK, Barrett SM, Dunstan GA (1994) Org Geochem 21:407
38. Rowland SJ, Robson JN (1990) Mar Environ Res 30:191
39. Versteegh GJM, Bosch HJ, de Leeuw JW (1997) Org Geochem 27:1
40. Volkman JK, Jeffrey SW, Nichols PD, Rogers GI, Garland CD (1989) J Exp Mar Biol Ecol 128:219
41. Bell MV, Dick JR, Pond DW (1997) Phytochemistry 45:303
42. Pagani M, Freeman KH, Arthur MA (2000) Geochim Cosmochim Acta 64:37
43. Kuypers MMM, Pancost RD, Sinninghe Damsté JS (1999) Nature 399:342
44. Grice K, Schaeffer P, Schwark L, Maxwell JR (1997) Org Geochem 26:677
45. Eglinton G, Hamilton RJ, Raphael RA, Gonzalez AG (1962) Phytochemistry 1:89
46. Cranwell PA (1981) Org Geochem 3:79
47. Kolattukudy PE (1976) The chemistry and biochemistry of natural waxes. Elsevier, Amsterdam
48. Eglinton G, Hamilton RJ (1967) Science 156:1322
49. Versteegh GJM et al (2004) Geochim Cosmochim Acta 68:411
50. Sarkanen KV, Ludwig CH (1971) Lignins. Wiley, New York
51. Hedges JI, Mann DC (1979) Geochim Cosmochim Acta 43:1809
52. Goñi MA, Yunker MB, Macdonald RW, Eglinton TI (2000) Mar Chem 71:23
53. Huang YS, Freeman KH, Eglinton TI, Street-Perrott FA (1999) Geology 27:471
54. Goñi MA, Teixeira MJ, Perkey DW (2003) Estuar Coast Shelf Sci 57:1023
55. Ourisson G, Rohmer M, Poralla K (1987) Annu Rev Microbiol 41:301
56. Rohmer M, Bisseret P, Neunlist S (1992) In: Moldowan JM, Albrecht P, Philp RP (eds) Biological markers in sediments and petroleum. Prentice-Hall, New Jersey, p 1–17
57. Summons RE, Jahnke LL, Hope JM, Logan GA (1999) Nature 400:554
58. Thiel V, Blumenberg M, Pape T, Seifert R, Michaelis W (2003) Org Geochem 34:81
59. Sinninghe Damsté JS, Rijpstra WIC, Schouten S, Fuerst JA, Jetten MSM, Strous M (2004) Org Geochem 35:561
60. Zundel M, Rohmer M (1985) Eur J Biochem 150:23
61. Gelpi E, Schneider H, Mann J, Oro J (1970) Phytochemistry 9:603
62. Shiea J, Brassell SC, Ward DM (1990) Org Geochem 15:223
63. Liaaen-Jensen S (1978) In: Clayton RK, Sistrom WR (eds) Photosynthetic bacteria. Plenum, New York, p 233–247
64. Koopmans MP, Schouten S, Kohnen MEL, Sinninghe Damsté JS (1996) Geochim Cosmochim Acta 60:4873
65. De Rosa M, Gambacorta A (1988) Prog Lipid Res 27:153
66. Koga Y, Morii H, Akagawa-Matsushita M, Ohga M (1998) Biosci Biotechnol Biochem 62:230

67. Schouten S, Hopmans EC, Pancost RD, Sinninghe Damsté JS (2000) Proc Natl Acad Sci USA 97:14421
68. Hoefs MJL, Schouten S, de Leeuw JW, King LL, Wakeham SG, Sinninghe Damsté JS (1997) Appl Environ Microbiol 63:3090
69. DeLong EF, King LL, Massana R, Cittone H, Murray A, Schleper C, Wakeham SG (1998) Appl Environ Microbiol 64:1133
70. King LL, Pease TK, Wakeham SG (1998) Org Geochem 28:677
71. Pearson A, Eglinton TI (2000) Org Geochem 31:1103
72. Pancost RD, van Geel B, Baas M, Sinninghe Damsté JS (2000) Geology 28:663
73. Holzer G, Oro J, Tornabene TG (1979) J Chromatogr 18:795
74. Risatti JB, Rowland SJ, Yon DA, Maxwell JR (1984) Org Geochem 6:93
75. Brassell SC, Wardroper AMK, Thomson ID, Maxwell JR, Eglinton G (1981) Nature 290:693
76. Schouten S, van der Maarel MJEC, Huber R, Sinninghe Damsté JS (1997) Org Geochem 26:409
77. Elvert M, Suess E, Whiticar MJ (1999) Naturwissenschaften 86:295
78. Bian LQ, Hinrichs KU, Xie TM, Brassell SC, Iversen H, Fossing H, Jorgensen BB, Hayes JM (2001) Geochem Geophys Geosys 2:2000.GC000112
79. Weiss RF (1974) Mar Chem 2:203
80. Mook WG, Bommerson JC, Staberman WH (1974) 22:169
81. Romanek CS, Grossman EL, Morse JW (1992) Geochim Cosmochim Acta 56:419
82. Zhang J, Quay PD, Wilbur DO (1995) Geochim Cosmochim Acta 59:107
83. Halas S, Szaran J, Niezgoda H (1997) Geochim Cosmochim Acta 61:2691
84. Zeebe RE, Wolf-Gladrow D (2001) CO_2 in seawater: equilibrium, kinetics, isotopes. Elsevier, Amsterdam.
85. Hayes JM, Strauss H, Kaufman AJ (1999) Chem Geol 161:103
86. Scholle PA, Arthur MA (1980) Am Assoc Pet Geol Bull 64:67
87. Dickens GR, O'Neil JR, Rea DK, Owen RM (1995) Paleoceanography 10:965
88. Roeske CA, O'Leary MH (1984) Biochemistry 23:6275
89. Raven JA, Johnston AM (1991) Limnol Oceanogr 36:1701
90. Bidigare RR et al (1997) Global Biogeochem Cycles 11:279
91. Popp BN, Laws EA, Bidigare RR, Dore JE, Hanson KL, Wakeham SG (1998) Geochim Cosmochim Acta 62:69
92. Farquhar GD, Richards PA (1984) Aust J Plant Physiol 11:539
93. Sharkey TD, Berry JA (1985) Carbon isotope fractionation of algae as influenced by an inducible CO_2 concentration mechanism. In: Lucas WJ, Berry JA (eds) Inorganic carbon uptake by aquatic photosynthetic organisms. American Society of Plant Physiologists, Rockville, p 389–401
94. Falkowski PG (1991) J Plankton Res 13:21
95. Farquhar GD, O'Leary MH, Berry JA (1982) Aust J Plant Physiol 13:281
96. Laws EA, Bidigare RR, Popp BN (1997) Limnol Oceanogr 42:1552
97. Cassar N, Laws EA, Popp BN, Bidigare RR (2002) Limnol Oceanogr 47:1192
98. Rau GH, Riebesell U, Wolf-Gladrow D (1996) Mar Ecol Prog Ser 133:275
99. Verity PG, Robertson CY, Tronzo CR, Andrews MG, Nelson JR, Sieracki ME (1993) Limnol Oceanogr 37:1434
100. Popp BN, Hanson KL, Dore JE, Bidigare RR, Laws EA, Wakeham SG (1999) Controls on the carbon isotopic composition of phytoplankton: Paleoceanographic perspectives. In: Abrantes F, Mix A (eds) Reconstructing ocean history: a window into the future. Plenum, New York, p 381–398

101. Goericke R, Montoya JP, Fry B (1994) In: Lajtha K, Michener B (eds) Stable isotopes in ecology. Blackwell, Oxford, p 187–221
102. Burns BD, Beardall J (1988) J Exp Mar Biol Ecol 107:75
103. Tortell PD, Reinfelder JR, Morel FMM (1997) Nature 390:243
104. Laws EA, Popp BN, Bidigare RR, Riebesell U, Burkhardt S, Wakeham SG (2001) Geochem Geophys Geosys 2:2000.GC000057
105. Riebesell U, Revill AT, Holdsworth DG, Volkman JK (2000) Geochim Cosmochim Acta 64:4179
106. Francois R, Altabet MA, Goericke R, McCorkle DC, Brunet C, Poisson A (1993) Global Biogeochem Cycles 7:627
107. Friedli H, Lotscher H, Oeschger H, Siegenthaler U, Stauffer B (1986) Nature 324:237
108. Broadmeadow MSJ, Griffiths H (1993) Carbon isotope discrimination and the coupling of CO_2 fluxes within forest canopies. In: Ehleringer JR, Hall AE, Farquhar GD (eds) Stable isotopes and plant carbon–water relations. Academic, San Diego, p 109–129
109. Price GD, McKenzie JE, Pilcher JR, Hoper ST (1997) The Holocene 7:229
110. O'Leary MH (1981) Phytochemistry 20:552
111. Madhavan S, Treichel I, O'Leary MH (1991) Bot Acta 104:292
112. Meinzer FC, Rundel PW, Goldstein G, Sharifi MR (1992) Oecologia 91:305
113. White JWC, Ciais P, Figge RA, Kenny R, Markgraf V (1994) Nature 367:153
114. Sauer M, Maurer S, Matyssek R, Landolt W, Gunthardt-Georg MS, Siegenthaler U (1995) Oecologia 103:397
115. Lockheart MJ, van Bergen PF, Evershed RP (1997) Org Geochem 26:137
116. Libby LM (1972) J Geophys Res 77:4310
117. Libby LM, Pandolfi LJ, Payton PH, Marshall III J, Becker B, Giertz-Sienbenlist V (1976) Nature 261:284
118. Christeller JT, Lang WA, Troughton JH (1976) Plant Physiol 57:580
119. Farmer JG (1979) Nature 279:229
120. Leavitt SW, Long A (1983) Isotope Geosci 1:169
121. Smith BN, Oliver J, McMillan C (1976) Bot Gaz 137:99
122. Troughton JH, Gard KA (1975) Planta 123:185
123. Chikaraishi Y, Naraoka H (2003) Phytochemistry 63:361
124. Marino BD, McElroy MB (1991) Nature 349:127
125. Ehleringer JR, Cerling TE, Helliker BR (1997) Oecologia 112:285
126. Whiticar MJ (1999) Chem Geol 161:291
127. Irwin H, Curtis C, Coleman M (1977) Nature 269:209
128. Elvert M, Suess E, Greinert J, Whiticar MJ (2001) Org Geochem 31:1175
129. Pancost RD, Sinninghe Damsté JS (2003) Chem Geol 195:29
130. Abelson PH, Hoering TC (1961) Proc Natl Acad Sci USA 47:623
131. Deines P (1980) The isotopic composition of reduced organic carbon. In: Fritz P, Fontes JC (eds) Handbook of environmental isotope geochemistry. Elsevier, Amsterdam
132. van der Meer MTJ, Schouten S, van Dongen B, Rijpstra WIC, Fuchs G, Sinninghe Damsté JS, de Leeuw J, Ward DM (2001) J Biol Chem 276:10971
133. van Dongen B, Schouten S, Sinninghe Damsté JS (2001) Rapid Commun Mass Spectrom 15:496
134. Pancost RD, Baas M, van Geel B, Sinninghe Damsté JS (2003) The Holocene 13:921
135. Huang YS, Bol R, Harkness DD, Ineson P, Eglinton G (1996) Org Geochem 24:273
136. Ruby EG, Jannasch HW, Deuser WG (1987) Appl Environ Microbiol 53:1940
137. House CH, Schopf JW, Stetter KO (2003) Org Geochem 34:345

138. Sirevåg R, Buchanan BB, Berry JA, Throughton JH (1977) Arch Microbiol 112:35
139. Quandt I, Gottshalk G, Ziegler H, Stichler W (1977) FEMS Microbiol Lett 1:125
140. Holo H, Sirevåg R (1986) Arch Microbiol 145:173
141. Strauß G, Fuchs G (1993) Eur J Biochem 215:633
142. Menendez C, Bauer Z, Huber H, Gad'on N, Stetter KO, Fuchs G (1999) J Bacteriol 181:1088
143. Preuß A, Schauder R, Fuchs G, Stichler W (1989) Z Naturforsch C 44:397
144. Van der Meer MTJ, Schouten S, Rijpstra WIC, Fuchs G, Sinninghe Damsté JS (2001) FEMS Microbiol Lett 196:67
145. Pearson A, McNichol AP, Benitez-Nelson BC, Hayes JM, Eglinton TI (2001) Geochim Cosmochim Acta 65:3123
146. Kuypers MMM, Blokker P, Erbacher J, Kinkel H, Pancost RD, Schouten S, Sinninghe Damsté JS (2001) Science 293:92
147. Summons RE, Franzmann PD, Nichols PD (1998) Org Geochem 28:465
148. Schouten S, Rijpstra WIC, Kok M, Hopmans EC, Summons RE, Volkman JK, Sinninghe Damsté JS (2001) Geochim Cosmochim Acta 65:1629
149. Jahnke LL, Summons RE, Hope JM, Des Marais DJ (1999) Geochim Cosmochim Acta 63:79
150. Hinrichs K-U, Hayes JM, Sylva SP, Brewer PG, DeLong EF (1999) Nature 398:802
151. Pancost RD, Sinninghe Damsté JS, de Lint S, van der Maarel MJEC, Gottschal JC (2000) The MEDINAUT Shipboard Scientific Party. Appl Environ Microbiol 66:1126
152. Thiel V, Peckmann J, Richnow HH, Luth U, Reitner J, Michaelis W (2001) Mar Chem 73:97
153. Teece MA, Fogel ML, Dollhopf ME, Nealson KH (1999) Org Geochem 30:1571
154. Zhang CL, Li Y, Ye E, Fong J, Peacock AD, Blunt E, Fang J, Lovley D, White DC (2003) Chem Geol 195:17
155. Monson KD, Hayes JM (1982) Geochim Cosmochim Acta 46:139
156. DeNiro MJ, Epstein S (1977) Science 197:261
157. Blair N, Leu A, NuAmerican Society of Plant Physiologists, Rockvillenoz E, Olsen J, Kwong E, Des Marais D (1985) Appl Environ Microbiol 50:996
158. Rohmer M, Sutter B, Sahm H (1989) J Chem Soc Chem Commun 19:1471
159. Rohmer M, Knani M, Simonin P, Sutter B, Sahm H (1993) Biochem J 295:517
160. Horbach S, Sahm H, Welle R (1993) FEMS Microbiol Lett 111:135
161. Rohmer M, Seeman M, Horbach S, Bringer-Meyer S, Sahm H (1996) J Am Chem Soc 118:2564
162. Lichtenthaler HK, Rohmer M, Schwender J (1997) Physiol Plant 101:643
163. Disch A, Schwender J, Muller C, Lichtenthaler HK, Rohmer M (1998) Biochem J 333:381
164. Schouten S, Breteler W, Blokker P, Schogt N, Rijpstra WIC, Grice K, Baas M, Sinninghe Damsté JS (1998) Geochim Cosmochim Acta 62:1397
165. Pancost RD, Freeman KH, Wakeham SG (1999) Org Geochem 30:319
166. Popp BN, Trull T, Kenig F, Wakeham SG, Rust TM, Tilbrook B, Griffiths FB, Wright SW, Marchant HJ, Bidigare RR, Laws EA (1999) Global Biogeochem Cycles 13:827
167. Collister JW, Rieley G, Stern B, Eglinton G, Fry B (1994) Org Geochem 21:619
168. Goñi MA, Ruttenberg KC, Eglinton TI (1997) Nature 389:275
169. Summons RE, Jahnke LL, Roksandik Z (1994) Geochim Cosmochim Acta 58:2853
170. van der Meer MTJ, Schouten S, Sinninghe Damsté JS (1998) Org Geochem 28:527
171. Sakata S, Hayes JM, McTaggart AR, Evans RA, Leckrone KJ, Togasaki RK (1997) Geochim Cosmochim Acta 61:5379

172. Bidigare RR et al (1999) Global Biogeochem Cycles 13:251
173. Rau GH, Takahashi T, Des Marais DJ, Repeta DJ, Martin JH (1992) Geochim Cosmochim Acta 56:1413
174. Jasper JP, Mix AC, Prahl FG, Hayes JM (1994) Paleoceanography 6:781
175. Jasper JP, Hayes JM (1990) Nature 347:462
176. Andersen N, Müller PJ, Kirst G, Schneider RR (1999) The $\delta^{13}C$ signal in $C_{37:2}$ alkenones as a proxy for reconstructing Late Quaternary pCO_2 in surface waters from the South Atlantic. In: Fischer G, Wefer G (eds) Proxies in paleoceanography: examples from the South Atlantic. Springer, Berlin Heidelberg New York, p 469–488
177. Pagani M, Freeman KH, Arthur MA (1999) Science 285:876
178. Nimer NA, Dixon GK, Merrett MJ (1992) New Phytol 120:153
179. Nimer NA, Guan Q, Merrett MJ (1994) New Phytol 126:601
180. Nimer NA, Merrett MJ (1996) New Phytol 133:383
181. Rost B, Zondervan I, Riebesell U (2002) Limnol Oceanogr 47:120
182. Pagani M, Freeman KH, Ohkouchi N, Caldeira K (2002) Paleoceanography 17:1069
183. Ohkouchi N, Kawamura K, Kawahata H, Okada H (1999) Global Biogeochem Cycles 13:695
184. Tsubota H, Ishizaka J, Nishimura A, Watanabe YW (1999) J Oceanogr 55:645
185. Blair NE, Aller RC (1995) Geochim Cosmochim Acta 59:3707
186. Borowski WS, Paull CK, Ussler W III (1996) Geology 24:655
187. Burns SJ (1998) Geochim Cosmochim Acta 62:797
188. Iverson N, Jørgensen BB (1985) Limnol Oceanogr 30:944
189. Reeburgh WS (1980) Earth Planet Sci Lett 46:345
190. Reeburgh WS (1996) Soft spots in the global methane budget. In: Lidstrom ME, Tabita FR (eds) Microbial growth on C_1 compounds. Kluwer, Dordrecht, p 334–342
191. Boetius A, Ravenschlag K, Schubert CJ, Rickert D, Widdel F, Gieseke A, Amann R, Jørgensen BB, Witte U, Pfannkuche (2000) Nature 407:623
192. Zhang CLL, Li YL, Wall JD, Larsen L, Sassen R, Huang YS, Wang Y, Peacock A, White DC, Horita J, Cole DR (2002) Geology 30:239
193. Zhang CL, Pancost RD, Sassen R, Qian Y, Macko SA (2003) Org Geochem 34:827
194. Thiel V, Peckmann J, Seifert R, Wehrung P, Reitner J, Michaelis W (1999) Geochim Cosmochim Acta 63:3959
195. Schouten S, Wakeham SG, Sinninghe Damsté JS (2001) Org Geochem 32:1277
196. Wakeham SG, Lewis CM, Hopmans EC, Schouten S, Sinninghe Damsté JS (2003) Geochim Cosmochim Acta 67:1359
197. Teske A, Hinrichs K-U, Edgcomb V, Gomez AD, Kysela D, Sylva SP, Sogin ML, Jannasch HW (2002) Appl Environ Microbiol 68:1994
198. Orphan VJ, House CH, Hinrichs KU, McKeegan KD, DeLong EF (2001) Science 293:484
199. Hinrichs K-U, Summons RE, Orphan V, Sylva SP, Hayes JM (2000) Org Geochem 31:1685
200. Pancost RD, Hopmans EC, Sinninghe Damsté JS (2001) The MEDINAUT Shipboard Scientific Party. Geochim Cosmochim Acta 65:1611
201. Pancost RD, Bouloubassi I, Aloisi G, Sinninghe Damsté JS (2001) The MEDINAUT Shipboard Scientific Party. Org Geochem 32:695
202. Werne JP, Baas M, Sinninghe Damsté JS (2002) Limnol Oceanogr 47:1694
203. Elvert M, Boetius A, Knittel K, Jorgensen BB (2003) Geomicrobiol J 20:403
204. Valentine DL, Reeburgh WS (2000) Environ Microbiol 25:477
205. Hoehler TM, Alperin MJ, Albert DB, Martens CS (1994) Geochim Cosmochim Acta 43:739

206. Peckmann J, Thiel V, Michaelis W, Clari P, Gaillard C, Martire L, Reitner J (1999) Int J Earth Sci 88:60
207. Goedert JL, Thiel V, Schmale O, Rau WW, Michaelis W, Peckmann J (2003) Facies 48:223
208. Peckmann J, Thiel V, Michaelis W, Clari P, Gaillard C, Martire L, Reitner J (1999) Int J Earth Sci 88:60
209. Boschker HTS, Nold SC, Wellsbury P, Bos D, de Graaf W, Pel R, Parkes RJ, Cappenberg TE (1998) Nature 392:801
210. Bull ID, Parekh NR, Hall GH, Ineson P, Evershed RP (2000) Nature 405:175
211. Laws EA, Popp BN, Bidigare RR, Kennicutt MC, Macko SA (1995) Geochim Cosmochim Acta 59:1131
212. Laws EA, Landry MR, Barber RT, Campbell L, Dickson ML, Marra J (2000) Deep-Sea Res II 47:1339
213. Prahl FG, Muehlhausen LA, Zahnle DL (1988) Geochim Cosmochim Acta 52:2303

Isotopic Tracers of the Marine Nitrogen Cycle: Present and Past

Mark A. Altabet

School for Marine Science and Technology, University of Massachusetts,
706 Rodney French Blvd, New Bedford, MA 02744-1221, USA
maltabet@umassd.edu

1	Overview	252
2	Dominant Marine N Processes and their Isotopic Effects	253
2.1	N_2 fixation	254
2.2	Denitrification	258
2.3	Autotrophic Assimilation	263
2.4	Remineralization	266
2.5	Nitrification	267
3	A Geochemical Approach to Marine N Cycling	268
4	Combined N Inventory Balance and Influence on Marine Biogeochemistry	270
4.1	What Controls Average Oceanic $\delta^{15}N$?	271
5	Paleo-Reconstruction of Marine N Cycle Processes	274
5.1	Signal Transfer to and Preservation in Sediments	274
5.2	Reconstruction of Water Column Denitrification	278
5.3	A Composite Denitrification Record	284
5.4	Toward Reconstruction of the Complete Oceanic N Budget	285
5.5	Oceanic Wide Changes in $\delta^{15}N$ Revisited	287
6	Summary and Synthesis	288
	References	289

Abstract The oceanic nitrogen cycle consists of a web of microbially mediated transformations driven in part by the large range in possible nitrogen oxidation states. Many of these transformations have corresponding isotope fractionation effects, usually leaving the product depleted in ^{15}N ($\delta^{15}N$). Due to the complexity of the nitrogen cycle, observed patterns of isotopic ratio could be expected to defy explanation. However in reality, a few geographically separated processes dominate the larger spatial and temporal scales in the open ocean. These are (1) NO_3^- assimilation by phytoplankton, (2) N_2 fixation, and (3) denitrification. The latter two have particular importance as the principal source and sink, respectively, of combined nitrogen to the ocean. As such, they together control the oceanic inventory for combined nitrogen which in turn is a factor controlling marine plant production and organic matter flux from the surface to the ocean's interior. Taking into account the effective isotopic fractionation effects for N_2 fixation and denitrification, the modern average $\delta^{15}N$ for the ocean is a potentially important constraint on the modern marine nitrogen budget. Past variation in these processes can be reconstructed on time scales from decades to millions of years from sediment cores with good preservation of organic matter. In particular, temporally well resolved δ^{15}N records show large variations in the three major water column denitrification regions in response to climate variations. Collectively, these variations in denitrification likely produced significant changes in the oceanic combined nitrogen inventory which appears to be confirmed by global-scale changes in δ^{15}N across the last deglaciation.

Keywords Nitrogen biogeochemistry · Denitrification · N_2 fixation · Stable isotope · Paleo-record

Abbreviations

‰	per mil
δ	"delta" convention for expressing natural variation in isotopic ratio
$\delta^{15}NO_3^-$	δ^{15}N of NO_3^-
ε	isotope fractionation factor
bp	before present
ACE	Antarctic climate event
DIN	dissolved inorganic nitrogen
D-O	Dansgaard–Oeschger event
ETNP	Eastern Tropical North Pacific
ETSP	Eastern Tropical South Pacific
f	fraction of remaining substrate
HNLC	high nutrient, low chlorophyll
kyr	kilo-year (10^3)
ka	kilo-annum (10^3)
MIS	marine isotope stage
Myr	mega-year (10^6)
N' or N^*	NO_3^- concentration anomaly
OM	organic matter
OMZ	oxygen minimum zone
POM	particulate organic matter
SOM	sedimentary organic matter
Tg	Teragram (10^{12})

1
Overview

Nitrogen is a major component of biomass and the organic matter (OM) derived from it. Throughout much of the ocean, the availability of usable nitrogen is a dominant control of autotrophic production and OM fluxes. This control has ramifications for global biogeochemistry, in general, including influence on the oceanic C cycle and its forcing of atmospheric CO_2. Marine N biogeochemistry itself has a number of microbially mediated components which in combination regulate the combined N content of the ocean. In this Chapter, I will review these components and their N isotope fractionation effects. The use of the N isotope ratio as a natural tracer for marine N cycle processes then will be discussed, concluding with applications from the sediment record for past changes in relation to climate changes.

For convenience, the small but robust natural variations in $^{15}N : ^{14}N$ ratio are typically reported using the "delta" notation employed for other stable isotope systems:

$$\delta^{15}N(‰) = (^{15}N : ^{14}N_{sample} - ^{15}N : ^{14}N_{standard}) / ^{15}N : ^{14}N_{standard} \times 1000 \tag{1}$$

The accepted standard is atmospheric N_2 which by definition has a $\delta^{15}N$ of 0‰. In the oceans, combined N in its various forms can range between −10 and +15‰ with few examples of more extreme values. External analytical precision of environmental materials using modern techniques and instrumentation can be as good as ±0.1‰. The potential for N transformation processes to alter isotopic ratios is measured by the isotopic fractionation factor. Current literature typically refers to ε values which can be defined as:

$$\varepsilon(‰) = \delta^{15}N_{substrate} - \delta^{15}N_{product} \text{ or } \delta^{15}N_{product} = \delta^{15}N_{substrate} - \varepsilon \tag{2}$$

where there is no significant depletion of the substrate. Typically, ε values are between 0‰ (no isotope fractionation) and 40‰ for microbially mediated N transformation processes. That is, the product will usually have a lower $\delta^{15}N$ value than the substrate.

In a closed system with depletion of substrate and matching accumulation of product over time, the Rayleigh equations describe the change in isotope ratio as a function of the fraction of remaining substrate (f):

$$\delta^{15}N_{substrate}(f) = \delta^{15}N_{substrate}(f=1) - \varepsilon \times \ln[f] \tag{3}$$

$$\delta^{15}N_{product}(f) = \delta^{15}N_{substrate}(f=1) + \varepsilon \times f/[1-f] \times \ln[f] \tag{4}$$

Modified Rayleigh equations are more appropriate to an open system akin to a chemostat with continual replenishment of substrate matched by removal of product:

$$\delta^{15}N_{substrate}(f) = \delta^{15}N_{substrate}(f=1) + \varepsilon \times [1-f] \tag{5}$$

$$\delta^{15}N_{product}(f) = \delta^{15}N_{substrate}(f=1) - \varepsilon \times f \tag{6}$$

2
Dominant Marine N Processes and their Isotopic Effects

Nitrogen is found in nature at a variety of oxidation states giving rise to a large number of potential biogeochemical transformations and chemical forms. This large potential list is reduced by the environmental conditions

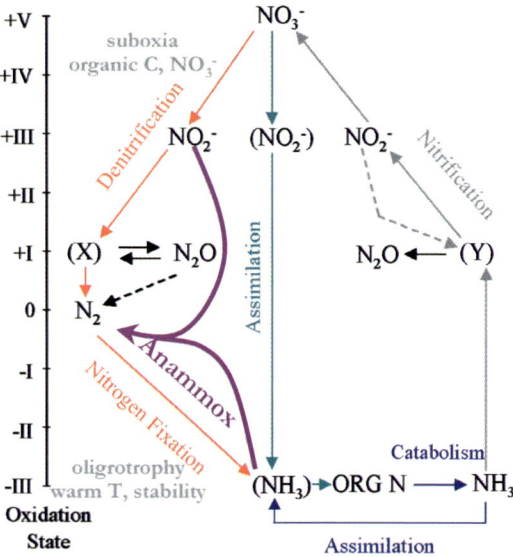

Fig. 1 Schematic representation of major nitrogen cycle paths based on an original Figure in [1]. The large range in nitrogen redox state is shown and is a factor in determining the circumstances under which any particular pathway is favorable. The recently discovered anammox pathway has been added

prevailing in today's ocean as well as the available biochemistries and the energy yield/requirement for particular transformations. While all significant transformations are biologically mediated, it is illustrative to present these with respect to redox state (Fig. 1 [1]). In the following discussion, the focus is on those processes with large-scale significance either with respect to oceanic N balance and/or creation of observed patterns in concentration and $\delta^{15}N$.

2.1
N$_2$ fixation

In the form of dinitrogen gas (N$_2$), nitrogen is an abundant element on the Earth's surface. It makes up $\sim 80\%$ of the atmosphere by volume and is a similarly dominant component of dissolved gases in seawater. If this form of nitrogen could be directly used by all autotrophs, nitrogen availability would exert little or no control on marine (or any other) biogeochemistry and there would be little sense in writing this chapter. In reality, breaking the triple bond holding N$_2$ together is only mildly thermodynamically favorable ($\Delta G = -43$ kJ/mol N) with apparent large kinetic barriers and few groups of organisms possess the biochemical machinery, specifically the nitrogenase enzyme, to do so. Those capable of N$_2$ fixation are exclusively prokaryotes and N$_2$ fixing autotrophs fall within the cyanobacteria (blue-green algae). In the ocean, the pelagic *Trichodesmium* species appear to be the dominant N_2 fixing organisms [2] followed by diatom endosymbionts (e.g. *Richelia*; see review by [3]).

The nitrogenase enzyme is both O_2 sensitive and has a high requirement for Fe [4]. The relative demand as compared to non-N_2 fixers, though, has been sharply revised downward to 4-fold [5]. Given the very low Fe concentrations found in the well-oxygenated open ocean of today and well-conserved structure of its active site across phylogenetic groupings, it has been suggested that nitrogenase evolved early in the Earth's history before oxygenation of the atmosphere several billion years ago [6]. O_2 sensitivity poses a problem for oxygenetic photoautotrophs with one solution being the formation of specialized non-photosynthetic, N_2 fixing cells referred to as heterocysts which have thickened cell walls to inhibit the inward diffusion of O_2. Despite its high O_2 environment in the near-surface open ocean, *Trichodesmium* lacks heterocysts and in fact couples photosynthetic energy production to N_2 fixation. How O_2 inhibition is overcome is not well understood.

The high requirement for Fe has further suggested an interaction between Fe and N biogeochemistry [6]. Since Fe is chiefly supplied by eolian dust input, proximity to these sources is thought to be a control on *Trichodesmium* population size and activity. Indeed, recent studies in the subtropical/tropical North Atlantic show *Trichodesmium* to be relatively Fe replete in this region which receives a large flux of Fe in the form of Saharan dust [7]. However, co-limitation with phosphorous has been reported [8]. Given the wide spatial variation in Fe delivery to the ocean's surface, it has been estimated that in 75% of the ocean's area, N_2 fixation is potentially limited by Fe [9].

N_2 fixers have a clear ecological advantage when the supply of combined N to photoautotrophs is very low. Hence, *Trichodesmium* blooms are typically found in oligotrophic parts of the ocean where high insolation, water column stability, and warm temperatures further favor this group of organisms. Assuming there is sufficient Fe, an apparent paradox is that low available combined N also implies a low flux of P. If *Trichodesmium* were constrained by the average (Redfield) ratio of N to P found in both phytoplankton and inorganic sources in deep water (16 : 1; [10]), little N_2 fixation could be accomplished before P limitation would set in. One possible solution would be faster recycling in near-surface waters of P as compared to combined N. However, recent field studies demonstrate that *Trichodesmium* N : P can be several fold higher than the Redfield ratio [7].

Since N_2 fixation contributes locally to regionally to the combined N inventory without any corresponding input of P, remineralization of relatively N rich OM results in subsurface NO_3^- : PO_4^{-3} ratios higher than the Redfield ratio. There are several formulations in the literature for quantifying the NO_3^- anomaly relative to PO_4^{-3} including the now well-known "N*" parameter devised by Gruber and Sarmiento [11] and later modified by Deutsch et al. [12]. I refer to my personal preference of N' for its simplicity and lack of assumptions regarding the N : P ratio of decaying organic mater:

$$N' = NO_3^- + NO_2^- - 16 \times PO_4^{-3} \tag{7}$$

Regardless of the specific formulation, geochemical approaches using NO_3^- anomalies dramatically changed views regarding the magnitude of oceanic N_2 fixation such that it is now seen as clearly the predominant source of combined N to the ocean. Prior estimates were based on combinations of *Trichodesmium* abundance estimates and N_2 rate measurements made using incubation techniques (e.g. [2]). In hindsight, prior studies likely underestimated oceanic N_2 fixation due to undersampling of *Trichodesmium* activity that is highly variable in time and space. Geochemical approaches have the advantage of integrating over the time and space scales of the circulation and renewal of the subsurface waters in which the NO_3^- anomalies accumulate—years to decades and hundreds to thousands of km. As expected, the most positive N* or N' values are found in subsurface waters (100 to 1000 m) underlying regions such as the Sargasso Sea where *Trichodesmium* blooms are known to regularly occur. However, estimation of pelagic marine N_2 fixation is far from settled in that recent work has suggested geochemical measures are closer to the prior biological ones due to smaller than assumed areal extent of the N* anomaly [13]. Large interannual variability in positive N* in the subsurface waters of the Sargasso Sea have also been observed [14] suggesting that temporal aliasing is also a concern with the geochemical approach.

Given the conservation of nitrogenase and presumably its catalytic mechanism and kinetics thereof, it is not surprising that fairly similar isotopic fractionation effects are observed for different phylogenetic groupings of N_2 fixers (e.g. [15–18]). Despite the energy required to break the triple bond in N_2, little discrimination against ^{15}N is observed, making the intracellular product NH_4^+ and subsequent organic N, similar in isotopic composition to the N_2 substrate. Regarding oceanic N_2 fixation, *Trichodesmium* isotopic composition can be measured directly to assess its impact on marine isotope biogeochemistry. Averages are between −1 and −2‰ [19, 20] consistent with small ε values of 2 to 3‰ given that the $\delta^{15}N$ of dissolved N_2 is about 0.7‰ higher than atmospheric N_2 (0‰ by definition) due to equilibrium isotopic fractionation between dissolved and gas phase N_2.

Average marine NO_3^- has a $\delta^{15}N$ value of $\sim 4.7‰$ [21, 22]. The contributions from N_2 fixation that produce positive NO_3^- anomalies (N') should also, by contributing ^{15}N depleted N, reduce NO_3^- $\delta^{15}N$ values ($\delta^{15}NO_3^-$) below 4.7‰. This is clearly seen in the subtropical N. Atlantic where the region of positive N' between 100 and 1000 m corresponds to $\delta^{15}NO_3^-$ values as low as 1‰ (Fig. 2). These results reflect influence from N_2 fixation in overlying surface waters which produces OM with higher than average N : P and $\delta^{15}N$ lower than 4.7‰. When this OM sinks below 100 m and degrades to its inorganic constituents, positive N' and lowered $\delta^{15}N$ NO_3^- values are produced. The region below 1000 m with slightly negative N' has $\delta^{15}NO_3^-$ values indistinguishable from the oceanic average. This is sensible since almost all sinking organic particles are degraded above 1000 m.

If positive N′ quantitatively reflects the contribution from N_2 fixation then:

$$\delta^{15}NO_3^- = [N' \times -2 + (NO_3^- + NO_2^- - N') \times 4.7]/(NO_3^- + NO_2^-) \quad (8)$$

but rearranging for convenient plotting:

$$\delta^{15}NO_3^- = 4.7 - 6.7 \times N'/(NO_3^- + NO_2^-) \quad (9)$$

Plotting the data in Fig. 2 as $\delta^{15}NO_3^-$ vs. $N'/(NO_3^- + NO_2^-)$ (Fig. 3) shows points distributed about this expected trend confirming that N′ is a generally reliable indicator for contributions from N_2 fixation. Consequences for global N isotope balance and possible reconstruction of past N_2 fixation are discussed below. Very recent analytical improvements in measuring the $\delta^{15}NO_3^-$, particularly at concentrations below 2 μM, will likely reduce scatter in the data distribution [23]. These new techniques in which NO_3^- is converted to N_2O for mass spectrometric analysis preserves oxygen isotopic information which can be used to detect influence on $\delta^{15}N$ values by removal processes such as denitrification or phytoplankton NO_3^- assimilation [24].

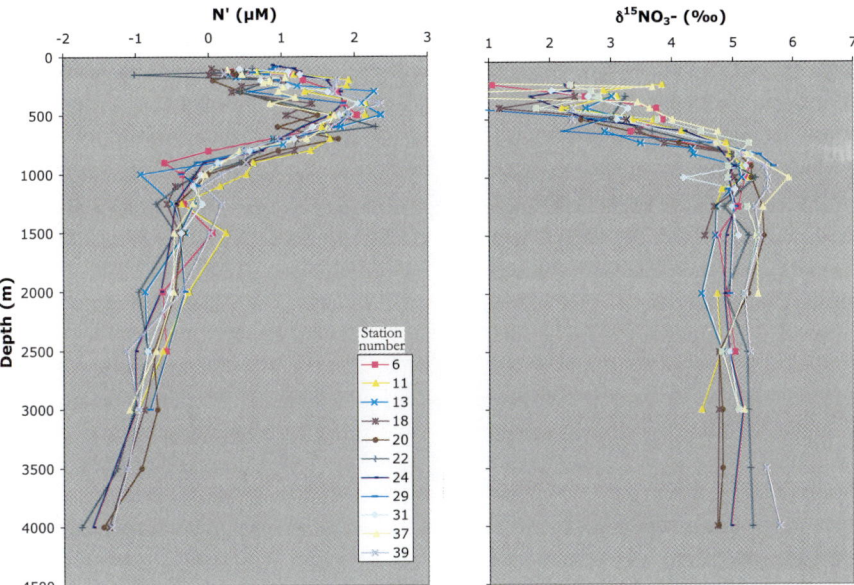

Fig. 2 Vertical profiles of N′ (see text) and $\delta^{15}NO_3^-$ from a series of stations in the Sargasso Sea (subtropical N. Pacific). The positive N′ values in the upper 1000 m are evidence for N_2 fixation in the overlaying surface waters. The relatively low $\delta^{15}NO_3^-$ values in this depth region are consistent with this perspective

2.2
Denitrification

If N_2 fixation is the entry point to the biosphere for combined N, then denitrification is the clear exit. In this case, its importance as a sink for combined N is incidental to its physiological significance. When O_2 falls to $< 5\,\mu M$, it becomes energy efficient for many heterotrophic bacteria to facultatively use the denitrification pathway to respire OM with NO_3^- as an electron acceptor [25]:

$$(CH_2O)_{106}(NH_3)_{16}H_3PO_4 + 84.8\,HNO_3$$
$$= 106\,CO_2 + 42.4\,N_2 + 16\,NH_3 + 148.4\,H_2O + H_3PO_4 \qquad (10)$$

The energy yield for this reaction ($\Delta G = -452\,kJ/mol\,C$) is only slightly less than for aerobic respiration ($\Delta G = -476\,kJ/mol\,C$). Only after NO_3^- is consumed are processes such as sulfate reduction with a much lower energy yield favored ($\Delta G = -82\,kJ/mol\,C$). Thus the requirements for denitrification are (1) a supply of NO_3^-, (2) suboxic conditions, and (3) a supply of OM. Not surprisingly #'s 2 and 3 often co-occur. Also note that N_2O is an intracellular intermediate (Fig. 1) such that denitrification potentially can result in either net production or consumption. However, N_2O is usually well below saturation concentrations (assuming equilibrium with the atmosphere) within active denitrification zones [26].

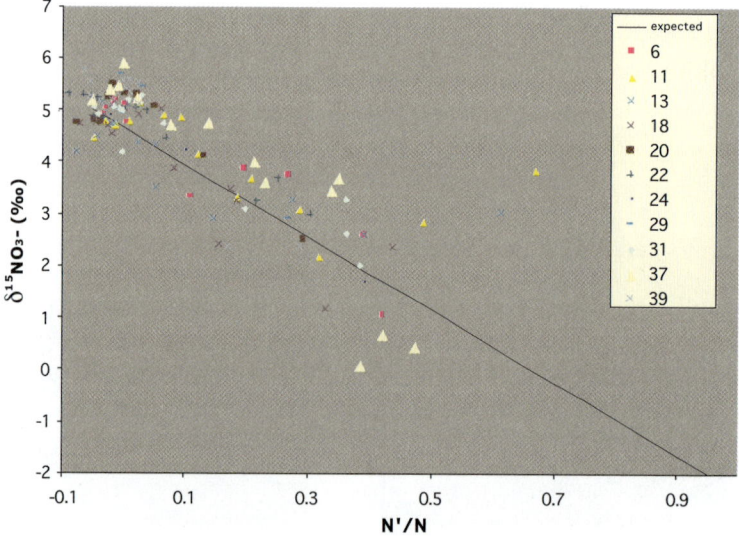

Fig. 3 Cross-plot of the data in Fig. 2. N' is normalized by the $NO_3^- + NO_2^-$ concentration (N'/N) since $\delta^{15}NO_3^-$ should be a function of the proportion of the total derived from N_2 fixation. The *line* represents an expected trend if the newly fixed nitrogen has a $\delta^{15}N$ of -2‰. *Numbers* refer to the station numbers in Fig. 2

The principal loci for denitrification in the ocean are regions with suboxic intermediate waters (water column denitrification) and continental margin sediments (sediment denitrification; e.g. [27]). Water column denitrification occurs primarily in three locations; the Arabian Sea [28], the Eastern Tropical N. Pacific (ETNP; [29]) off the western Mexican margin, and the Eastern Tropical S. Pacific (ETSP; [30]) off the Peruvian and northern Chilean margins. In each of these regions, suboxic conditions are vertically found between near the base of the euphotic zone (50 to 100 m) down to as deep as 500 to 1000 m (Fig. 4). Suboxia results from a combination of poor ventilation of intermediate waters and high downward flux of OM. Each of these regions experience coastal upwelling to the surface of nutrient-rich waters which in turn stimulate high rates of primary production. Both upwelling and intermediate water ventilation can be sensitive to climate change suggesting mechanisms for temporal variability in water column denitrification (see below). Although Eq. 10 indicates a build up of NH_4^+ should occur, this is not observed in any of the three water column denitrification regions. An alternative stoichiometry has been used in which organic N as well as NO_3^- is transformed to N_2 raising its effective yield by 17%;

$$(CH_2O)_{106}(NH_3)_{16}H_3PO_4 + 94.4\,HNO_3$$
$$= 106\,CO_2 + 55.2\,N_2 + 177.2\,H_2O + H_3PO_4 \qquad (11)$$

NO_2^- is an intermediate for denitrification (see Fig. 1) that can reach substantial concentrations (1 to 10 μM) in subsurface waters as a consequence of intense denitrification. The anammox pathway in which NH_4^+ and NO_2^- are combined to form N_2 by specialized chemosynthetic bacteria has been recently identified and may explain the lack of NH_4^+ buildup [31, 32].

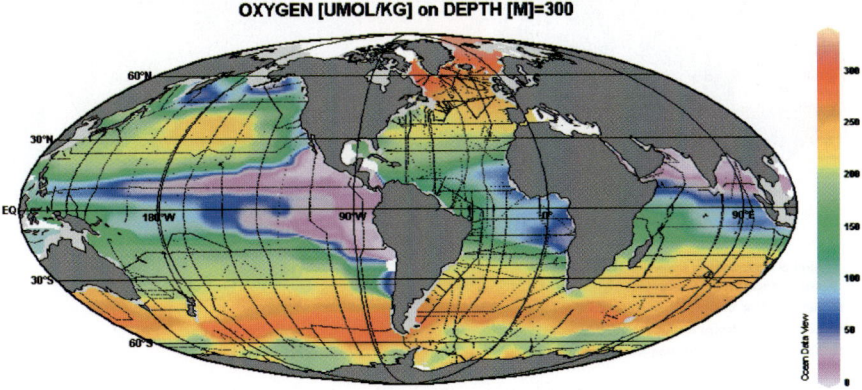

Fig. 4 Global distribution of O_2 at 300 m depth. This depth intersects the cores of the major water column O_2 minimum zones (OMZ's) off the Mexican Margin, the Peru Margin, and in the Arabian Sea

Anammox will be further discussed below. A limited number of measurements of N_2 anomaly relative to changes in NO_3^- have suggested that even Eq. 11 underestimates denitrification yield relative to NO_3^- removal by up to 50% [33].

While only a handful of N_2 measurements of sufficient precision exist for assessing water column denitrification, a relatively large nutrient database exists for calculation of N'. In the case of water column denitrification, the removal of NO_3^- results in negative anomalies relative to PO_4^{-3}. N' can reach $-20\,\mu M$, indicating that up to half of the initial NO_3^- has been removed by denitrification. Just as with N_2 fixation, maps of N' can provide a detailed picture of the horizontal and vertical extent of the influence of denitrification, if not the active process itself (Fig. 5). Combined with estimates of fluid motion and mixing, a regional denitrification rate can be calculated that is nevertheless dependent on assumptions of stoichiometry (e.g. [12]).

Sediment denitrification is biochemically equivalent to the water column process and occurs in the upper sediment column of inshore regions, the continental shelf, and continental slope. In contrast to water column denitrification, water column O_2 concentrations have secondary influence. Along almost all continental margins, OM inputs are sufficiently high to support respiration rates that result in suboxic conditions. In fact, in margin sediments with the highest denitrification rates, the suboxic zone is found vertically within the sediments as a narrow mm to cm scale band between the oxic sedi-

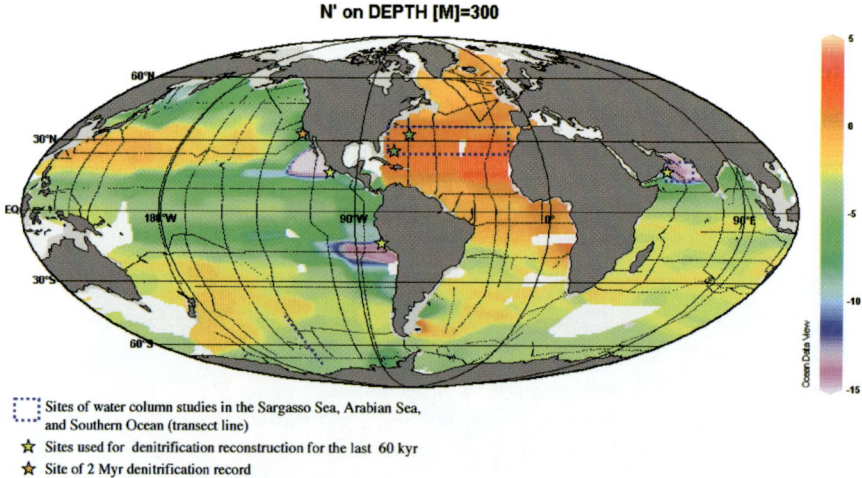

Fig. 5 Global distribution of N' at 300 m depth. The most negative values correspond to the cores of the OMZ's shown in Fig. 4 and are produced by denitrification. Positive values in the Sargasso Sea and other subtropical zones are the result of N_2 fixation. Site locations for data discussed in this chapter are as indicated

ment surface and an sulfate-reducing, anoxic region below. Until the work of Christensen and Devol [27, 34, 35], the downward diffusion of NO_3^- from the overlying water was thought to control the overall sediment denitrification rate and was used as its estimate. They discovered that in situ production of NO_3^- in the sediment oxic layer by nitrification could be the principal source for denitrification. Their resulting estimates put global marine sediment denitrification on par in importance with respect to water column denitrification (~ 100 Tg – N/yr). Codispoti et al. [33] and Brandes and Devol [36] have recently suggested even higher rates (> 200 Tg – N/yr). Unfortunately, sediment denitrification estimates are based on extrapolation of flux determinations from a limited number of individual sediment cores from a limited number of regions. This contrasts with water column denitrification in which each major region has been characterized using geochemical estimates.

Unlike N_2 fixation, denitrification strongly discriminates against ^{15}N (large isotope fractionation potential). The two most important implications are (1) locally, denitrification can produce NO_3 strongly enriched in ^{15}N and (2) as an important overall combined sink, denitrification raises oceanic average $\delta^{15}N$ above the average of the combined N sources (discussed further below). Whether measured in the laboratory or in the field, the inherent fractionation factor for denitrification (ε_{den}) is between 20 and 30‰ [37–43]. The clearest examples of denitrification produced isotope enrichment are for the regions of water column denitrification discussed above. While the oceanic average for $\delta^{15}NO_3^-$ is near 5‰, depths with the greatest intensity of denitrification reach $> 20‰$. In each of the ocean's three large water column denitrification regions, ^{15}N enrichment in NO_3^- is clearly associated with the core of the O_2 minimum zone (OMZ), and the most negative N' values. Denitrification also increases $\delta^{18}O$ in the residual NO_3^- and recent studies have shown the relationship with the increase in $\delta^{15}N$ is approximately 1 : 1 (D. Sigman, pers. comm.).

N' can be used to estimate the residual fraction NO_3^- remaining after denitrification permitting fits to the Rayleigh equations (Eqs. 3 and 5):

$$f = (NO_3^- + NO_2^-)/(NO_3^- + NO_2^- - N') \tag{12}$$

Using the Arabian Sea as an example, $\delta^{15}NO_3^-$ maxima within the OMZ vary between stations suggesting variable isotopic fractionation (Fig. 6). Nevertheless, when plotting all station data together, strong linear correlations are found between $\delta^{15}NO_3^-$ and its fractional removal as estimated by Eq. 12, (Fig. 7). Neither the linear fit nor the value for ε_{den} estimated from the slope appears sensitive to the choice of an open or closed system equation. ε_{den} is between 20 and 30‰ consistent with both laboratory and field observations. Similar results have been reported for the ETNP [43].

Although benthic denitrification employs the same biochemical pathways, the measured isotopic fractionation in continental margin sediments is a tenth or less as found in the water column [44, 45]. The cause for this

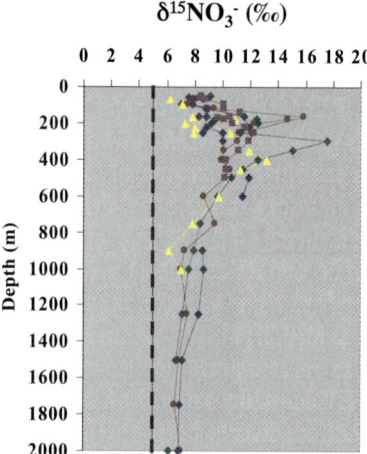

Fig. 6 Depth profiles for $\delta^{15}NO_3^-$ from several stations in the Arabian Sea denitrification zone. The maxima between 200 and 400 m are in the core of this OMZ and co-occur with the most negative N' values. The *vertical dashed line* represents average oceanic $\delta^{15}NO_3^-$ for reference

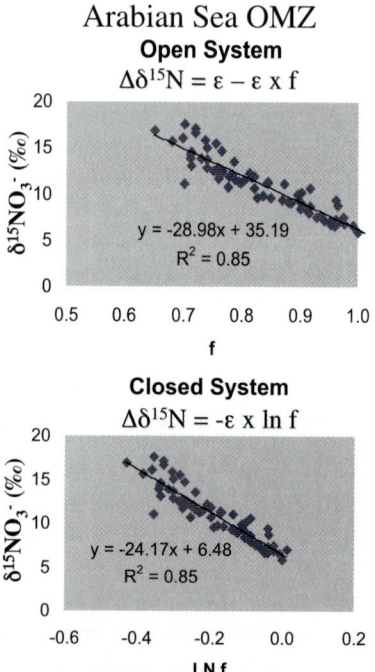

Fig. 7 Cross plots of the $\delta^{15}NO_3^-$ data in Fig. 6 with either the estimated fraction of NO_3^- remaining after denitrification (f) or the ln of f as prescribed by the open and closed system Rayleigh equations (Eqs. 3 and 5), respectively. The regression slopes estimate the fractionation factor (ε)

contrast in ε_{den} between these environments is an apparent distinction in the rate-limiting step. In the case of water column denitrification, the activity of nitrate reductase appears limiting allowing for a near-full expression of the enzymatic fractionation factor. In the case of the benthic denitrification, the sediment matrix impedes the transport of substrates, such that diffusion of NO_3^- is rate limiting. Since this process only very weakly discriminates (fractionates) between isotopes, observed ε_{den} is sharply reduced to values indistinguishable from 0‰.

2.3
Autotrophic Assimilation

While neither a source nor sink to marine combined N inventory, autotrophic assimilation can produce important transient or regional N isotopic signals. Analogous to denitrification, autotrophic consumption tends to discriminate against the heavy isotope such that partial nutrient drawdown leaves the remaining substrate isotopically enriched and the OM produced isotopically depleted. The most important sources of combined N for phytoplankton are NO_3^- and NH_4^+ and significant uptake isotopic effects (ε_u) have been observed during the uptake of either. For NO_3^-, laboratory experiments show ε_u to vary with taxa with some evidence of diatoms having higher values [46–51] as compared with other common marine phytoplankton species.

Aside from eutrophic estuaries/embayments and anoxic basins, NH_4^+ is usually found in the ocean as small, rapidly overturning pools. Any isotopic effects during uptake are not expressed at observable temporal/spatial scales. By contrast, large spatial/temporal variations in NO_3^- concentration are observed in the near-surface ocean which are forced by shifts in the balance between supply and biological consumption. For example, in most temperate oceanic waters, deep winter mixing followed by springtime thermal stratification and a phytoplankton bloom produces an annual cycle of high wintertime NO_3^- followed by springtime drawdown in near-surface waters.

In high-nutrient low chlorophyll regions (HNLC; e.g. Subarctic Pacific, Equatorial Pacific, Southern Ocean), this cycle is muted to varying degrees as a result of Fe limitation of phytoplankton growth (e.g. [52]). In the polar Southern Ocean, summertime NO_3^- drawdown is only between 10 to 20% of the wintertime maximum surface concentration. Northward advection across the polar front into the Subantarctic zone followed by further NO_3^- drawdown produces a large meridional concentration gradient. Since the Subantarctic zone is in an important location for intermediate water formation, incomplete nutrient utilization has important consequences for the chemical composition of these waters that ultimately upwell in the subtropics and tropics [53]. A qualitatively similar phenomenon is observed in the Equatorial Pacific where upwelling to the surface of nutrient-rich waters occurs at the equator and more intensely to the east. Here too, Fe limitation prevents complete local utilization of NO_3^-, but continued utilization with advection of surface waters poleward as well as westward produces a well-known concentration gradient, e.g. [54].

Because these surface NO_3^- gradients are created by phytoplankton drawdown, there are corresponding and inverse variations in the $\delta^{15}NO_3^-$. The best example is from the S. Ocean where near-surface $\delta^{15}NO_3^-$ values can reach 15‰ as $[NO_3^-]$ approaches 0 μM in the vicinity of the subtropical front (Fig. 8; [55]). A Rayleigh-type relationship is observed in all sectors though the estimated value for ε_u (4 to 6‰) may be reduced from actual values due to mixing and seasonal effects. During the JGOFS AESOPS program, there

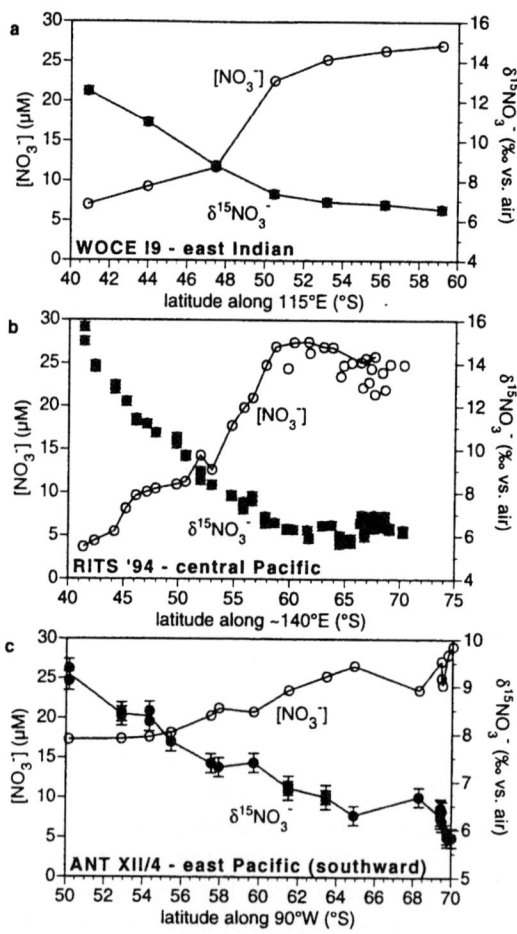

Fig. 8 Changing NO_3^- concentration and $\delta^{15}NO_3^-$ with latitude across the Southern Ocean frontal zone system. Increasing $\delta^{15}N$ with decreasing NO_3^- is the result of isotopic fractionation during phytoplankton uptake. Figure from [55]

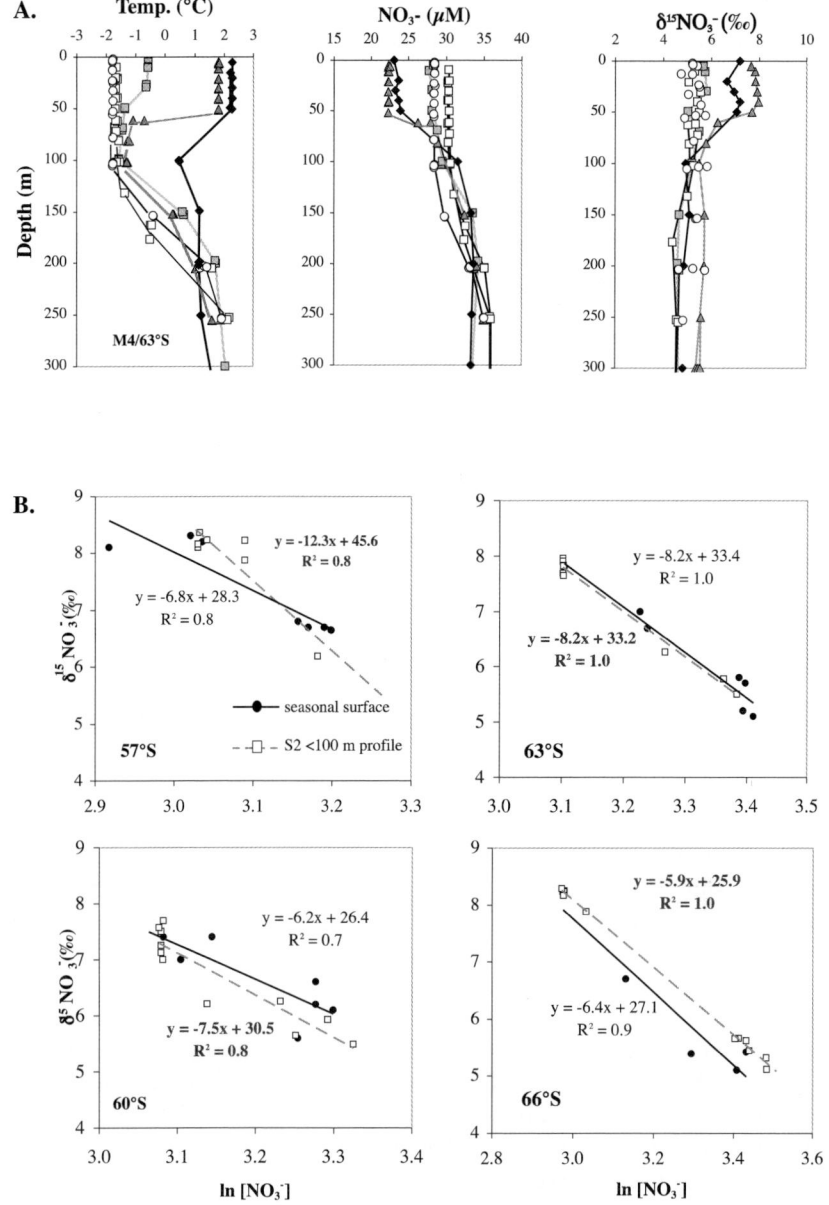

Fig. 9 (a) Example of seasonal changes in NO_3^- concentration and $\delta^{15}NO_3^-$ in the near-surface waters of the Southern Ocean due to partial removal by phytoplankton. The corresponding temperature profile marks the progression from well-mixed surface waters in austral winter to thermal stratification in spring. (b) Cross plots with ln $[NO_3^-]$, an approximation for ln f, for four stations along a N-S transect of the polar frontal region in the SW Pacific sector. Data from profiles taken at the time of maximal drawdown as well as seasonally distributed data are shown

was seasonal sampling at locations within both the subpolar and polar regions north of the Ross Sea. Surface data showed a similar $\delta^{15}NO_3^-$: $[NO_3^-]$ relationship with an obvious seasonal pattern of austral summer reduction in $[NO_3^-]$ and increase in $\delta^{15}NO_3^-$ [56]. Four stations spanning the polar front were studied in detail allowing for more rigorous analysis of quantitative relationships. ε_u was estimated and compared for both summertime vertical profiles of $[NO_3^-]$ and $\delta^{15}NO_3^-$ as well as mixed-layer seasonal timeseries (Fig. 9). For both, a similar range in ε_u of 6 to 8 was deduced with the exception of the depth-distributed data at one station. Sediment trap collections at these sites further showed that the annual average $\delta^{15}N$ for sinking POM at each site was consistent with (a) annual NO_3^- drawdown (f) and (b) estimated ε_u. For the low extent of nutrient drawdown observed, it was shown there was insignificant sensitivity to closed (Eq. 4) or open (Eq. 6) system equations. This showed the $\delta^{15}N$ of POM reaching the sediments to be sensitive to surface ocean NO_3^- utilization in HNLC regions.

2.4
Remineralization

Remineralization refers to all processes that transform organic N back into its inorganic constituents. Biochemically this usually involves the heterotrophic removal of the amino group from amino acids with subsequent production of ammonia (NH_3) which at seawater pH is found as ammonium (NH_4^+). While some metazoans subsequently detoxify ammonia by conversion to urea, etc., ammonia is the dominant release product of remineralization in marine systems. The biogeochemical significance of this process is two-fold: First, it is an obligatory step in nitrogen cycling resupplying pools of DIN. Second, it is the "other side of the coin" for organic nitrogen diagenesis—organic nitrogen preservation equates with incomplete remineralization. Conditions that enhance remineralization also enhance diagenesis and reduce preservation.

Evidence for N isotope effects during remineralization is equivocal. Significant isotopic discrimination is indicated by direct observation of low $\delta^{15}N$ for NH_4^+ excreted by zooplankton which is consistent with the well known and ubiquitous phenomenon of ^{15}N trophic enrichment [57–60]. The 3 to 4‰ higher $\delta^{15}N$ for a heterotrophic organism relative to its food is well explained as a mass balance with loss of ^{15}N depleted excreta. The trophic shift in $\delta^{15}N$ has been widely used to trace foodwebs particularly in conjunction with $\delta^{13}C$ which shows a relatively modest change with trophic exchange and is much more indicative of the primary producers at the base of the food web (e.g. C3 vs. C4 plants). Bulk $\delta^{15}N$ has been complemented with individual amino acid isotopic analysis. Large variations within microorganisms have been observed that appear to match position in the transamination pathway [15, 61]. For metazoans which may not have the ability to synthesize all required protein amino acids, the $\delta^{15}N$ of "essential" amino acids appears to retain the

δ^{15}N of the primary producers while others exhibit an amplified trophic enrichment. Thus, spectra of δ^{15}N variation among amino acids in zooplankton have been used to deduce trophic level [62].

Additional evidence for isotopic fractionation during remineralization include the universal observation of 5 to 10‰ [63, 64] increases in δ^{15}N for small, suspended particulate organic matter (POM) with depth below the euphotic zone in the open ocean. With the presumption that deep POM was ultimately derived from the overlying surface waters, the rise in δ^{15}N with decreasing concentrations was interpreted as reflecting progressive diagenesis with depth and removal of ^{15}N-depleted NH_4^+. Elevated δ^{15}N (3 to 5‰) for sea floor sediment N relative to sinking particle inputs in regions with poor OM preservation has been interpreted similarly [65].

In contrast, large, fast-sinking POM in the open ocean as sampled by a sediment trap either shows constant or even decreasing δ^{15}N with depth (1000s of m) and diminishing flux [56, 64–67]. The few direct measurements of porewater NH_4^+ δ^{15}N show values similar to sediment OM [68]. These measurements were made in the upper 50 cm of relatively OM-rich sediments in regions of the sediment column not influenced by nitrification. It has also been pointed out that even where sediment OM content is relatively high, most of the original input of organic N has been removed by remineralization but nevertheless sediment δ^{15}N matches well inputs as measured by sediment traps.

Laboratory studies of diagenetic isotopic effects do not resolve these ambiguities. Under oxic and anoxic conditions, δ^{15}N of the remaining OM can either increase or decrease [47, 69, 70]. These apparently contradictory results may be the result of microbial assimilation of NH_4^+ occurring simultaneously with production. Nevertheless, there appears to be a consensus view that NH_4^+ produced during microbial OM diagenesis is not significantly shifted in δ^{15}N away from its source [71, 72].

2.5
Nitrification

Nitrification is the last step completing the internal marine N cycle in which NH_4^+ is in two steps oxidized by obligative chemosynthetic bacteria back to NO_3^- (Fig. 1). In the first step, NH_4^+ is oxidized to NO_2^- and in the second NO_2^- is oxidized by a different class of bacteria (e.g. *Nitrobacter*) to NO_3^-. In most of the open ocean, the production of NH_4^+ limits nitrification such that rarely NH_4^+ or NO_2^- concentrations (< 0.2 μM) reach significant levels as compared to NO_3^- (1 to 40 μM below the euphotic zone). These intermediaries may build up where nitrification is inhibited by low O_2 as at the boundaries of water column suboxic/anoxic zones or where NH_4^+ production rates are very high such as in eutrophic estuaries.

Laboratory and field studies clearly show a substantial N isotope fractionation effect associated with each of the nitrification steps (15 to 40‰; [73–

75]). The distinction with the diagenesis/remineralization studies is that single strains were used thereby isolating the experiments to single N transformations. New and exciting work shows for the first time variations in ε as a function of phylogenetic grouping presumably due to differences in enzyme structure and/or regulation of microbial physiology [76]. This approach holds much promise for understanding part of the natural variations observed in isotopic fractionation for this and other transformations.

In most of the ocean, nitrification does not produce any observed variation in $\delta^{15}N$ since any remineralized NH_4^+ is rapidly and completely converted to NO_3^-. Only where there is only partial removal or a gradient in utilization in eutrophied coastal waters such as Delaware Bay [74], is there an effect observed through progressive ^{15}N enrichment in NH_4^+ with its consumption by nitrification. A similar effect may occur vertically in OM-rich sediments at the suboxic/oxic transition zone where NH_4^+ is diffusing upward from the anoxic layer and upon entering the near-sediment-surface oxic layer is subject to nitrification [72]. However, if diffusion is rate limiting, little isotopic fraction may occur as observed for sediment denitrification.

N_2O is a by-product of nitrification (Fig. 1) whose yield appears to be dependent on O_2 concentration [77, 78]. Though a very minor sink for combined marine N, this flux may be a large fraction of total global N_2O sources [79, 80]. Atmospheric N_2O acts as a minor greenhouse gas but through photochemical reactions producing NO is a major term controlling stratospheric O_3. The oceanic N cycle through N_2O production thus impacts the chemistry of the upper atmosphere. A number of studies have made use of the $\delta^{15}N$ and $\delta^{18}O$ of N_2O to better constrain its biogeochemistry. Where there is net depletion due to denitrification, ^{15}N and ^{18}O are correspondingly strongly enriched [81]. Recent studies have exploited the asymmetry of the N_2O molecule to examine isotopomer composition, that is the differential $^{15}N/^{14}N$ ratio of the central vs. end position N atom [82, 83].

3
A Geochemical Approach to Marine N Cycling

Availability of combined N is an important control on photosynthetic carbon fixation (primary production) in much of the ocean. Combined N appears to fulfill the definition of a limiting nutrient as first coined by Liebig in his "law of the minimum". Marine biogeochemists have in the last two decades found reality to be rather more complex than this simple view with respect to co-limitation by other macro and micro-nutrients. Moreover, related but distinct processes may have different suites of controlling factors (e.g. primary vs. export production). Nevertheless, it is clear that outside of HNLC regions practically all combined inorganic nitrogen available to the euphotic zone is utilized on an annual basis. This includes regions with perennial coastal upwelling centers that are persistently enriched near-surface in NO_3^- that is consumed with subsequent along-surface advection of the upwelled water.

As mentioned above, marine phytoplankton produces OM with an average C : N ratio of 6.6 and an average N : P ratio of 17 [10]. This Redfield stoichiometry is a central and powerful concept in marine biogeochemistry and the quantitative link between nitrogen with other elemental cycles. These relationships can be expressed as an equation for the interconversion between average marine OM and its inorganic precursors:

$$(CH_2O)_{106}(NH_3)_{16}H_3PO_4 + 138\,O_2$$
$$= 106\,CO_2 + 16\,HNO_3 + H_3PO_4 + 122\,H_2O \qquad(13)$$

To the left, OM production consumes NO_3^-, PO_4^{-3}, CO_2, and increases O_2 concentrations in the proportions prescribed. To the right, OM remineralization increases their concentrations by these proportions. An obvious exception is denitrification as discussed above. However, there is still much confusion in the field regarding the precision of the Redfield stoichiometry. A common misperception is that instead of reflecting average behavior, these ratios reflect the strict requirements of all phytoplankton. Therefore, N : P delivery ratios even slightly above 16 would reflect PO_4^{-3} limitation and delivery ratios below N limitation. In reality, there appears to be considerable biological plasticity with respect to nutrient stoichiometry and limitation. Culture and field studies show phytoplankton cell quotas can vary much more dramatically in response to P vs. N limitation (e.g. [84, 85]). This is likely due to the greater structural role that N-containing molecules have. An important confirmation of Redfield's seminal work was the observation that marine phytoplankton, even when nutrients were abundant, utilized NO_3^- and PO_4^{-3} at ratios near 16 : 1. However, more recent work shows considerable variability between taxonomic groups; [86] and that theoretically optimal stoichiometry can vary widely with life strategy [87]. Particularly noteworthy, as mentioned above, is the apparent high N : P ratio, 3 to 4-fold higher than Redfield, for the N_2 fixer *Trichodesmium* [88, 89].

These observations lead to the question of whether available nutrient stoichiometry drives phytoplankton (and OM) elemental ratios or visa versa. Redfield's seminal 1934 paper concluded with the inference that because N sources and sinks were largely controlled by ocean biology whereas P was largely controlled abiologically (source from riverine delivery, sink from sediment burial), the N cycle would adjust to variations in oceanic P balance in order to maintain observed stoichiometry. For example a drop in P delivery would increase N : P, but subsequent inhibition of N_2 fixation would restore it to Redfield values. Since then, many geochemists have viewed P as the ultimate limiting nutrient while biological oceanographers, based on physiological studies, have viewed N as limiting outside of HNLC regions. The observed variability in phytoplankton preferred N : P and in particular the high N : P (relative to Redfield) of *Trichodesmium* call into question the simplicity of the traditional geochemist's view. As discussed below there is now considerable evidence for past large variations in the ocean N cycle that would have forced fixed N inventory changes independent of P.

Despite these issues, it is clear that there is considerable coupling between oceanic elemental cycles. Of particular interest is the coupling between nutrient and carbon cycles. There is 60-fold more CO_2 dissolved in the ocean than in the atmosphere. It follows that on time-scales of exchange between the atmosphere and ocean ($\sim$ 1000 yr), oceanic processes determine atmospheric CO_2 content which in turn is a major influence on global climate in its role as a greenhouse gas (e.g. [90]). The 40% glacial to interglacial oscillation in atmospheric CO_2 observed in polar ice cores for the last 400 kyr is widely inferred to be driven by oceanic processes [91]. Numerous theories have been proposed over the last 20 years as to which oceanic phenomena produced these changes (see brief review in [53]). Extensive paleoceanographic study has led to many of these being discarded.

The surviving theories include two for which $\delta^{15}N$ data have provided important support. The first theory (or set of theories) focuses on the modern Southern Ocean as the region (1) through which the large volume of abyssal ocean waters is ventilated with the atmosphere and (2) with the largest area of HNLC surface waters [92–94]. Briefly, low glacial CO_2 may have been driven by increased downward export of organic C stimulated by an increase in Fe-rich dust [95]. Alternatively, increased near-surface stratification decreased ventilation of deep CO_2-rich waters. In either case, relative NO_3^- utilization would have increased in glacial polar waters and there is now substantial $\delta^{15}N$ data to support this hypothesis [96, 97]. The second theory involves increased glacial export productivity in all non-HNLC regions due to an increase in average oceanic nutrient concentration. It has been argued from residence time considerations that the response time of the oceanic P inventory is too long relative to the 10 000 yr period for the glacial to interglacial transition in CO_2 concentration [98]. The N residence time is on the order of several thousand years (see below), and is consistent with the oceanic N cycle being able to be perturbed with sufficient rapidity to contribute to the observed change in atmospheric CO_2.

4
Combined N Inventory Balance and Influence on Marine Biogeochemistry

Estimates for annual flux estimates for ocean N sources and sinks have recently been extensively reviewed [33]. The current consensus is that source terms are dominated by oceanic N_2 fixation with significant but more minor contributions from riverine and atmospheric inputs. Denitrification in the water column as well as continental margin sediments is the predominant sink with very minor contributions from sediment burial. Overall, for a steady state to exist, N_2 fixation must roughly balance denitrification. There are a number of possible feedbacks including the average NO_3^- content of the ocean itself. For example, increasing NO_3^- (excess source over sink) could increase denitrification through increased export productivity and reduced subsurface O_2. Increased average NO_3^- concentration may also inhibit N_2 fixation if subtropical oligotrophy was reduced. Decreasing NO_3^- could produce inverse effects. While such coupling must exist on a very-long time scale, a number of other independent factors can influence either denitrification or N_2 fixation resulting in partial decoupling on climatically significant time scales.

The magnitude of total sources or total sinks are on the order of 200 Tg/yr. Given that the current combined N inventory is 6×10^5 Tg (> 95% in the form of NO_3^-), the residence time (inventory/flux) for combined N in the ocean is about 3 kyr. Any imbalance between sources and sinks would be reflected in a change in inventory on this time scale which is comparable to the time scale for past changes in atmospheric CO_2. By contrast, P residence time is about 5 to 10-fold longer, such that it is much less likely that significant changes in P inventory occurred on deglacial timescales [98]. There is considerable uncertainty, though, in several of the N flux estimates. For example, it was only after integrative, geochemical approaches were taken to quantify average rates of very temporally and spatially variable N_2 fixation that its dominance was recognized as discussed above [11].

While Gruber and Sarmiento [11] assumed that the ocean's N cycle was in rough balance, Codispoti et al. [33] have suggested it is, *at present*, well out of balance by increasing prior estimates of both water column and sediment denitrification to ~ 450 Tg N/yr. The increase in water column denitrification is based on limited data showing the yield of N_2 relative to observed negative NO_3^- anomaly to be up to 4-fold higher than calculated based on Eq. 11. Sediment denitrification was also assumed to be higher than previously thought based on extension of the water column results and recent discovery of anammox and other pathways involving Mn [99].

4.1
What Controls Average Oceanic $\delta^{15}N$?

In parallel with an oceanic N budget, there must also be a ^{15}N budget that controls the average $\delta^{15}N$ of the ocean. If removal processes did not alter (e.g. did not fractionate), nitrogen isotopic ratio then average oceanic $\delta^{15}N$ would simply reflect the weighted average of the sources regardless of the overall magnitude of oceanic N fluxes and whether they were in balance. With N_2 fixation being the dominant term, average $\delta^{15}N$ would be near -1‰, unless the smaller contributions from riverine and atmospheric inputs are very enriched or depleted in ^{15}N. In reality, average oceanic δ^{15}N is near 5‰ [22] and is the result of isotopic fractionation during the predominant sink— denitrification. This value though is well below the one that would be estimated from the known range in ε_{den}; 20 to 30‰ assuming steady state;

$$\text{Flux}_{input} \times \delta^{15}N_{avg.\ input} = \text{Flux}_{denitrification} \times \delta^{15}N_{N2\ lost\ via\ denitrification} \tag{14}$$

since at steady state $\text{Flux}_{input} = \text{Flux}_{denitrification}$

$$\delta^{15}N_{avg.\ input} = \delta^{15}N_{N2\ lost\ via\ denitrification} \tag{15}$$

Eq. 2 applies such that $\delta^{15}N_{N2\ lost\ via\ denitrification} = \delta^{15}N_{avg.\ ocean} - \varepsilon_{denitrification}$

$$\delta^{15}N_{avg.\ input} = \delta^{15}N_{avg.\ ocean} - \varepsilon_{denitrification} \tag{16}$$

or $\delta^{15}N_{avg.\ ocean} = \delta^{15}N_{avg.\ input} + \varepsilon_{denitirification}$

Brandes and Devol [36] have pointed out that lower than expected average δ^{15}N can be accounted for by the nil effective ε associated with sediment denitrification. Average oceanic δ^{15}N at steady state is thus largely dependent on the relative contributions of water column and sedimentary denitrification:

$$\delta^{15}N_{avg.\ ocean} = \delta^{15}N_{avg.\ input} + f_{wc} \times \varepsilon_{wc\ denitirification} + (1 - f_{wc}) \times \varepsilon_{sed\ denitirification} \tag{17}$$

where f_{wc} is the fraction of total denitrification occurring in water column suboxic zones and $1 - f_{wc}$ the fraction in sediments. Considering $\varepsilon_{sed\ denitrification}$ is close to 0:

$$\delta^{15}N_{avg.\ ocean} \sim \delta^{15}N_{avg.\ input} + f_{wc} \times \varepsilon_{wc\ denitrification} \tag{18}$$

It had previously been thought that water column and sedimentary denitrification each made roughly equal contributions to oceanic N loss. Brandes and Devol conclude, though, that to achieve an average δ^{15}N of 5‰, sedimentary denitrification must dominate and account for about 80% of total oceanic denitrification. Since as mentioned above, flux estimates for water column denitrification are likely conservative, the implication is that sedimentary and thus total denitrification is substantially greater than previously thought as concluded by Codispoti et al. (2001). The resulting large imbalance, though, would indicate non-steady $\delta^{15}N_{avg.\ ocean}$.

The Brandes and Devol model assumes that use of $\varepsilon_{wc\ denitrification}$ values derived from cultures and/or from changes $\delta^{15}NO_3^-$ is appropriate for use in Eq. 18. However, there are at least three factors which could reduce the effective isotopic fractionation (difference in $\delta^{15}N_{N2\ lost\ via\ denitrification}$ and $\delta^{15}N_{avg.\ ocean}$) associated with water column denitrification. The first takes into account that within the water column denitrification zone there is substantial reduction of NO_3^- concentration such that Eq. 4 rather than Eq. 2

would be a more appropriate predictor of the $\delta^{15}N$ of N_2 lost under "closed" system conditions. Using a realistic profile from the Arabian Sea, the effective $\varepsilon_{denitrification}$ is reduced from 25 to 20‰. Second, the canonical denitrification equation shows that 16% of the N_2 produced is actually derived from OM which in water column denitrification regions has a $\delta^{15}N$ between 8 to 10‰ due to the elevated $\delta^{15}N$ of upwelled NO_3^-. The contribution of organic N conversion to N_2 further lowers the effective $\varepsilon_{wc\ denitrification}$ to 16‰. Finally, the nature of the physical flow system also influences effective isotopic fractionation. At one extreme, water parcels can be imagined as being isolated from each other as they pass through (advect) through the denitrification zone and the closed system approach would be most accurate. Alternatively at the other extreme, mixing between parcels greatly exceeds advection and open system assumptions apply. Applying the open system Eq. 6, would further reduce $\varepsilon_{wc\ denitrification}$ to 13‰. Applying all three conditions produces a steady state balance in which denitrification is close to being equally apportioned between water column and sedimentary denitrification (Fig. 10). The lower overall magnitude of sedimentary denitrification called for also makes more likely that a near-steady state in the ocean N budget exists.

The available means to estimate oceanic combined N flux have sufficient inherent uncertainty that the question of modern balance will not be answered through estimation of present-day fluxes. As discussed above, open ocean N_2 fixation is very patchy in time and space though integrative geochemical estimates based on intermediate water NO_3^- anomalies partially overcome this problem. Sedimentary denitrification so far has relied on ex-

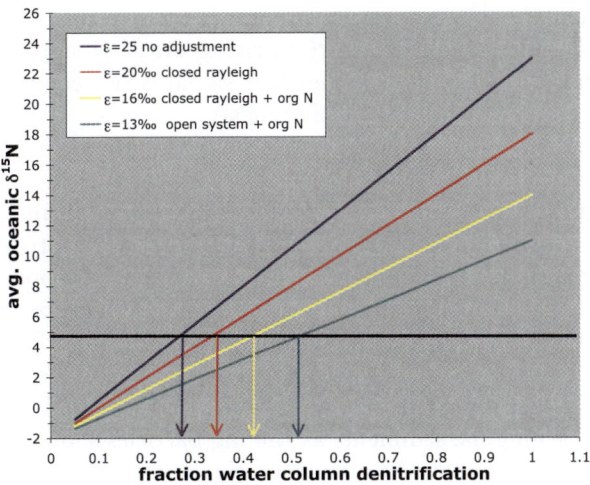

Fig. 10 Estimated relationships between the fraction of total ocean N source consumed by water column denitrification according to Eq. 18. Relationships assuming different values for the effective ε are shown (see text)

trapolation from a limited set of core incubation experiments. However, even if all modern estimates were "perfect" it is unlikely that they would be equivalent to average fluxes over the most recent 3 kyr or approximately 1 N residence time period. It is on this time scale that any net imbalance would have a significant impact on oceanic N inventory. For example, a 20% deficit sustained over the last 3 kyr would roughly result in a 20% decrease in N inventory and average oceanic NO_3^- concentration. Average oceanic $\delta^{15}N$ would similarly respond to unsteady conditions on this time scale. Altabet and Curry [100] showed theoretically that during periods of increasing N inventory $\delta^{15}N$ decreases during the transition regardless of whether the cause is an increase in N_2 fixation or a decrease in denitrification. Decreasing N inventory would correspondingly produce a maximum in $\delta^{15}N$ on the transition. When balance was reestablished, initial average $\delta^{15}N$ would be returned to even though the N inventory was higher or lower unless the relative proportion of sedimentary and water column denitrification had changed. The rest of this chapter focuses on $\delta^{15}N$ reconstruction using sediment cores with application to both the problems of near-modern balance as well as whether changes in N inventory were coupled to past global climate change.

5
Paleo-Reconstruction of Marine N Cycle Processes

5.1
Signal Transfer to and Preservation in Sediments

To use sedimentary $\delta^{15}N$ records for oceanic N cycle reconstruction, it is necessary to (1) understand the factor(s) driving the $\delta^{15}N$ signal reaching the sediments and (2) determine whether that signal is subsequently altered by sedimentary organic matter (SOM) diagenesis.

Dealing first with item #1, it has been long recognized there must be on average a balance in non-HNLC regions between the flux of "new" nitrogen into the euphotic zone where it is utilized by phytoplankton and the loss of N primarily in the form of fast-sinking particles [101]. New nitrogen is distinguished from nitrogen remineralized and reutilized by phytoplankton in the euphotic zone. It is supplied in the form of NO_3^- from deeper waters when they upwell into and/or mix with nutrient-depleted near-surface waters. Just as there is an on average balance between new N input and loss to the oceanic euphotic zone, there must also be an isotopic balance [102]. If so, then the $\delta^{15}N$ signal reaching the sediments as sinking particles should be equivalent to the $\delta^{15}N$ of NO_3^- supplied to the euphotic zone above. In HNLC regions, the same would apply except the $\delta^{15}N$ observed is lowered by the fractionation effect associated with partial NO_3^- utilization as in the Equatorial Pacific and Southern Ocean. A priori, this may not be realized in a $\delta^{15}N$ comparison between NO_3^- and sinking particles if: (1) local vertical processes were not dominant such that balance was only realized by integrating over very large horizontal scales or (2) possible secondary contributors to N loss such as mixing/advection of DON or excretion at depth by vertically migrating zooplankton had large isotope effects associated with them.

Whether sinking particles accurately transfer the NO_3^- isotopic signal to the sediments has been extensively tested by local comparison at a number of sites. At each, in addition to $\delta^{15}NO_3^-$ data, a sediment trap time series was available covering at least one annual cycle. The latter point is important in that many of the locations studied have seasonal increases in euphotic zone NO_3^- due to either seasonal upwelling or wintertime deep convective mixing. Isotopic fractionation due to partial phytoplankton NO_3^- utilization may be expressed on a seasonal basis, but not in the annual average $\delta^{15}N$ for sinking particles since eventually almost all near-surface NO_3^- is utilized at some point in the annual cycle. The sites examined also cover the > 10‰ range in $\delta^{15}N$ observed in the open ocean, from low $\delta^{15}N$ regions influenced by N_2 fixation (Sargasso Sea) to high $\delta^{15}N$, denitrification regions such as the Gulf of California (Fig. 11A). Overall, the comparison is excellent with all points falling on or near the 1 : 1 line. If HNLC regions were included (e.g. Southern Ocean or Equatorial Pacific), these points would fall to the right of the 1 : 1 line in Fig. 11A. In summary, it is clear that local isotopic balance between the new nitrogen source in the form of NO_3^- and loss as sinking particles is realized and that a reliable and well-identified isotopic signature is transferred to the sediments.

This comparison is now extended to the surficial sediments. While the isotopic signature reaching the sediment reflects near-surface processes, early diagenetic alteration may alter it even in recent sediments. Even where sedimentary SOM preservation is considered excellent, only a few percent of the input is preserved. While the isotopic effects of diagenesis *per se* remain ambiguous, SOM may also have altered $\delta^{15}N$ relative to inputs if bacterial metabolism produces new organic nitrogen from NH_4^+ or NO_3^- with $\delta^{15}N$ distinct from sinking particles. The comparison with surficial sediment $\delta^{15}N$, though, shows these concerns are not actualized when SOM preservation is good to excellent. For these sites in which %N in surficial sediment is relatively high (> 0.1% by weight) as a result of a combination of high export production and/or inhibited diagenesis resulting from low bottom water O_2, the 1 : 1 comparison between SOM $\delta^{15}N$ and $\delta^{15}NO_3^-$ is just as robust if not more so as between $\delta^{15}NO_3^-$ and sinking particle $\delta^{15}N$ (Fig. 11B). In contrast, deep-ocean sites such as the Sargasso Sea with poor SOM preservation have $\delta^{15}N$ values elevated by 3–6‰ relative to sinking particles. This "diagenetic effect" is clearly qualitatively linked to reduced preservation in bottom environments that receive less OM input, have high O_2, are well bioturbated, and have lower sediment accumulation rates. However, the mechanisms producing this effect are equally unknown as for the ^{15}N enrichment of deep,

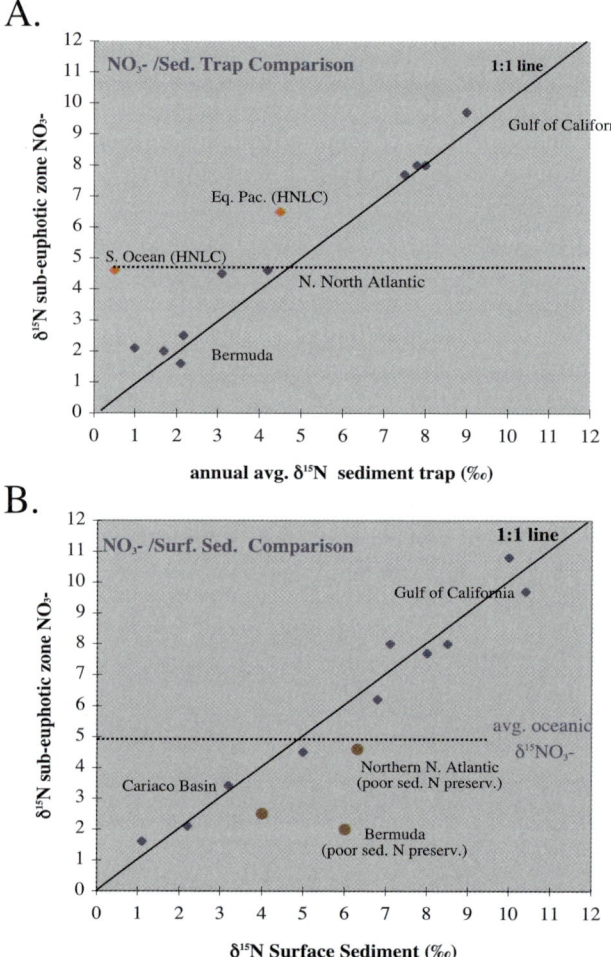

Fig. 11 A Sub-euphotic zone $\delta^{15}NO_3^-$ vs. annual average $\delta^{15}N$ for sinking particle sampled by sediment trap. B Sub-euphotic zone $\delta^{15}NO_3^-$ vs. surface sediment $\delta^{15}N$. For both panels, a 1:1 line as well as a *horizontal dashed line* representing the oceanic average $\delta^{15}NO_3^-$ are shown for reference. Note the departure from the 1:1 *line in panel B* for sediments with poor N preservation. Data sources other than those already cited are found in [128–143]

suspended particles. While it may be tempting to assume that the diagenetic increase in $\delta^{15}N$ is constant downcore at any particular site, we have no means so far to independently determine this.

Reliable and detailed $\delta^{15}N$ paleo-reconstructions are thus, to date, restricted to sediments with good to excellent preservation of OM with one important exception. In the polar Southern Ocean, sediments are typically rich in biogenic opal consisting primarily of diatom frustule microfossils. Like all organisms which produce biominerals, diatoms use organic templates to direct their growth which are then permanently incorporated in the mineral matrix. The organic template with a typically high protein content is relatively protected from diagenetic alteration after the organism's death. Biogenic opal has a number of advantages as a reservoir of protected OM in the sediments. It has significantly lower density than the other major components of deep-sea marine sediments, clays and calcites, such that it can be readily separated using density centrifugations. Its chemical resistance permits the use of strong oxidants such as hot perchloric acid for rigorous surface cleaning of extraneous OM. Sigman et al. [97] showed that lowered apparent N content from cleaned diatom frustules corresponded to lower $\delta^{15}N$ and a less "noisy" downcore record. As discussed above past changes in $\delta^{15}N$ in the Southern Ocean are most likely associated with changes in NO_3^- drawdown by phytoplankton and results to date indicate greater drawdown of near-surface NO_3^- in the polar Southern Ocean during the last glacial period [97, 103]. Recently it has been shown, though, that the presence of atmospheric N_2 "dissolved" in frustules may pose a contamination artifact that would bias $\delta^{15}N$ results [144], (Altabet unpublished). Potentially other biominerals may be exploited as repositories of unaltered OM for isotopic reconstruction and there has been continued effort to make use of the preserved calcite tests of foraminifera [100] and coccoliths produced by coccolithophorids.

Isolation of specific biomarker molecules from sediments for C and H isotope analysis has proven very successful in both avoiding the effects of diagenesis and to isolate the source of the isotopic signature. In contrast, there has been comparatively little progress in N-specific biomarker isotope analysis. The reasons for this are both technical and phenomenological. N is found in much lower abundance in molecules as compared to C and H requiring much larger quantities to be extracted and purified. N-containing biomarker molecules are also typically more polar and less volatile (e.g. peptides and amino acids) making them more difficult to purify and separate using now standard GC interfaces to isotope ratio mass spectrometers (IRMS). Extra care in the GC-IRMS interface needs to be taken to ensure that N_2 is the only product of combustion (no N oxides) and that there is no contamination from atmospheric leaks. Additionally, common N-containing biomarkers are not specific; all organisms produce amino acids and nucleic acids. An important exception is chlorophyll and its derivatives (chlorins) which are produced only by photosynthetic organisms and contain 4 N atoms per molecule. Despite the difficulties of off-line preparation of sufficient material for isotopic analysis, $\delta^{15}N$ chlorin data have been used to show that the relatively high $\delta^{15}N$ of Mediterranean low OM sediment sequences is largely due to diagenesis [104, 105]. Efforts to develop a GC-IRMS technique for these low volatility molecules have to date not been successful. Neither have amino acid $\delta^{15}N$ analysis been applied to sediments, though their organic geochemistry in recent sediments has been examined using $\delta^{13}C$ [106].

5.2
Reconstruction of Water Column Denitrification

The most successful application of downcore δ^{15}N reconstruction has been in identifying and understanding past changes in water column denitrification on time scales from 10^2 to 10^6 years. Water column denitrification is regionally isolatable, produces large isotopic signals as shown above, and is strongly coupled to past climate change. Suboxic intermediate waters overlying margin sediments with high OM input ensure both fidelity of the preserved δ^{15}N signal and the availability of cores with high accumulation rates for good to excellent temporal resolution. In these sediments, N content is high enough that only ten's of mg of material are necessary for isotopic analysis. Reconstruction of other marine biogeochemical processes such as near-surface nutrient utilization have been made, but instead of a broad review, we focus here on denitrification for detailed illustration.

Denitrification records based on sediment δ^{15}N have been published for the Arabian Sea [42, 107, 108] and the ETNP [109–111]. In the latter case, sites along the western American margin are included since upwelled waters are derived from the California Undercurrent which in turn carries denitrification-influenced water northward from the ETNP. Current studies now also include the Peru Margin (Higginson et al., submitted), the third of the ocean's major water column denitrification zones.

Variation in Arabian Sea denitrification occurs on a variety of climatologically relevant time scales. The longest record from the Owen Ridge off Oman reaches back nearly 1 Myr from the present [42]. Variation is primarily at the Milankovitch periods of Earth' orbital variation (23, 41, and 100 kyr) which are also found in the glacial cycles of Pleistocene climate. In general, lower δ^{15}N and denitrification is found during cold periods of glacial maxima. Nearby cores from the Oman Margin have 10-fold higher accumulation rates and provide better temporal resolution (Fig. 12). A detailed study of the last 60 kyr at centennial resolution reveals strong millennial variability during MIS 3 (30 to 60 kyr BP), with transitions from δ^{15}N minima to maxima occurring relatively abruptly, on the order of a few hundred years or less [108]. The pattern of variation over this period is remarkably similar to temperature changes recorded in Greenland ice cores and known as Dansgaard–Oeschger events after their discoverers (e.g. [112]). Again, cold events are associated with δ^{15}N minima and visa versa. It is apparent that broad-scale (perhaps global), rapid climate change is strongly coupled to Arabian Sea denitrification intensity.

We can make use of nature's "past experiments" to probe the causes of changing denitrification intensity by analysis of synoptic proxies of relevant processes. A leading hypothesis linking N. Hemisphere climate to Arabian Sea denitrification is that climate change modulates the strength of the summer South Asian monsoon and upwelling favorable winds off Oman. Corresponding oscillation in productivity influences subsurface suboxia and thus denitrification intensity in the Arabian Sea. Consistent with this hypothesis, a site on the nearby Owen Ridge has records for wind strength (sediment grain size) and productivity (%N) which are positively correlated with $\delta^{15}N$ at the orbital periodicities. For the Oman Margin sites, further support is found in positive relationships between $\delta^{15}N$, %N, and chlorin concentration.

The high accumulation rate of ODP site 723 (15 cm/kyr) and the low material requirements for key analyses permit an extremely well-resolved exploration of the relationship between climate-linked denitrification and its putative proximal forcings [113]. A 20 cm section of ODP Site 723A corres-

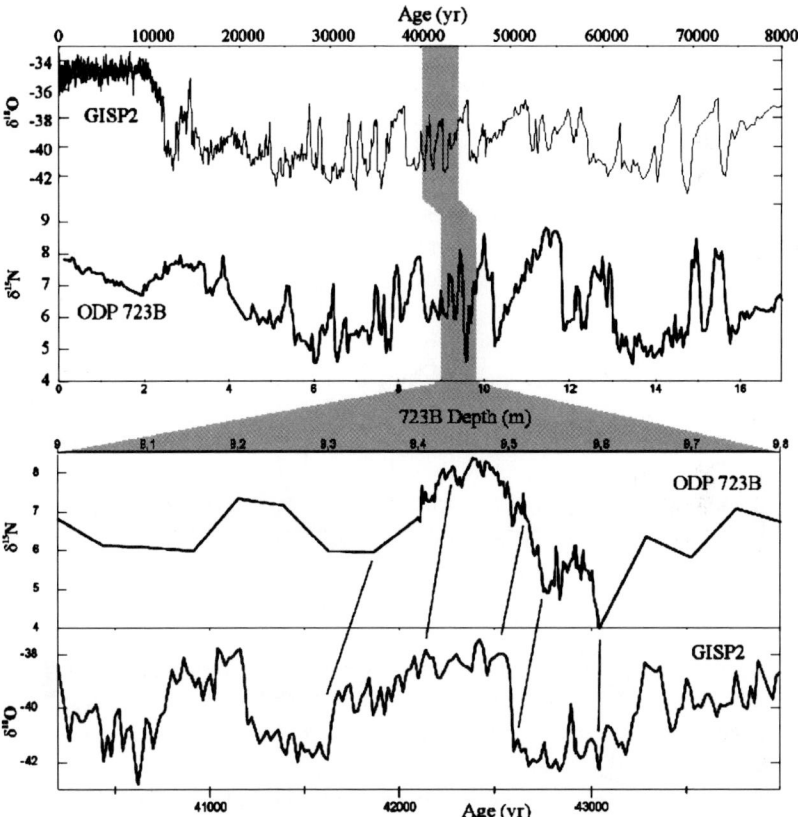

Fig. 12 *Upper Panel*. Comparison between Greenland ice core record of climate (ice $\delta^{18}O$ reflects local temperature) over the last 80 kyr and Arabian Sea denitrification ($\delta^{15}N$). ODP Site 723B is located on the Oman Margin at about 800 m depth at the lower boundary of the OMZ. MIS 3 is the period between 30 and 60 kyr before present and the sub-millennial scale rapid D-O events are reflected in both records. *Lower Panel*. The approximate 1 kyr period examined at very high temporal resolution through 2 mm sectioning of the sediment core. This section corresponds to D-O event #11

ponding to the D-O event #11 was sectioned at 2 mm intervals to achieve decadal resolution (Fig. 12). δ^{15}N results show a very well-resolved transition over 100 to 200 years from little or no denitrification during the cold-stadial period to near-maximal denitrification during the warm, interstadial period. At the point of this transition, there is a simultaneous rise in productivity proxies; OM accumulation and Ba (Fig. 13). Higher export productivity and therefore increased OM transport to the sediments should result in higher sediment respiration as confirmed by reduced $CaCO_3$ preservation. For productivity to force denitrification, it must produce suboxia in intermediate waters. Low O_2 is known to increase V and Mo and reduce Mn in sediments (e.g. [114]). These changes are also observed coincident with the increase in productivity and denitrification (Fig. 13). Other sediment proxies can be

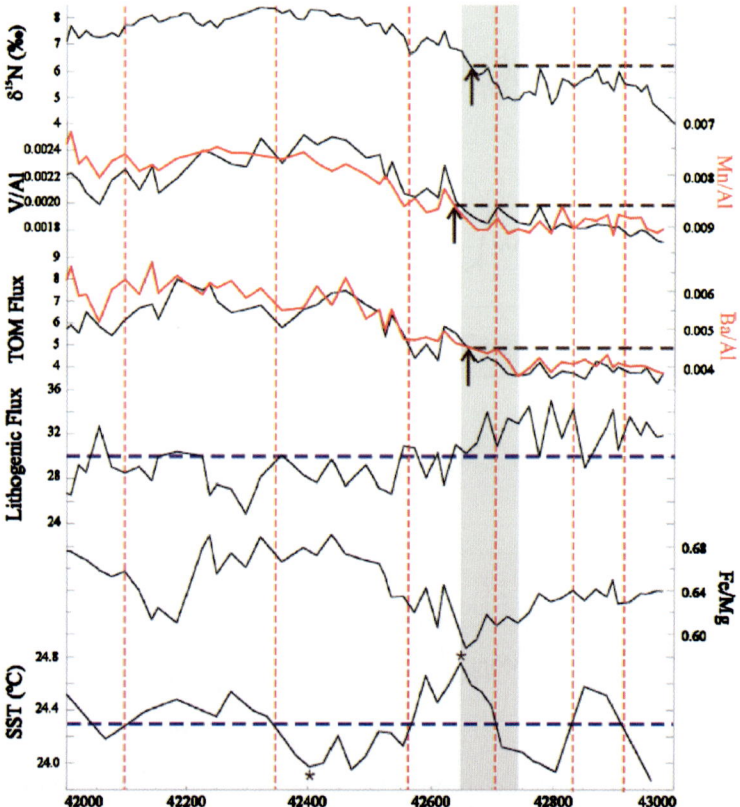

Fig. 13 Multi-proxy, very high resolution reconstruction of D-O event #11 from ODP Site 723B in the Arabian Sea. The increase in denitrification starting about 42 700 yrs bp is accompanied by decreased oxygenation of the overlying water column (V/Al increases, Mn/Al decreases) and increased productivity (increasing Ba/Al and TOM flux). Changing Fe/Mg also marks changing terrestrial aridity in eolian dust source regions

exploited to probe the cause and effect linkages between climate change and marine biogeochemical processes such as denitrification. One example is dust-born proxies for terrestrial conditions transported by winds to the study region. Here the Fe/Mg ratio varies such that the stadial period of low Arabian Sea denitrification is also marked by more arid conditions on nearby continental land masses. The strength of the S. Asian monsoon may be modulated at these suborbital timescales by changes in land cover and corresponding ability to absorb sunlight.

In sum, intermediate water suboxia and denitrification are regionally forced by changes in southern Asia monsoon intensity. It is likely that productivity and subsurface respiration are the linking processes. While remote changes in water mass ventilation can be ruled out, coincident changes in local ventilation of intermediate waters cannot, without studies of past changes in the hydrography of these waters.

Studies in the ETNP have shown similar linkages between past-climate change and denitrification with glacial periods marked by lowered levels. To date there is no evidence of D-O event forcing as strong as observed in the Arabian Sea. This observation is likely related to the greater potential for remote forcing of ventilation in this eastern boundary current. It has been argued that coincidence with productivity proxies demonstrates local forcing of suboxia and denitrification [109]. More recent work indicates significant phase lags between the two [111]. It has also been argued that changes in the proportion and ventilation of the source of intermediate water in the ETNP is the major proximal control on ETNP intermediate water oxygenation [115]. One source is "old" equatorial underwater which is remotely ventilated in the Southern Ocean. Another is Pacific Intermediate Water ventilated in the Subarctic NW Pacific. If true, denitrification in the ETNP could be modulated principally by high latitude processes as opposed to proximal subtropical processes as in the Arabian Sea.

We have examined denitrification variations for nearly the last 2 Myr years from ODP Site 1012 at the northern boundary of the ETNP (Fig. 14). The significance of this portion of the latest Pleistocene is that it encompasses the mid-Pleistocene transition from 41-kyr period dominance to 100-kyr dominance in glacial/interglacial variations. The cause of this transition is the subject of intensive research particularly since the 100 kyr period is the weakest with respect to solar forcing despite its dominance over the latest 0.5 Myr (e.g. [116]). Our data show large variations in δ^{15}N and denitrification throughout this time period with no long-term trend. Spectral analysis demonstrates the importance of the 100, 41, and 23 kyr periods over the last 0.6 Myr (Fig. 14 inset). However the earliest 0.6 Myr shows no significant 100 kyr periodicity but a strong 41 kyr period along with a harmonic near double this period. Clearly the climate-denitrification link extends to these Myr timescales.

The Peru Margin is the third and last of the major water column denitrification regions to be subjected to paleoceanographic reconstruction. A priori

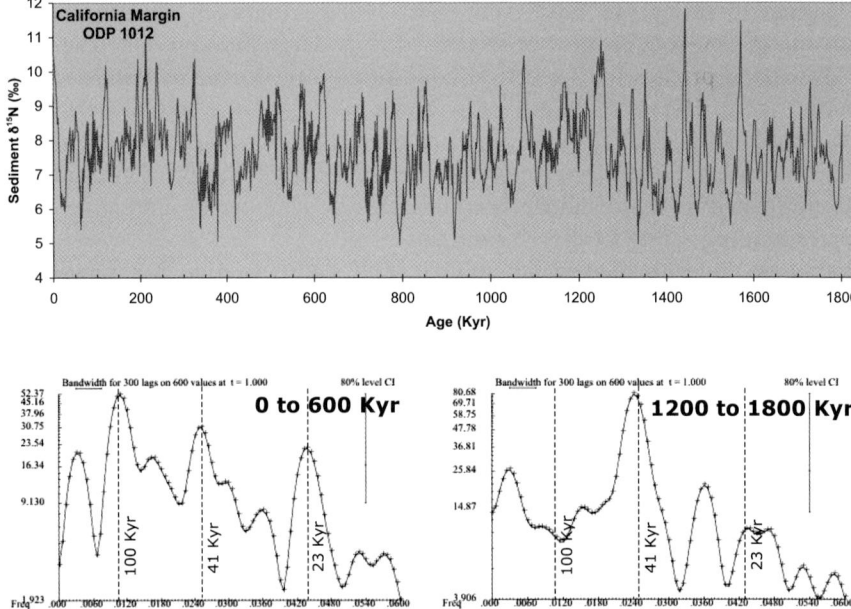

Fig. 14 Denitrification record for the ETNP over the last 1.8 Myr. ODP site 1012 is located on the southern California margin. Though this site is not properly within the ETNP, a coastal undercurrent that is the source of upwelled waters carries NO_3^- whose isotopic composition is influenced by ETNP denitrification. The inset shows spectral analysis results showing the change in major mode of variability from 41 to 100 kyr bp at the mid-Pleistocene transition about 600 kyr ago [Liu Z, Altabet MA, Herbert TD (in press) Geophys Res Lett]

we may expect differences in climate response as compared to the Arabian Sea and ETNP since this region is only one of the three located in the Southern Hemisphere. As elsewhere, margin sediments overlaid by low O_2 present the best targets due to excellent OM preservation and high accumulation. As in the ETNP, laminations and banding are indicative of periods of sufficiently low O_2 to exclude bioturbating organisms and present the potential for ultra-high temporal resolution studies. However, the Peru margin has presented a number of challenges for paleoceanographic research, the foremost of which are (1) difficulties in dating and age model construction due to paucity of foram tests preserved in the sediments and (2) the presence of hiatuses in most cores marked by phosphorite and/or course grain layers. We have largely overcome these problems by employing for nearby Peru margin sites an innovative dating technique based on ^{14}C dating of extracted and purified algal biomarker compounds [117]. While no single core available to us continuously spans the last 20 ka, the regional nature of the $\delta^{15}N$ signal allows for overlapping sections to be correlated to produce a composite

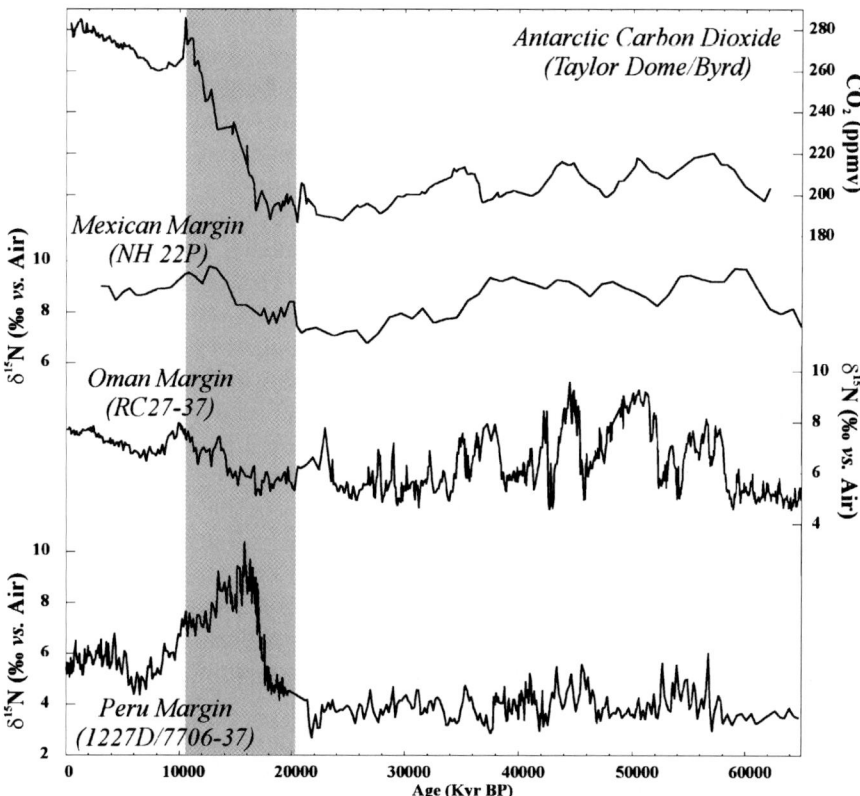

Fig. 15 Comparison of $\delta^{15}N$ records from the three major water column denitrification zones covering the last 60 kyr. Note the relatively low $\delta^{15}N$ values and absence of D-O events for the Peru Margin record during MIS 3 (30 to 60 ka bp) and the early and sharp rise during the early part of the last deglacial period (14 to 18 ka bp). The ice core record for atmospheric CO_2 is shown for [145]. The Mexican Margin data is from [109] and the Oman Margin data is from [108]

$\delta^{15}N$ record (Fig. 15). The chronology of the composite high-resolution $\delta^{15}N$ record is verified by comparison with a low-temporal resolution record from a low accumulation rate core from off margin with conventional dating based on foraminiferal ^{14}C and $\delta^{18}O$. We have extrapolated our chronology beyond our last dated interval near 25 ka to give a 1st order view of MIS 3 (25 to 60 ka before present).

Not surprisingly, large variations in $\delta^{15}N$ and hence denitrification are observed over the last 50 ka. As in the Arabian Sea and ETNP, upon deglaciation there is a dramatic ramp up in denitrification. However, the history of Peru denitrification departs from its northern hemisphere counterparts in several striking respects. First, the deglacial peak in denitrification occurs very early at between 15 and 16 ka, when most of the planet was still very much experiencing glacial conditions. We suspect that this is the result of Southern Hemisphere forcing of intermediate water ventilation. An important source for the intermediate waters of the Peru OMZ is Subantarctic Mode Water (SAMW) which is formed in the Subantarctic section of the Southern Ocean and a number a climate records from this and the polar region even further to the south suggest early initiation of the deglacial sequence (e.g. [118]). Second, MIS 3 (30 to 60 ka) shows overall low denitrification similar to LGM levels except for modest maxima that may be related to Antarctic Climate Events (ACE's) as observed in Antarctic ice cores. This again is in contrast to the Northern Hemisphere records which show near-Holocene values with D-O event variability further supporting distinct Southern Hemisphere forcing. Third, there is substantial late Holocene variability in denitrification with excursions in $\delta^{15}N$ almost as large as upon deglaciation occurring over centuries or less. Of the three denitrification regions, Peru is most impacted by ENSO (El Niño/La Niña) variability, and while individual ENSO events cannot be observed at our temporal resolution, changes in the overall frequency and amplitude would be. During El Niño, isopycnal surfaces are depressed off Peru such that upwelling brings nutrient-poorer water to the surface with lowered productivity and the OMZ resides more deeply. It would thus be expected that sinking particles would have reduced $\delta^{15}N$ during El Niño events, and the minima in the late Holocene record may correspond to periods where these are more frequent and more intense.

5.3
A Composite Denitrification Record

With good to high quality $\delta^{15}N$ records now available from each of the major water column denitrification zones, a first attempt can be made to assess the combined influence of these regions over the last 60 kyr. This is done by first normalizing the variability in each record selected about its average value and producing a 3-kyr smoothed composite of the three assuming each contributes one third of total water column denitrification for the time period considered. The result is found in Fig. 16 which shows low overall denitrification in MIS 3 and the LGM, a rapid rise to high late deglaciation and early Holocene values, and moderate levels during the mid- and late Holocene. When compared to atmospheric CO_2 as observed in ice cores, there are striking similarities including the 4 maxima of moderate intensity during MIS 3. Importantly, the deglacial rise in denitrification leads CO_2 by the approximate residence time for combined N in the ocean. That is, if changing denitrification is influencing atmospheric CO_2, it would do so by changing the oceanic N inventory with a lag on the order of its residence time of about 3 kyr.

The changes in water column denitrification described were proximally forced by changes in the intensity and extent of OMZ's. OMZ modulation in turn was forced by changes in organic carbon flux and/or intermediate water

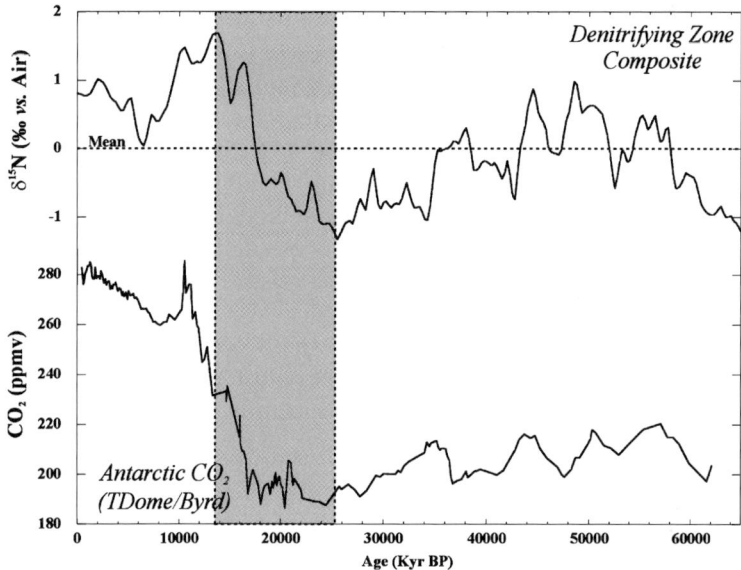

Fig. 16 Comparison of global CO_2 record [145] and composite global water column denitrification record. (see text for details)

ventilation and there has been discussion in the literature of the sensitivities of these to climate change phenomenon. Another consequence of OMZ modulation is likely change in oceanic N_2O flux to the atmosphere. As mentioned above, N_2O is an intermediate in the denitrification pathway as well as a by-product of nitrification. Its yield relative to NO_3^- production is negatively correlated with O_2 concentration such that the highest oceanic N_2O concentrations are found in oceanic OMZ's prior to the onset of denitrification [119]. It has been hypothesized that expansion/contraction of oceanic OMZ's would also correspond to changes in oceanic N_2O flux to the atmosphere. A correlation between atmospheric N_2O as recorded in ice cores and Arabian Sea denitrification has been cited as evidence [120]. N_2O also acts as a powerful greenhouse gas and would add to the proposed climate change effect from CO_2 changes brought about by changes in oceanic N inventory.

5.4
Toward Reconstruction of the Complete Oceanic N Budget

Sediment $\delta^{15}N$ records from the three water column denitrification zones have provided well-resolved histories for each of these regions. Though important, they still provide only an incomplete understanding of both past changes in marine N cycle and its modern dynamics. A wish list in this respect would include reconstruction of sedimentary denitrification, N_2 fixation, and average state of the marine N cycle.

Unfortunately, it is unlikely that $\delta^{15}N$ can be used to directly reconstruct sediment denitrification. It leaves little or know isotopic imprint on oceanic $\delta^{15}NO_3^-$ and it is widely distributed throughout all sediments with suboxic layers, particularly along continental margins. There is also no evidence for sedimentary denitrification influencing the $\delta^{15}N$ of sedimentary organic N which is the source for coupled nitrification–denitrification. Fortunately, inability to reconstruct sedimentary denitrification does not eliminate any hope of sufficiently complete marine N cycle reconstructions. Knowing water column denitrification, N_2 fixation, as well as relative changes in N inventory would be sufficient for this.

Any thorough reconstruction of the ocean's N cycle, though, does require assessment of any past changes in N_2 fixation. Just as denitrification imparts a regional "heavy" N isotopic signature to subsurface NO_3^- and ultimately to the underlying sediments, N_2 fixation should impart a "light" isotopic signature to the sediments below. Above, we showed evidence for the modern ocean of decreased $\delta^{15}NO_3^-$ which in turn influences the $\delta^{15}N$ of sinking particles. Unlike denitrification zones, N_2 fixation is found principally in oligotrophic subtropical gyres where the underlying sediments typically have poorly preserved OM as a result of low OM flux, great depth, and high bottom water O_2. Accordingly, sediment $\delta^{15}N$ has been modified upward by diagenesis. While $\delta^{15}N$ records of modest temporal resolution can be obtained in regions such as the Sargasso Sea (Fig. 17), we have little confidence in assuming the upward diagenetic shift has been constant over time. Nevertheless, the lower $\delta^{15}N$ values during the LGM suggest that N_2 fixation did not decrease to compensate for reduced denitrification. In fact, the broad features in this record are seen in a number of records from outside of both denitrifying and HNLC zones suggesting that a global response is recorded instead of local changes.

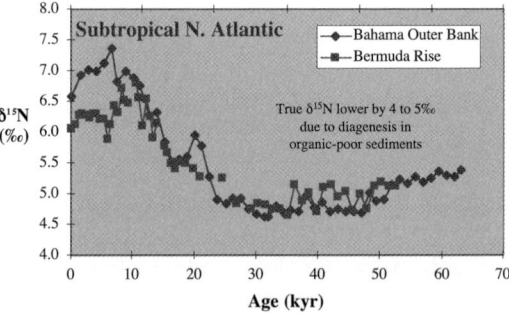

Fig. 17 Sediment $\delta^{15}N$ records from two locations in the Sargasso Sea. Organic N is only poorly preserved at these deep-sea, oligotrophic sites and $\delta^{15}N$ values are shifted upward by 4 to 6‰ from the primary signal. Nevertheless, the observation of lower glacial values is indicative that N_2 fixation in this region was not lower than at present

There are a number of avenues to overcome the diagenetic problem in N_2 fixation as well as other regions with poor sedimentary OM preservation. Just as in the Southern Ocean diatom microfossils were exploited as reservoirs of preserved OM [97], calcareous microfossils which are more abundant in N_2 fixation regions may prove to have similar utility. Assessments of diagenetic state independent of $\delta^{15}N$ may be used to "correct" downcore $\delta^{15}N$ data for variations in diagenetic influence.

5.5
Oceanic Wide Changes in $\delta^{15}N$ Revisited

Due to good OM preservation and not being located in either an HNLC nor denitrification region, sediment records for $\delta^{15}N$ from the South China Sea have been cited as evidence for little change in ocean average $\delta^{15}N$ over the last glacial cycle [121]. This has led to the conclusion that either N_2 fixation changes compensated for any change in denitrification and/or there was no change in the proportioning of denitrification between the sediment and the water column [36]. However, subsequent work in this area has shown that significant variations occurred in a nearby site and assigned them to possible changes in water mass source due to sea level change [122]. Another approach is to consider all available records from regions that are neither directly influenced by water column denitrification nor are within HNLC regions regardless of OM preservation to detect common trends (Table 1). While each of these records could individually be questioned as to the influence of local processes or diagenesis, in aggregate a consistent picture is apparent in which there is a 2 to 3‰ increase in $\delta^{15}N$ from the LGM to early Holocene and a subsequent 1 to 2‰ decrease to the present. Qualitatively, this supports a global oceanic impact from changes in water column denitrification; increasing (as a fraction of N_2 fixation) upon deglaciation with subsequent decrease perhaps due to a "slow" feedback mechanism. If Fig. 10 is used to estimate the magnitude of these changes, water column denitrification as a portion of the total N flux varied by up to ±15% between the LGM and present. Alternatively, N_2 fixation could have decreased upon deglaciation but this scenario is not supported by any of the water column denitrification records nor the putative $\delta^{15}N$ record from the Sargasso Sea (Fig. 17). These changes would have resulted in significant variations in oceanic combined N inventory. For example a 15% increase in denitrification relative to N_2 fixation over 5 kyr would decrease total ocean inventory by about 25%. Of course, this perspective assumes no insidious compensation by opposite and equal changes in sedimentary denitrification. In fact, it has been argued that sedimentary denitrification would only have increased upon deglaciation and the increased shelf sediment rise produced by sea level rise [27]. As discussed above, the major parameters needed for near complete reconstruction of past variations in the oceanic N budget are potentially obtainable.

Table 1 Summary of $\delta^{15}N$(‰) changes at oceanic sites which are neither within HNLC or denitrifying regions. Overall these results are indicative of change in global average marine $\delta^{15}N$. *Upper panel* from sites with a relatively high sediment accumulation rate (> 10 cm/kyr) and good N preservation. The *bottom panel* for sites with a low accumulation rate (< 5 cm/kyr) and sufficiently poor N preservation that diagenetic effects are apparent. Early to late Holocene refers to the period from approximately 10 ka b.p. to present. LGM (last glacial maximum) to early Holocene refers to the period from 20 to 10 kyr b.p.

High-resolution, good N preservation sites	Early to Late Holocene	
Cariaco Basin *	– 3	
S. China Sea [122]	– 1	
Gulf of Maine *	– 2	
Delmarva Slope *	– 0.8	
Carolina Slope *	– 2	

Low-resolution, poor N preservation sites	Early to Late Holocene	LGM to Early Holocene
Bermuda Rise +	– 1	2.5
Bahama Outer Bank +	– 1	3
Benguela [123]	– 0.5	2
Arctic Ocean [124]	– 1	3
N.W. Africa [125]	– 1	3
W. Equatorial Pacific [126]	– 1.5	3.5
Bering Sea [127]	– 1.5	1.5

* Altabet unpublished data, + Fig. 17

6
Summary and Synthesis

The major microbial transformation pathways within the marine N cycle have been well known for almost a century. Perception of a quasi-balanced biogeochemical system between sources and sinks was first made by Karl Brandt in 1902. Within the last three decades, serious and credible efforts have been made to quantify global ocean N fluxes and that with more detailed study estimates have almost been steadily revised upward. Particularly important were the respective discoveries of the importance of sedimentary denitrification and the dominance of oceanic N_2 fixation as a combined N source. In the last decade and a half, advances have been made in the application of geochemical tracers; objective analysis of NO_3^- anomalies and application of isotope natural abundance. It is not clear if there is modern balance between sources (primarily N_2 fixation) and losses (sedimentary and water denitrification) but total throughput is on the order of 200 to 400 Tg/yr with a residence time for marine combined N < 3000 yrs. This is relatively short compared to other major biogeochemical cycles such as for P or Si. The N cycle is further distinguished by having the major source for combined N within the ocean and all major fluxes controlled by biological transformations. P and Si sources to the ocean, in contrast, are in primarily the form of riverine fluxes. Paleoceanographic reconstructions have clearly demonstrated large variations in water column denitrification coupled to climate change on time scales from hundreds to millions of years. In particular, increasing denitrification across the last deglaciation likely caused a significant change in the oceanic N inventory that in turn may have influenced atmospheric CO_2 and N_2O.

Important challenges remain in understanding the present past oceanic N cycle. It appears surprising that anammox as a potentially important pathway has only been recently recognized. Related problems include a reexamination of the stoichiometry of NO_3^- removal to N_2 production in open ocean denitrification. Greater certainty in sedimentary denitrification may be achieved by applying complementary geochemical approaches analogous to the applications made to N_2 fixation and water column denitrification. Perhaps just as important as establishing well-constrained flux estimates is achieving a realistic understanding of the controls and feedbacks within the oceanic N cycle that in turn determine total inventory when in quasi-steady state and the system's capacity for imbalance and transition to perhaps other states. Such mechanistic understanding must be based on further study in the modern of the factors to which denitrification and N_2 fixation are in reality most sensitive as well as paleoceanographic reconstructions to determine which states have been realized and their forcings.

Acknowledgements The author thanks Peng Feng and Rehka Singh for technical assistance in generating much of the data presented, Matt Higginson for thoughtful discussions and data analysis, Joe Montoya, Dave Murray, and Tim Herbert for on-going collaborations. The US NSF has provided the bulk of the support for the author's research group.

References

1. Liu KK (1979) Geochemistry of inorganic nitrogen compounds in two marine environments; the Santa Barbara Basin and the ocean off Peru. Doctoral Thesis, University of California, Los Angeles
2. Capone DG, Zehr JP, Paerl HW, Bergman B, Carpenter EJ (1997) Science 276:1221
3. Karl D, Michaels A, Bergman B, Capone D, Carpenter E, Letelier R, Lipschultz F, Paerl H, Stal L (2002) Biogeochemistry 57–58:47
4. Raven JA (1988) New Phytol 109:279
5. Kustka A, Sanudo-Wilhelmy S, Carpenter EJ, Capone DG, Raven JA (2003) J Phycol 39:12
6. Falkowski PG (1997) Nature 387:272

7. Sanudo-Wilhelmy SA, Kustka AB, Gobler CJ, Hutchins DA, Yang M, Lwiza K, Burns J, Capone DG, Raven JA, Carpenter EJ (2001) Nature 411:66
8. Mills MM, Ridame C, Davey M, La Roche J, Geider RJ (2004) Nature 429:292
9. Berman-Frank I, Cullen JT, Shaked Y, Sherrell RM, Falkowski PG (2001) Limnol Oceanogr 46:1249
10. Redfield A (1934) James Johnstone Memorial Volume, p 176
11. Gruber N, Sarmiento JL (1997) Glob Biogeochem Cycle 11:235
12. Deutsch C, Gruber N, Key RM, Sarmiento JL, Ganachaud A (2001) Global Biogeochem Cycle 15:483
13. Hansell DA, Bates NR, Olson DB (2004) Mar Chem 84:243
14. Bates NR, Hansell DA (2004) Mar Chem 84:225
15. Macko SA, Fogel ML, Hare PE, Hoering TC (1987) Chem Geol 65:79
16. Hoering T, Ford HT (1960) J Am Chem Soc 82:376
17. Delwiche CC, Zinke PJ, Johnson CM, Virginia RA (1979) Bot Gaz 140 (Suppl):s65
18. Mariotti A, Mariotti F, Amarger N, Pizelle G, Ngambi J-M, Champigny M-L, Moyse A (1980) Physiol Veg 18:163
19. Minagawa M, Wada E (1986) Mar Chem 19:245
20. Carpenter EJ, Harvey HR, Fry B, Capone DG (1997) Deep-Sea Res I 44:27
21. Sigman DM, Altabet MA, McCorkle DC, Francois R, Fischer G (2000) J Geophys Res-Oceans 105:19599
22. Sigman DM, Altabet MA, Michener R, McCorkle DC, Fry B, Holmes RM (1997) Mar Chem 57:227
23. Sigman DM, Casciotti KL, Andreani M, Barford C, Galanter M, Bohlke JK (2001) Anal Chem 73:4145
24. Casciotti KL, Sigman DM, Hastings MG, Bohlke JK, Hilkert A (2002) Anal Chem 74:4905
25. Richards FA (1965) In: Riley JP, Skirrow G (eds) Chemical Oceanography. Academic Press, London, p 611
26. Cohen Y, Gordon LI (1978) Deep-Sea Res 25:509
27. Christensen JP, Murray JW, Devol AH, Codispoti LA (1987) Glob Biogeochem Cycle 1:97
28. Deuser WG, Ross EH, Mlodzinska ZJ (1978) Deep Sea Res 25:431
29. Codispoti LA, Richards FA (1976) Limnol Oceanogr 21:379
30. Codispoti LA, Packard TT (1980) J Mar Res 38:453
31. Kuypers MMM, Sliekers AO, Lavik G, Schmid M, Jorgensen BB, Kuenen JG, Sinninghe Damsté JS, Strous M, Jetten MSM (2003) Nature 422:608
32. Dalsgaard T, Canfield DE, Petersen J, Thamdrup B, Acuna-Gonzalez J (2003) Nature 422:606
33. Codispoti L, Brandes J, Christensen J, Devol A, Naqvi S, Paerl H, Yoshinari T (2001) In: Gili J, Pretus J, Packard T (eds) A Marine Science Odyssey into the 21st Century (Scientia Marina), vol 65 (Suppl 2), p 85
34. Christensen JP, Smethie JW, Devol AH (1987) Deep-Sea Res 34:1027
35. Devol AH (1991) Nature 349:319
36. Brandes JA, Devol AH (2002) Glob Biogeochem Cycle 16:67
37. Miyake Y, Wada E (1971) Rec Oceanog Works Japan 11:1
38. Wellman RP, Cook FD, Krouse HR (1968) Science 161:269
39. Cline JD, Kaplan IR (1975) Mar Chem 3:271
40. Barford CC, Montoya JP, Altabet MA, Mitchell R (1999) Appl Environ Microbiol 65:989
41. Brandes JA, Devol AH, Yoshinari T, Jayakumar DA, Naqvi SWA (1998) Limnol Oceanogr 43:1680

42. Altabet MA, Murray DW, Prell WL (1999) Paleoceanography 14:732
43. Voss M, Dippner JW, Montoya JP (2001) Deep Sea Res II 48:1905
44. Brandes JA, Devol AH (1997) Geochim Cosmochim Acta 61:1793
45. Sigman DM, Robinson R, Knapp AN, Geen AV, McCorkle DC, Brandes JA, Thunell RC (2003) Geochem Geophys Geosyst 4:1
46. Wada E, Hattori A (1978) Geomicrobio J 1:85
47. Wada E (1980) In: Goldberg ED, Horibe Y (eds) Isotope Marine Chemistry. Uchida Rokakuho, Tokyo, p 375
48. Montoya JP, McCarthy JJ (1995) J Plank Res 17:439
49. Waser NA, Yin K, Yu Z, Tada K, Harrison PJ, Turpin DH, Calvert SE (1998) Mar Ecol Prog Ser 169:29
50. Needoba JA, Sigman DM, Harrison PJ (2004) J Phycol 40:517
51. Needoba JA, Harrison PJ (2004) J Phycol 40:505
52. Martin JH, Gordon RM, Fitzwater S, Broenkow WW (1989) Deep-Sea Res 36:649
53. Sigman DM, Boyle EA (2000) Nature 407:859
54. Chavez FP, Buck KR, Coale KH, Martin JH, DiTullio GR, Welschmeyer NA, Jacobson AC, Barber RT (1991) Limnol Oceanogr 36:1816
55. Sigman DM, Altabet MA, McCorkle DC, Francois R, Fischer G (1999) Glob Biogeochem Cycle 13:1149
56. Altabet MA, Francois R (2001) Deep-Sea Res II 48:4247
57. Checkley DM Jr, Entzeroth LC (1985) J Plank Res 7:553
58. Checkley DM Jr, Miller CA (1989) Deep-Sea Res 36:1449
59. DeNiro MJ, Epstein S (1981) Geochim Cosmochim Acta 45:341
60. Minagawa M, Wada E (1984) Geochim Cosmochim Acta 48:1135
61. Macko SA, Fogel Estep ML, Engel MH, Hare PE (1986) Geochim Cosmochim Acta 50:2143
62. McClelland JW, Montoya JP (2002) Ecology 83:2173
63. Saino T, Hattori A (1985) In: Sigleo AC, Hattori A (eds) Marine and Estuarine Geochemistry. Lewis Publishers, Chelsea, MI, p 697
64. Altabet MA (1988) Deep-Sea Res 35:535
65. Altabet MA (2001) Limnol Oceanogr 46:368
66. Altabet MA (1989) Limnol Oceanogr 24:1185
67. Altabet MA, Deuser WG, Honjo S, Stienen C (1991) Nature 354:136
68. Sweeney RE, Kaplan IR (1980) Mar Chem 9:81
69. Macko, Estep (1984) Org Geochem 6:787
70. Lehmann MF, Bernasconi SM, Barbieri A, McKenzie JA (2002) Geochim Cosmochim Acta 66:3573
71. Kendall C (1998) In: Kendall C, McDonnell JJ (eds) Isotope Tracers in Catchment Hydrology. Elsevier, Amsterdam, p 519
72. Lehmann MF, Sigman DM, Berelson WM (2004) Mar Chem 88:1
73. Yoshida N (1988) Nature 335:528
74. Cifuentes LA, Fogel ML, Pennock JR, Sharp JH (1989) Geochim Cosmochim Acta 53:2713
75. Horrigan SG, Montoya JP, Nevins JL, McCarthy JJ (1990) Est Coast Shelf Sci 30:393
76. Casciotti KL, Sigman DM, Ward BB (2003) 20:335
77. Goreau TJ, Kaplan WA, Wofsy SC, McElroy MB, Valois FW, Watson SW (1980) Appl Env Microbiol 40:526
78. Lipschultz F, Zafiriou OC, Wofsy SC, McElroy MB, Valois FW, Watson SW (1981) Nature 294:41
79. Nevison CD, Weiss RF, Erickson DJ (1995) J Geophys Res—Oceans 100:15809

80. Suntharalingam P, Sarmiento JL, Toggweiler JR (2000) Glob Biogeochem Cycle 14:353
81. Yoshinari T, Altabet MA, Naqvi SWA, Codispoti L, Jayakumar A, Kuhland M, Devol A (1997) Mar Chem 56:253
82. Popp BN, Westley MB, Toyoda S, Miwa T, Dore JE, Yoshida N, Rust TM, Sansone FJ, Russ ME, Ostrom NE, Ostrom PH (2002) Glob Biogeochem Cycle 16:12
83. Yoshida N, Toyoda S (2000) Nature 405:330
84. Goldman JC (1979) Microb Ecol 5:153
85. Goldman JC, McCarthy JJ (1978) Limnol Oceanogr 23:695
86. Quigg A, Finkel ZV, Irwin AJ, Rosenthal Y, Ho T, Reinfelder JR, Schofield O, Morel FMM, Falkowski PG (2003) Nature 425:291
87. Klausmeier CA, Litchman E, Daufresne T, Levin SA (2004) Nature 429:171
88. White AE, Letelier RM, Spitz YH (2003) J Phycol 39:59
89. Villareal T, Carpenter E (2003) Microb Ecol 45:1
90. Broecker WS (1982) Prog Oceanogr 11:151
91. Petit JR, Jouzel J, Raynaud D, Barkov NI, Barnola JM, Basile I, Beders M, Chappellaz J, Davis M, Delaygue G, Delmotte D, Kotlyakov VM, Legrand M, Lipenkov VY, Lorius C, Pepin L, Ritz C, Saltzman E, Stievenard M (1999) Nature 399:429
92. Knox F, McElroy M (1984) J Geophys Res 89:4629
93. Sarmiento J, Toggweiler R (1984) Nature 308:621
94. Sieganthaler, Wenk (1984) Nature 308:624
95. Martin JH (1990) Paleoceanogr 5:1
96. Francois R, Altabet MA, Yu E-F, Sigman D, Bacon MP, Frank M, Bohrmann G, Barielle G, Labeyrie L (1997) Nature 389:929
97. Sigman DM, Altabet MA, Francois R, McCorkle DC, Gaillard JF (1999) Paleoceanogr 14:118
98. McElroy M (1983) Nature 302:328
99. Anschutz P, Sundby B, Lefrancois L, Luther GW, Mucci A (2000) Geochim Cosmochim Acta 64:751
100. Altabet MA, Curry WB (1989) Glob Biogeochem Cycle 3:107
101. Eppley RW, Peterson BJ (1979) Nature 282:677
102. Altabet MA, McCarthy JJ (1985) Deep-Sea Res 32:755
103. Crosta X, Shemesh A (2002) Palaeoceanogr 17:10
104. Sachs JP, Repeta DJ, Goericke R (1999) Geochim Cosmochim Acta 63:1431
105. Sachs JP, Repeta DJ (1999) Science 286:2485
106. Keil RG, Fogel ML (2001) Limnol Oceanogr 46:14
107. Altabet MA, Francois R, Murray DW, Prell WL (1995) Nature 373:506
108. Altabet M, Higginson M, Murray DW (2002) Nature 415:159
109. Ganeshram RS, Pedersen TF, Calvert SE, Murray JW (1995) Nature 376:755
110. Ganeshram RS, Pedersen TF, Calvert SE, McNeill GW, Fontugne MR (2000) Paleoceanogr 15:36
111. Kienast SS, Calvert SE, Pedersen TF (2002) Paleoceanogr 17:1055
112. Dansgaard W, Clausen HB, Gundestrup N, Hammer CU, Johnsen SF, Kristinsdottir PM, Reeh N (1982) Science 218:1273
113. Higginson MJ, Altabet MA, Murray DW, Murray RW, Herbert TD (2004) Geochim Cosmochim Acta 68:3807–3826, DOI: 10.1016/j.gca.2004.03.015
114. Morford JL, Emerson S (1999) Geochim Cosmochim Acta 63:1735
115. Hendy IL, Kennett JP (2003) Quat Sci Rev 22:673
116. Raymo ME, Nisancioglu K (2003) Paleoceanogr. 18:1011
117. Ohkouchi N, Eglinton TI, Keigwin LD, Hayes JM (2002) Science 298:1224

118. Kanfoush SL, Hodell DA, Charles CD, Janecek TR, Rack FR (2003) Palaeogeogr Palaeoclim Palaeoecol 182:329
119. Codispoti LA, Elkins JW, Yoshinari T, Friederich GE, Sakamoto CM, Packard TT (1992) In: Desai BN (ed) Oceanography of the Indian Ocean. Oxford and IBH Publishing Co., New Dehli, p 271
120. Suthhof A, Ittekkot V, Gaye-Haake B (2001) Glob Biogeochem Cycle 15:637
121. Kienast M (2000) Paleoceanogr 15:244
122. Higginson MJ, Maxwell JR, Altabet MA (2003) Mar Geol 201:223
123. Holmes ME, Schneider RR, Muller PJ, Segl M, Wefer G (1997) Paleoceanogr 12:604
124. Schubert CJ, Stein R, Calvert SE (2001) Paleoceanogr 16:199
125. Martinez P, Bertrand P, Calvert S, Pedersen T, Shimmield G, Lallier-Verges E, Fontugne M (2000) J Mar Res 58:809
126. Nakatsuka T, Harada N, Matsumoto E, Handa N, Oba T, Ikehara M, Matsuoka H, Kimoto K (1995) Geophys Res Lett 22:2525
127. Nakatsuka T, Watanabe K, Handa N, Matsumoto E, Wada E (1995) Paleoceanogr 10:1047
128. Nelson JR, Beers JR, Eppley RW, Jackson GA, McCarthy JJ, Soutar A (1987) Cont Shelf Res 7:307
129. Haug GH, Pedersen TF, Sigman DM, Calvert SE, Nielsen B, Peterson LC (1998) Paleoceanogr 13:427
130. Altabet MA, Pilskaln C, Thunell R, Pride C, Sigman D, Chavez F, Francois R (1999) Deep-Sea Res 46:655
131. Voss M, Altabet MA, Bodungen BV (1996) Deep-Sea Res II 43:33
132. Velinsky DJ, Fogel MF (1999) Mar Chem 67:3
133. Holmes ME, Muller PJ, Schneider RR, Segl M, Patzold J, Wefer G (1996) Mar Geol 134:1
134. Thunell RC, Kepple AB (2004) Glob Biogeochem Cycle 18:1001
135. Nakatsuka T, Handa N (1997) J Oceanogr 53:105
136. Pantoja S, Repeta DJ, Sachs JP, Sigman DM (2002) Deep-Sea Res I 49:1609
137. Antia AN, Maassen J, Herman P, Voss M, Scholten J, Groom S, Miller P (2001) Deep-Sea Res II 48:14
138. Kerherve P, Minagawa M, Heussner S, Monaco A (2001) Oceanol Acta 24S:S77
139. Karl D, Letelier R, Tupas L, Dore J, Christian J, Hebel D (1997) Nature 388:533
140. Fry B, Jannasch HW, Molyneaux SJ, Wirsen CO, Muramoto JA, King S (1991) Deep-Sea Res A 38:S1003
141. Liu K-K, Kaplan IR (1989) Limnol Oceanogr 34:820
142. Smith SL, Henrichs SM, Rho T (2002) Deep-Sea Res II 49:6031
143. Wong GTF, Chung S-W, Shiah F-K, Chen C-C, Wen L-S, Liu K-K (2002) Geophys Res Lett, 29:1029
144. Robinson RS, Brunelle BG, Sigman DM (2004) Paleoceanogr 19:3001
145. Indermuhle A, Monnin E, Stauffer B, Stocker TF, Wahlen M (2000) Geophys Res Lett 27:735

Degradation and Preservation of Organic Matter in Marine Sediments

Stuart G. Wakeham[1] (✉) · Elizabeth A. Canuel[2]

[1] Skidaway Institute of Oceanography, 10 Ocean Science Circle, Savannah, GA 31411, USA
stuart@skio.peachnet.edu

[2] Virginia Institute of Marine Science, The College of William and Mary, P.O. Box 1346, Gloucester Point, VA 23062, USA
ecanuel@vims.edu

1	Introduction	296
2	Substrate Character	297
3	Matrix Effects	301
4	Oxygen/Redox Control of Aerobic and Anaerobic Degradation	306
5	Sediment Mixing Regime	312
6	Concluding Remarks	316
	References	318

Abstract Organic matter that is deposited in aquatic sediments is subject to an intense diagenetic reactor that determines how much organic carbon is eventually preserved in sediments. The balance between organic matter degradation and preservation has immense consequences for the global carbon and oxygen cycles. A diverse set of hypotheses regarding the controls on organic matter degradation/preservation have received considerable attention over the past decade, most often revolving around the relative roles of bottom water and pore water oxygen and the rate of organic matter delivery to the sediments. These overriding hypotheses have in turn spawned numerous other hypotheses on specific topics. In this review, we discuss four important controls that impact on the degradation and subsequent preservation of organic matter in aquatic sediments. Our focus areas are: (1) the chemical nature of the organic substrate; (2) the potential influence of matrix on preservation; (3) the role of redox effects in degradation; and (4) the effects of physical mixing of sediments. Although we have divided our discussion under these headings, it will immediately become apparent that these subsections are at best arbitrary and that the four factors are indeed intimately related.

Keywords Diagenesis · Organic carbon · Organic carbon preservation · Organic carbon degradation · Redox oscillation · Co-metabolism

1
Introduction

Aquatic sediments serve as an intense reactor through which organic matter moves from the overlying water column toward sedimentary rocks [1]. The reactions taking place are largely mediated by sedimentary microorganisms that efficiently degrade $\sim 99\%$ of the organic matter that rains down onto the water/sediment interface in open ocean settings. Ultimately, only $\sim 1\%$ of this organic rain is preserved in underlying deep-ocean sediments to become part of the sedimentary record. Burial efficiencies in continental margin sediments may be substantially greater, and in some cases up to 40% of the input flux may be preserved. The consequences of this efficient reactor are profound. The organic matter that is eventually preserved is the source of fossil fuels and provides insight into the Earth's history. The balance between loss by remineralization, preservation by burial, and weathering of uplifted kerogen-containing sedimentary rocks inextricably links the global carbon, oxygen, and sulfur cycles [2-4]. Achieving a better understanding of the fate of organic matter during early diagenesis is also of practical importance, because of the use of biomarkers and other proxies in paleoenvironmental studies and the reconstruction of past environmental changes.

It is these global implications that drive the need to understand the biogeochemical processes that determine the character and quantity of organic matter that is either degraded or preserved. Over the past decade, theories regarding the dominant control(s) on organic matter preservation in marine sediments have revolved around the competing roles of water column production and organic matter delivery to the sediment versus bottom water oxygen content [5-12]. Related factors include organic matter source, molecular character and selective preservation of recalcitrant molecules [13-15], sediment accumulation rate [16], effects of bioturbation [17], oscillating redox conditions [18, 19], oxygen exposure time [20, 21], microbial dynamics [22-24], sorptive preservation on mineral surfaces [4, 25, 26], and protective encapsulation within macromolecular organic matrices [27, 28].

In this chapter, we review four important controls on organic matter degradation and preservation in marine sediments, building on concepts developed in the past and using new results to refine the ideas and theories that have been put forward. These four focus areas are: (1) substrate character; (2) matrix effects; (3) redox controls; and (4) sediment mixing regime. Although we have divided our discussion under these four headings, they are by no means the only important factors that may come into play nor are they mutually exclusive. In fact, as we will show, these four factors are closely interrelated.

2
Substrate Character

The classic "multi-G" model of Berner and coworkers for organic matter degradation in sediments [29-31] describes sedimentary organic matter as composed of many fractions, each with different susceptibilities to degradation. Implicit in this model is that each type of organic matter degrades independently of both other types of organic matter and the overall metabolic activity of the sediment. Middelburg [32] and Boudreau and Ruddick [33] have developed "continuous multi-G" models that have a continuous spectrum of G-types and a continuum of rate constants. In a further refinement, Canfield [34] described a "pseudo-G" model in which the metabolic activity of the sediment was controlled by degradation of the most labile fraction. In this model, two or more types of organic matter are present, and once the most labile fraction is consumed, the next most reactive form controls the overall metabolic activity of the sediment.

Degradation of the more refractory components is linked to the decay of the labile components, and high overall metabolic activity enhances the decomposition of refractory organic matter. This linked "co-metabolism" results from a relationship between the degradation of refractory organic matter and sediment metabolic activity [23, 35], where some metabolic activity in highly microbially active sediments is channeled into the oxidation of compounds that on their own would be resistant to decay. Ultimately, the key to both aerobic and anaerobic decomposition is the nature of the organic substrate [34]. In fact, Canuel and Martens [36] developed an approach for determining in situ decomposition rates by following the behavior of individual compounds within parcels of sediment of known age, rather than relying on down-core profiles obtained from a sediment core. Their analysis provides convincing evidence that organic matter reactivity changes with time (and burial), as apparent decomposition rates were substantially higher at the sediment surface than in deeper horizons (Fig. 1).

That there is a continuum of reactivities for organic matter comes as no surprise to organic geochemists. Polysaccharide components of vascular plants have been shown to be degraded two to five times faster than lignin components of vascular plants [37]. Similarly, in a study of the comparative geochemistries of lignins and carbohydrates in an anoxic fjord, Hamilton and Hedges [38] showed that neutral sugars were consistently the most reactive class. Among early studies of lipid biomarker distributions in sediments were observations that compounds displayed a range of stabilities. For instance, Cranwell [39] reported that reductions in abundance for various lipids indicated an order of stability: n-alkanes > alkan-2-ones > sterols > n-alkanoic acids > n-alkanols > n-alkenoic acids, and that within classes, shorter-chained components apparently were lost more rapidly than longer-chained ones. In part, variability in degradation rate may be due to mo-

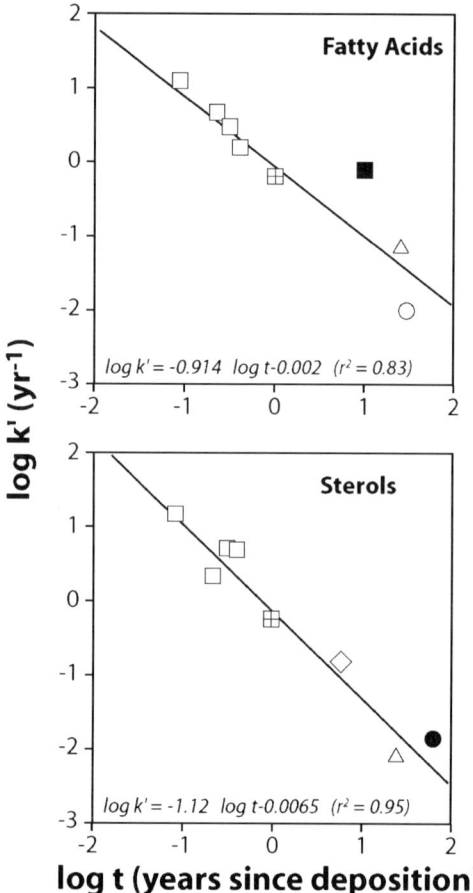

Fig. 1 Rate constants (k') for total fatty acids and total sterols for surficial sediments as a function of time since deposition. Key: ◇ Peru; •, ○ Buzzards Bay; △ Black Sea; □, ⊞ Cape Lookout Bight. After Canuel and Martens [36]

lecular structural features, i.e., short-chain lipids are more reactive than long chain lipids, unsaturated bonds are more reactive than saturated ones, and numerous recent studies confirm these trends in lipid reactivity [36, 40–44].

Arnosti [45–47] has studied polysaccharide hydrolysis and demonstrated that rates of extracellular enzyme hydrolysis vary considerably as a function of polysaccharide substrate (Fig. 2), with differences resulting from a mismatch between substrate structure and extracellular enzyme availability and activity for hydrolysis steps. Similarly, among major biochemical classes, such as amino acids and carbohydrates, differential degradation is common. Harvey et al. [48], for example, conducted laboratory experiments to evaluate the decomposition of algal organic matter and found carbohydrates to be more reactive than protein under oxic conditions, but the reverse under anoxic con-

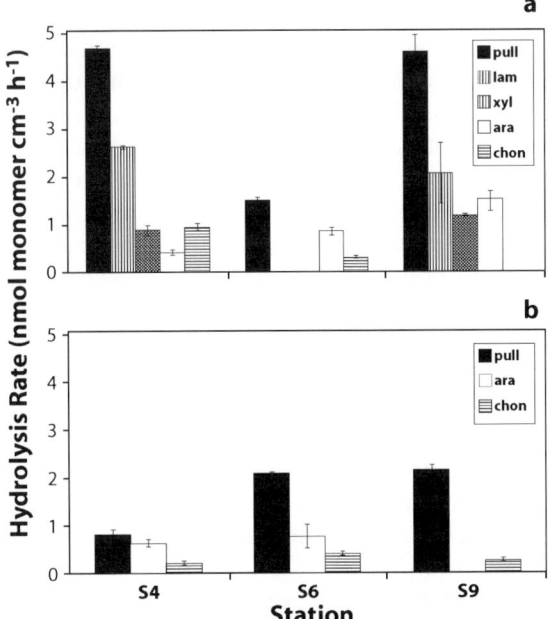

Fig. 2 a Hydrolysis of polysaccharides in homogenized surface (0–3 cm: S4; 0–1 cm: S6 and S9) sediments. **b** Hydrolysis of substrates in homogenized subsurface (3–6 cm) sediments. Pull – pullulan; lam – laminarin; xyl – xylan; ara – arabinogalactan; chon – chondritin sulfate. Error bars are for triplicate incubations. After Arnosti and Holmer [47]

ditions. Lignin is remarkably stable in sediments because there are relatively few enzymes produced by aquatic organisms that are capable of hydrolyzing the lignin macromolecule [49–51], but it can in fact be degraded under both oxic and anoxic conditions (review by Gough et al. [52]). These diverse studies point out that highly disparate views of organic matter degradation may arise, depending on what substrates are being examined and the environment in which they are being studied.

Despite numerous laboratory simulations and measurements in natural settings showing that individual (or specific) organic matter classes and compounds behave differently toward degradation, susceptibility to diagenetic alteration is clearly not related to molecular structure alone. Concentrations of compounds susceptible to degradation and total organic carbon often never drop to zero in sediments; organic molecules of identical structure often occur in both labile and relatively refractory forms (e.g., extractable and bound). It is thus likely that environmental conditions and/or protective matrices must be involved in determining the fate of organic matter. These physically protected forms may be relatively rare in fresh, undegraded organic material, and/or they may be concentrated in geochemical samples as the bulk and more labile organic substrates are extensively and preferentially degraded.

Since long carbon chain lipids tend to be derived largely from terrigenous vascular plant tissues whereas short-chained compounds generally originate from algae and bacteria, and since allochthonous compounds appear more refractory during diagenesis than autochthonous lipids, carbon chain length has been widely used to distinguish between allochthonous and autochthonous sources. Terrigenous compounds are generally considered to be more refractory, and thus better preserved, than algal compounds, based on changes in relative abundance in sediments [11, 36, 44, 53–55]. But is reactivity a function of molecular structure, or is it a function of differential packaging?

In an experiment by Reiley et al. [56], lipids of the vascular plant, *Fagus salvatica*, were found to be more resistant than lipids of the alga, *Isochrysis galbana*. Potentially, the difference observed by Reiley and coworkers may be due to differences in cellular and structural materials in their susceptibility to degradation, and/or different cellular matrices for vascular plants and this alga. As long as bacterial lipids remain associated with the membranes of bacterial cells, their constituent fatty acids are protected from degradation, but once the cells die and are subject to disruption, autolysis and further decomposition are rapid [57, 58]. Individual lipids common to two marine phytoplankton, the diatom *Thalassiosira weissflogii* and the cyanobacterium *Synechococcus* sp., showed different patterns of decay in a decomposition experiment, suggesting that factors other than molecular structure might be active [41]. Structural polysaccharides are less subject to diagenetic decomposition and are thus preferentially preserved compared to quickly degraded storage polysaccharides [59]. In addition, when allochthonous materials are delivered to aquatic environments, they may be sorbed to clays or sediment particles, providing additional protection from degradation (see below).

Over a decade ago, Tegelaar et al. [13] reappraised the processes involved in the formation of kerogen. In the condensation/humification scenario [60], simple biochemicals, generated by hydrolysis of complex substances, abiotically condense to produce complex assemblages (Fig. 3) that are difficult to define structurally [15]. Recent evidence [61] continues to indicate that some refractory sedimentary organic matter with a melanoidin-type structure is indeed formed by a degradation–recondensation of products derived mainly from polysaccharide and proteinaceous material. Intermolecular incorporation of inorganic sulfur with functionalized lipids [62] leads to complex and biologically resistant macromolecular material; this mechanism is still the subject of intense research [63]. An alternate theory, the preferential preservation mechanism, relies on the preservation of abundant hydrolysis-resistant biomacromolecules that are now known to be present in vascular plants and some algae [13, 14, 64] and that can be traced into sediments and kerogens [65, 66]. Hydrolysis-resistant biomacromolecules including algaenans, sporopollenins, cutans, suberans, and lignin, among others (see review by de Leeuw and Largeau [14]) are highly cross-linked and highly aliphatic in nature, often being associated with cell wall and/or structural organelles

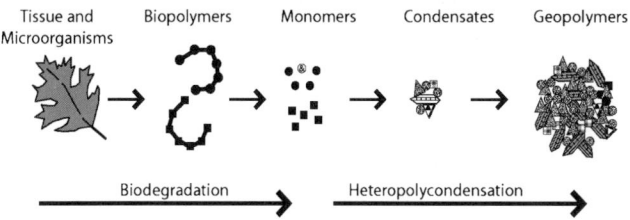

Fig. 3 Illustrations of the conventional biodegradation/repolymerization model (*top*) and the alternate biodegradation/sorption model for organic matter diagenesis and preservation (*bottom*). After Hedges and Keil [111]

of plants. Preferential preservation of these biomacromolecules results from their surviving microbial decomposition during early diagenesis (Table 1).

3
Matrix Effects

Organic matter is associated with mineral particles [67–69] and this association slows decomposition [70–72]. Distributions of "free" (released by solvent extraction) and "bound" (released by some hydrolysis step) compounds are often different [73–75], suggesting that this association is not equal for all molecular structures. Suess [76] observed a correlation between mineral surface area and organic carbon content of calcite-rich marine sedimentary particles. He found that the organic carbon (OC) loading per unit of surface

Table 1 Inventory of presently known biomacromolecules, their occurrence in extant organisms, and their potential for survival during sedimentation and diagenesis[a]

Biomacromolecules	Occurrence	"Preservation potential"[b]
Starch	Vascular plants; some algae; bacteria	–
Glycogen	Animals	–
Fructans	Vascular plants; algae; bacteria	–
Laminarans	Mainly brown algae; some other algae and fungi	–
Poly-β-hydroxyalkanoates (PHA)	Eubacteria	–
Cellulose	Vascular plants; some fungi	–/+
Xylans	Vascular plants; some algae	–/+
Pectins	Vascular plants	–/+
Mannans	Vascular plants; fungi; algae	–/+
Galactans	Vascular plants; algae	–/+
Mucilages	Vascular plants; (seeds)	+
Gums	Vascular plants	+
Alginic acids	Brown algae	–/+
Fungal glucans	Fungi	+
Dextrans	Eubacteria; fungi	+
Xanthans	Eubacteria	+
Chitin	Anthropods; copepods; crustacea; fungi; algae	+
Glycosaminoglycans	Mammals; some fish; Eubacteria	–/+
Proteins	All organisms	–/+
Extensin	Vascular plants; algae	–/+
Mureins	Eubacteria	+
Teichoic acids	Gram-positive Eubacteria	+
Teichuronic acids	Gram-positive Eubacteria	+
Lipoteichoic acids (LTA)	Gram-positive Eubacteria	+
Bacterial lipopolysaccharides (LPS)	Gram-negative Eubacteria	++
DNA, RNA	All organisms	–
Glycolipids	Plants; algae; Eubacteria	+/++
Polyisoprenols (rubber and gutta)	Vascular plants	+
Polyprenols and dolichols	Vascular plants; bacteria; animals	+
Resinous polyterpenoids	Vascular plants	+/++

[a] After de Leeuw and Largeau [14]
[b] Preservation potential ranges from – (extensive degradation under depositional conditions) to ++++ (no degradation under any depositional conditions)

Table 1 continued

Biomacromolecules	Occurrence	"Preservation potential"[b]
Cutins, suberins	Vascular plants	+/++
Lignins	Vascular plants	++++
Tannins	Vascular plants; algae	+++/++++
Sporopollenins	Vascular plants	+++
Algaenans	Algae	+++
Cutans	Vascular plants	++++
Suberans	Vascular plants	++++
Cyanobacterial sheaths	Cyanobacteria	+

[a] After de Leeuw and Largeau [14]
[b] Preservation potential ranges from – (extensive degradation under depositional conditions) to ++++ (no degradation under any depositional conditions)

area was similar to that for single layers of protein associated with interfaces, and suggested that the calcite-rich sediments under study consisted of highly irregular particle surfaces. The significance of OC–mineral associations was extended by Mayer [25, 26], who reported a widespread relationship between OC concentration and mineral surface area that approximated a monolayer of adsorbed OC on mineral surfaces (a "monolayer equivalent"). This led to the hypothesis that there was a surface area control on the stabilization and burial of OC in sediments, especially those on continental shelves. Further work led to a refined hypothesis, that OC saturates adsorption sites within small pores ("mesopores") on mineral surfaces that are small enough to exclude hydrolytic enzymes and hence protect otherwise intrinsically labile organic matter against biological attack. Several subsequent studies [77–81] support aspects of these hypotheses.

Mayer's "sorptive preservation" hypotheses [25, 26] have been the subject of considerable testing. Hedges and Keil [4] synthesized the early evidence in favor of the sorptive preservation hypothesis (Fig. 3). However, recent microscopic analyses [81] showed that organic matter distributions on mineral surfaces were patchy, discrete, and discontinuous rather than the continuous distributions that the monolayer equivalent hypothesis would imply. This study revealed that the vast bulk of OC in sediments is not in direct contact with the mineral surface, and that more attention needs to be directed toward understanding the relationships between mineralogy and surface area [82]. Bock and Mayer [83] determined pore size distributions of organic–clay aggregates and found most surface area to be within small mesopores (< 10 nm in width) that consist of interparticle slitlike spaces between clay grains rather than intraparticle dissolution features. The implication of this observation is that the formation of these aggregates involves an organic "glue" rather than a physical adsorption. Observations such as these have led Mayer [84] to recast the sorptive preservation hypothesis, recognizing that most organic matter is not adsorbed in a monolayer and that mineral surfaces are largely uncoated, with the result that most sediments in the ocean are actually naked aluminosilicate surfaces.

There is general agreement that a continuum of reactivity exists based on the chemical structure of the organic substrate, and that this continuum can be altered by interactions with minerals which can stabilize labile organic matter [85], leading to the well-established correlation between organic carbon content and mineral surface area (Fig. 4). But what are the molecular implications of organic–mineral associations? Compositional differences between sediment size and density fractions are well known [77, 79, 80], but more work is needed in order to characterize the relative lability of specific organic substances associated with these different fractions. Organic matter that is incorporated into silicate and carbonate tests during biological deposition of these minerals is better preserved than cellular organic matter: mineral-bound amino acids are well protected from diagenesis and remain relatively unaltered chemically [86, 87] compared to cellular amino acids. In a study of opal-rich Southern Ocean sediments, Ingalls et al. [88] showed that the proportion of silica-bound amino acids increased significantly with increasing depth in the sediments, reaching > 50% of total hydrolyzable amino

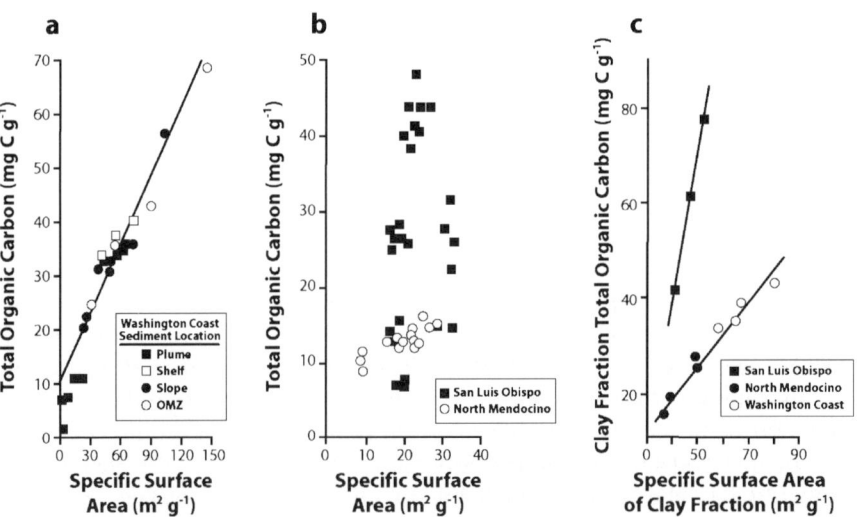

Fig. 4 Correlation between mineral surface area and total organic carbon content of marine sediments collected from **a** the Washington continental margin and **b** the California continental margin. **c** The relationship between the total organic carbon content of the clay fraction and surface area of the clay fraction of the Washington and California continental margin sediments. After Baldock and Skjemstad [85]

acids compared to negligible amounts in the diatom-rich plankton in overlying surface waters. Compositions also changed between mineral-bound and nonmineral-bound amino acids with increased depth.

Mineral associations are apparently not a prerequisite for organic matter preservation. While proteinaceous material can be preserved by interactions with mineral surfaces [89, 90], there is mounting evidence that proteinaceous material survives in systems where minerals are absent or in low abundance. This observation led Knicker and Hatcher [27] to study organic matter diagenesis in Mangrove Lake, Bermuda, an environment characterized by sapropelic sediments with low mineral content. Using ^{15}N NMR, Knicker and Hatcher [27] determined that since amide N was the dominant form of N in diagenetically altered 4000-year-old sapropel, and that there was little contribution from heterocyclic N, the refractory organic N could not derive from heterocyclic aromatic N compounds formed via condensation and polymerization of monomeric or oligomeric hydrolysis products of bacterial degradation. Rather, this organic N must survive via some interaction with other refractory macromolecular organic matter, whereby proteins become sandwiched or "encapsulated" between highly aliphatic macromolecular layers during diagenesis.

In follow-up studies, Harvey and coworkers [91–93] provided further support for the encapsulation hypothesis. Zang et al. [91] conducted dual-labeling experiments using ^{13}C and ^{15}N to follow the degradation of *Botryococcus braunii*, a prolific producer of biopolymeric algaenan. They found that biologically labile proteins and carbohydrates were preferentially lost during the time course of the experiment, but proteinaceous material remained the major form of organic N even after 200 days. Again, there was no evidence of formation of heterocyclic N compounds via depolymerization–recondensation reactions. Nguyen and Harvey [92] reported that noncovalent associations, such as hydrophobic interactions and hydrogen bonding, of protein could enhance preservation by stabilizing structures that are resistant to degradation. Nguyen et al. [93] used pyrolysis gas chromatography–mass spectrometry and ^{13}C NMR to show the preferential loss of intracellular material during degradation of *B. braunii* coupled with preservation of cell wall material. Furthermore, there were significant differences in degradation rates as a function of phytoplankton species (Fig. 5), implying distinctly different cell wall matrices. The highly aliphatic macromolecular fraction was refractory and the intrinsically labile proteinaceous material was protected against degradation.

The consideration of matrix effects on organic matter preservation now comes full circle when considered along with condensation, and in fact Collins et al. [94] contended that the two mechanisms work in concert with one another. The condensation pathway for kerogen formation [60] is based on the condensation of labile biomolecules, but condensation of organic compounds from solution is thermodynamically unfavorable. On the other hand, the process of selective adsorption represents a mechanism for concentrating the

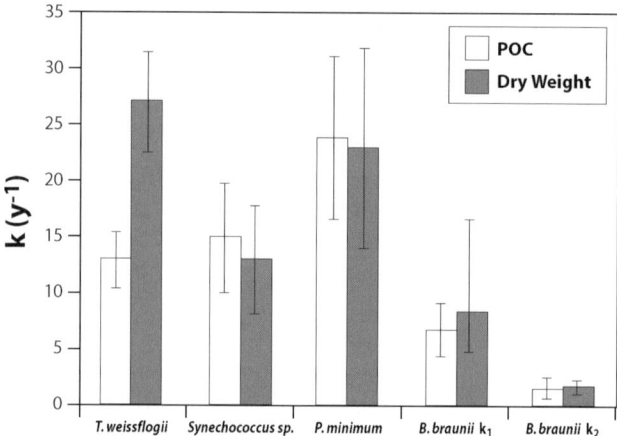

Fig. 5 Comparison of first-order decay constants (k) for particulate organic carbon (POC) and dry weight during oxic degradation of four phytoplankton species: *Thalassiosira weissflogii* (diatom), *Synechococcus* sp. (cyanobacterium), *Prorocentrum minimum* (dinoflagellate), and *Botryococcus braunii* (green alga). After Nguyen et al. [93]

labile organic compounds as mineral coatings, protects them from biodegradation, and provides a monolayer template that can irreversibly incorporate labile organic matter into an evolving macromolecule. Condensation reactions between adsorbed compounds leads to the formation of strongly bound macromolecules, which could lead to unexpected preservation of labile compounds in an organomineral phase that effectively transfers the compounds into the realm of the "molecularly uncharacterized component" [15].

4
Oxygen/Redox Control of Aerobic and Anaerobic Degradation

Considerable effort over the past decade has gone toward determining the relative efficiency of microbial decomposition operating through aerobic or anaerobic metabolic pathways. Traditionally, aerobic oxidation has been accepted as being more important because energy yields for aerobic decomposition are generally greater than those for anaerobic decomposition reactions. Oxygen serves two functions in organic matter degradation [95]: that of terminal electron acceptor during oxidation of organic carbon, and as a reactant in the oxygenase-catalyzed primary attack on substrate molecules. The first function may be transferred to other oxidized compounds (sulfate, nitrate) in the absence of oxygen, but there is no equivalent to O_2 that can fulfill its function as a reactant. Limitations in the ability of anaerobes to hydrolyze certain structurally complex compound types result in slower rates of decomposition in anoxic zones. This may also be due to the fact that the organic matter has already been partially decomposed by aerobic bacteria, and the anaerobes are exposed to the "leftovers" that tend to contain higher proportions of refractory, residual material. Thus, substrate lability and apparent decomposition rates usually decrease rapidly with depth in the sediment.

In addition, aerobic metabolism is generally more direct than anaerobic metabolism. Aerobic decomposition involves diverse enzymes, many of which are specific to individual types of organic functional groups, and each substrate is often rapidly and completely metabolized to CO_2 and biomass by a single microorganism [34]. Anaerobic heterotrophs, in contrast, are unable to degrade most polymeric compounds [96, 97] and must rely on slow hydrolytic and fermentative bacteria using various oxidized inorganic compounds (NO_3^-, Fe, Mn, and SO_4^{2-}) as electron acceptors for the supply of metabolizable low molecular weight substrates. The result is that anaerobic degradation often involves a consortium of cooperating organisms.

There is considerable molecular evidence from laboratory simulation experiments and natural settings addressing redox effects on organic matter degradation/preservation (Table 2). These studies have shown that the residence times for organic compounds present in marine sediments can vary as a result of environmental conditions such as bioturbation, physical mixing, and the presence or absence of oxygen and other electron acceptors. Some evidence indicates significant differences in degradation rates when comparing oxic and anoxic experimental conditions [43, 98–101], and field measurements also suggest an oxygen effect [42, 53, 55]. Other observations suggest that anoxic decomposition may not be intrinsically slower than oxic decomposition [22, 102], at least for simple substrates and at the onset of diagenesis. But, as discussed above, anaerobes often must deal with the more refractory substrates that have already survived attack by aerobes.

Teece et al. [100] reported that initial rates of degradation of lipids from *Emiliania huxleyi* were rapid for both oxic and anoxic conditions, but that rates for anaerobic decomposition slowed significantly, and initial rates of decay poorly reflected the overall extent of degradation. One implication of this study is that experiments of anaerobic decomposition need to be carried out over longer timescales to more accurately assess the long-term effect of anaerobic decomposition. Furthermore, Teece et al. [100] observed that since decomposition patterns were different under the two anoxic conditions examined, sulfate reduction and methanogenesis, the specific anaerobic pathways involved also needed to be considered rather than simply the overall anoxic state. A recent study by Lehmann et al. [101] tracking changes in $\delta^{13}C$ and $\delta^{15}N$ in incubation experiments showed similar rates of decomposition for reactive organic matter under oxic and anoxic conditions, again supporting the idea that aerobic and anaerobic metabolic pathways are capable of degrading labile components. However, this study showed that once the labile component was degraded, the proportion of organic matter resistant to degradation was lower under oxic than anoxic conditions.

Table 2 Comparison of turnover times for organic compounds in marine sediments

Location	Bottom water	Sediment interface	τ (days)	Refs.
Bioturbated				
Carteau Bay, Mediterranean Sea (surface sediments)	Oxic	Oxic	*n*-alkenes: 23–28 *n*-alkyl diols: 55–100 sterols: 40–59 acids: 21–76	[43]
Long Island Sound, NY (surface sediments)	Oxic	Oxic	sterols: 24–50 acids: 11–17 acids: 8–50 TOC (reactive): 51 TOC (nonreactive): 365	[118] [98] [31]
York River, VA (33–84 cm; ~ 30–40 years in age)	Oxic	Oxic	sterols: 28×10^3 acids: 22×10^3 TN: 243×10^3 TOC: 166×10^3	[55]
Physically mixed				
York River, VA (41–54 cm; ~ 30–40 years in age)	Oxic	Oxic	sterols: 20×10^3 acids: 21×10^3 to 30×10^3 TN: 54×10^3 TOC: 64×10^3	[55]

Table 2 continued

Location	Bottom water	Sediment interface	τ (days)	Refs.
Nonbioturbated				
Cape Lookout Bight, NC (surface sediments)	Oxic	Dysoxic/anoxic	sterols: 11–170 acids: 11–17 n-alkanes: 14–45	[36]
Cape Lookout Bight, NC (0–1 m)			TOC: 33–379 TN: 68–456 acids: 250–2500	[119] [120] [40]
Chesapeake Bay (0–3 m)	Oxic/seasonally hypoxic or anoxic	Oxic/seasonally hypoxic or anoxic	TOC: 14×10^3 to 33×10^3	[121]
Peru upwelling (surface sediments)	Dysoxic	Dysoxic	br alkene: 2300 n-alkanols: 2100–6100 sterols: 680–2400	[122]
Black Sea (surface sediments)	Anoxic	Anoxic	br alkene: 8900 n-alkyl diols: 26 000–33 000 sterols: 16 000–26 000 acids: 6500–14 000	[42]

Because the most energetically favorable metabolic pathways for bacteria involve oxygen as the electron acceptor, it follows that organic carbon degradation (and preservation) in sediments is strongly controlled by the average time that organic matter is exposed to oxygen, or the oxygen exposure time, OET [20, 21]. In a transect across the Washington continental margin, slope, and adjacent abyssal plain, measurements of the penetration of O_2 into surface sediment along with sediment accumulation rates allowed calculation of the oxygen exposure time [21]. There was a marked increase in OET in the further offshore sediments that corresponded to decreased OC concentrations and OC/surface area ratios and increased molecular indications of organic matter degradation. Studies of deep-sea turbidites characterized by an oxidation front (the Madeira abyssal plain (MAP) turbidites) clearly implicate molecular oxygen as a key agent in organic matter degradation [103–105]. As the MAP turbidites were laid down, the slumping sediment was thoroughly mixed mineralogically and presumably chemically. Following deposition, the upper half-meter of the turbidite was exposed to bottom water oxygen before being "capped" by the next deposit that returned the turbidite to a suboxic state. There was marked degradation—orders of magnitude reduced concentrations—of organic carbon and biomarkers in the oxidized zone compared to the unoxidized zone, producing sharp organic gradients across the redox front. In addition, there were significantly different apparent degradation rates for individual compounds and classes across the oxidation fronts. Prolonged exposure of sedimentary organic matter to oxygen not only led to greater alteration of that organic matter than occurred for sulfate reduction only, but also resulted in marked changes in absolute and relative distributions of biomarkers. Further evidence for the importance of both oxygen exposure and variable effects on different molecular structures comes from the work of Sinninghe Damsté et al. [11], who investigated the biomarker record in sediments of the Arabian Sea that had been exposed to varying oxygen exposure times (Fig. 6). Under anoxic conditions, a much larger fraction of biomarker flux accumulated than under oxic conditions, and it was apparent that different biomarkers were subject to differences in the degree of degradation, and hence variable preservation. Compositional changes such as these could substantially compromise our ability to use biomarkers for paleoenvironmental reconstructions.

Given that oxygen plays a significant role in organic matter degradation/preservation, and since intrinsic molecular character also plays a role, Hedges and Keil [4] built on previous work (references cited above) showing that organic matter does not degrade as a single pool but rather as a composite of multiple rates among the various classes of organic components. In the oxygen-sensitive organic matter model of Hedges and Keil [4] there are (at least) three forms of organic matter in sediments. Materials such as charcoal are totally refractory. An oxygen-sensitive fraction, lignin for example, degrades slowly in the presence of oxygen, but not at all under anoxic con-

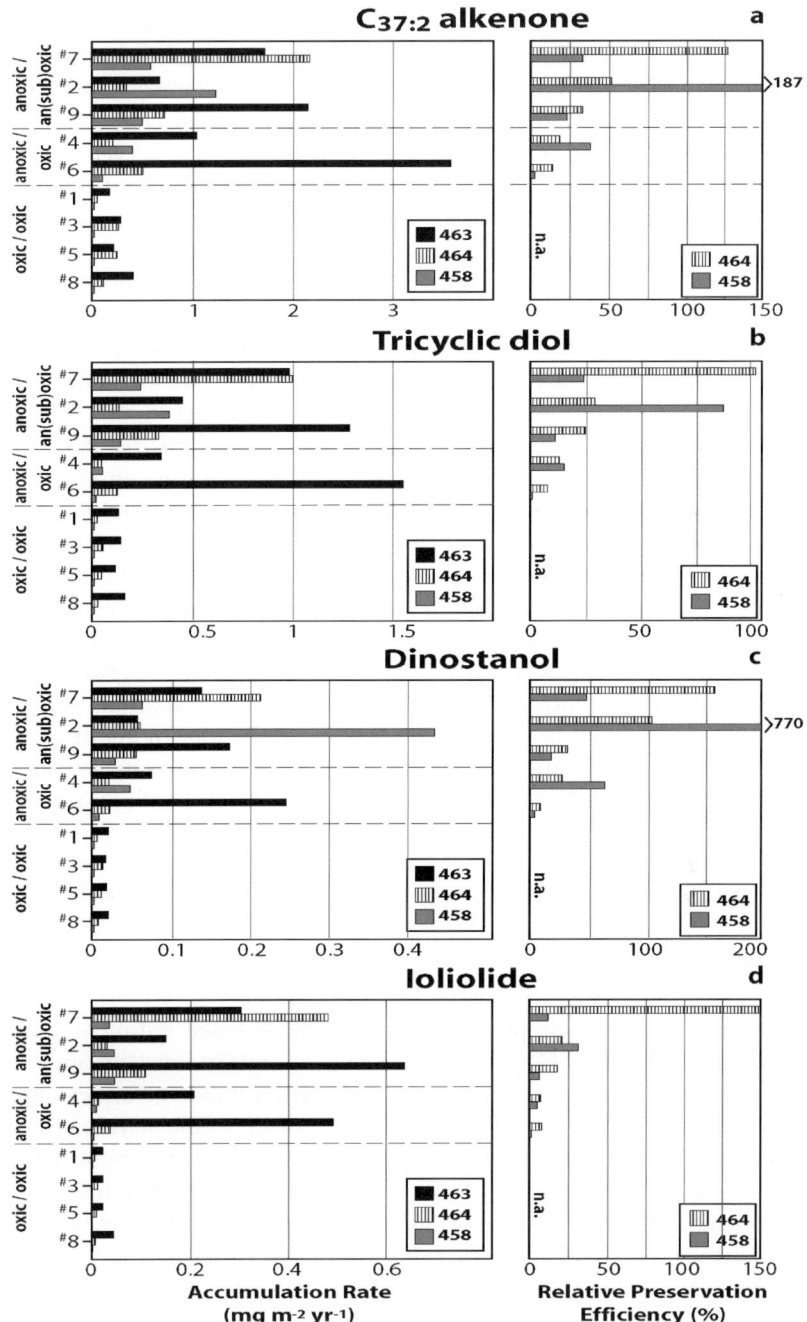

Fig. 6 Accumulation rates and preservation efficiencies for biomarkers in sediment cores from the Arabian Sea, grouped according to the degree of oxygen depletion in bottom and pore waters at the time of sediment deposition. After Sinninghe Damsté et al. [11]

ditions. Hydrolyzable molecules such as proteins and polysaccharides are readily degraded regardless of conditions. Some combination of remineralization of these three types of OC, albeit on very different timescales, is additive in proportion to their abundance, generating the profiles observed in sediments. Note that the oxygen-sensitive organic matter model relies on molecular properties as a key determinant of organic degradation rates, whereas the sorption model is largely based on sediment surface area, although the two are certainly not mutually exclusive.

Hulthe et al. [106] have provided a synthesis of concepts developed by, among others, Mayer [25, 26], Kristensen et al. [95], and Hedges and Keil [4], in terms of mineral association and oxygen effects. Hulthe et al. [106] suggest that aerobic and anaerobic degradation rates are fundamentally similar for fresh organic matter because the organic matter has yet to attach to mineral grains; that is, both aerobes and anaerobes are equally adept at leaching and hydrolyzing organic matter. As time and depth increase in sediments, a greater fraction of the organic matter becomes associated with mineral surfaces, and anaerobic bacteria have increasing difficulty with its hydrolysis. Over time, apparent rates of aerobic decomposition become faster than rates for anaerobic decomposition because aerobes have the capability of producing stronger oxidizing agents, such as H_2O_2, which penetrate into mesopores where enzymes cannot reach. As sorption increases, aerobic degradation becomes progressively more effective than anaerobic degradation. However, some mechanism, such as physical mixing, bioturbation, or redox oscillations, is required to continually reintroduce dissolved O_2 into subsurface sediments that have higher proportions of protectively sorbed organic material.

5
Sediment Mixing Regime

Most organic matter decomposition occurs in bioturbated sediments underlying oxygenated waters [18]. Whereas many laboratory investigations have been designed to study decomposition under strictly oxic or anoxic conditions, extrapolation to natural environments can be problematic if there is heterogeneity in the bioturbated zone. Mobile benthos burrow and irrigate sediment and then often move on, allowing oxygen to penetrate into the upper sediments, but then as the oxygen is consumed by aerobic processes, the sediments may return to an anoxic state. These redox oscillations also lead to oscillations in aerobe–anaerobe communities and decomposition processes that may be distinct from the aerobic and anaerobic end-members. As the size of the pool of organic carbon that is preserved in sediments depends on the delivery rate, the rate of degradation, the nature of the organic carbon available, and the length of time the substrates are exposed to a particular degradation mechanism, processes that prolong exposure to oxygen will enhance degradation and decrease preservation, in keeping with the oxygen exposure time hypothesis. The overall effect of bioturbation in sediments is suggested in a study by van der Weijden et al. [8], who found that the depth of bioturbation significantly influenced the accumulation of organic carbon in sediments across the oxygen minimum zone in the Arabian Sea (Fig. 7).

The abundance of aerobic infauna capable of bioturbation thus may be an important control on carbon decomposition and subsequent preservation, because they may provide a mechanism for renewing the oxygen content of pore waters and for transporting organic material up into the oxygenated zone where material resistant to anaerobic degradation is exposed to aerobic metabolism or "primes" co-metabolism. Laboratory experiments have now shown that decomposition rate constants for algal lipids [99] and chlorophyll *a* [107] are proportional to the abundance of subsurface-deposit feeders, such as *Yoldia limatula* (Fig. 8). In an earlier study where Bianchi et al. [108] studied the effects of macrofauna on the degradation of chloropigments, the conversion of chlorophyll *a* to phaeophorbide *a* was enhanced in sediments containing either the bivalve, *Macoma balthica* (surface-deposit feeder), or the polychaete, *Leitoscoloplos fragilis,* compared to controls with no macrofauna. In addition to the physical mixing of sediments, deposit-feeding animals can regulate the dynamics of bacterial growth. In the experiments conducted by Sun et al. [99], *Yoldia* grazing on bacteria limited the accumulation of bacterial fatty acids in the sediments, and presumably this

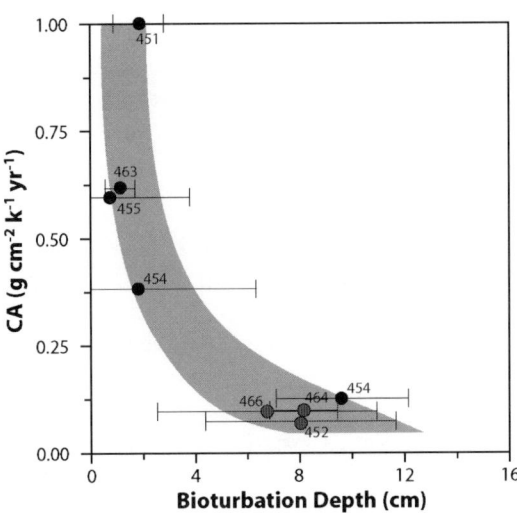

Fig. 7 Accumulation rates of organic carbon vs bioturbation depth estimated from differences in ^{14}C ages between organic carbon and foraminiferal carbonates for sediments crossing the oxygen minimum zone of the Arabian Sea. After van der Weijden et al. [8]

reduced the extent of bacterial decomposition. This observation supports the hypothesis of Lee [22] that predation on bacterial biomass, either by protozoa as suggested by Lee or by macrofauna [99, 107], would reduce bacterial decomposition and enhance carbon preservation. Aller [18] suggested that in a similar manner, periodic catastrophic death of a portion of a microbial population due to rapid redox change might significantly alter the net efficiency of remineralization.

Redox oscillation in nature is undoubtedly a highly variable event both temporally and spatially. Sun et al. [109] conducted microcosm experiments using ^{13}C- and ^{15}N-labeled algae to assess the effects of the frequency of oxic–anoxic oscillation on the rates and pathways of degradation of algal lipids in surface sediments. These experiments, carried out in the absence of bioturbating macrofauna, clearly show that the degradation of lipids is faster when redox oscillation is more frequent and, as a result, exposure to oxygen diffusing into the sediments from the overlying bottom waters is longer in duration (Fig. 9). Fatty acid analysis also indicated that redox oscillations strongly affected net synthesis of bacterial biomass, and that turnover of this biomass was faster under continuously or occasionally oxic conditions than under continuously anoxic conditions.

In addition to bioturbation, physical mixing, such as during large tidal excursions and storm events, can alter the redox environment of sediments or redistribute organic matter into different redox zones. Such physical mix-

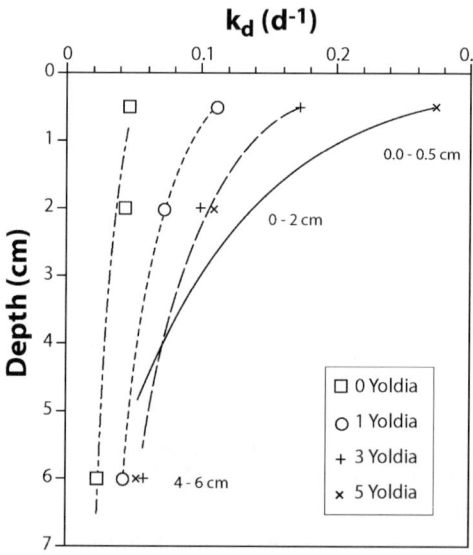

Fig. 8 Relationships between the decay constant (k_d) of chlorophyll *a* and depth interval for laboratory experiments involving varying abundances of *Yoldia*. After Ingalls et al. [107]

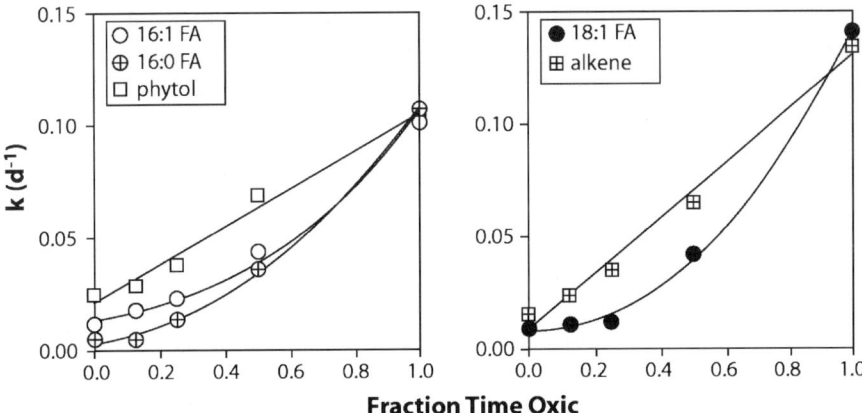

Fig. 9 Relationships between degradation constants for cell-associated ^{13}C-labeled algal lipids and the fraction of time of exposure to oxygen in redox oscillation experiments. After Sun et al. [99]

ing is more important in highly dynamic estuarine, deltaic, and coastal areas than in quiescent lagoonal or deep-sea environments. In these active depositional environments, even the relatively refractory terrestrial component may be susceptible to rapid degradation [110, 111]. Aller [112] provided an explanation for this scenario: mobile deltaic and continental shelf muds, driven by fluvial energy, estuarine circulation, tidal energy, coastal upwelling, and wind-driven waves. In environments such as the continental shelves off the Amazon River and the Fly River, Papua New Guinea, muds act as fluidized-bed reactors due to exposure to repetitive redox successions, the availability of electron acceptors, and the supply of planktonic organic carbon. Together, these conditions result in efficient remineralization of both labile marine and refractory terrestrial material under oxic and suboxic conditions. Sorbed organic matter may be released as readily degradable dissolved organic carbon [78] and the introduction of new labile substrates likely stimulates cometabolism. Bioturbation is not a prerequisite for the intense mixing of these highly mobile muds, and in fact biomass of benthic communities tends to be significantly reduced compared to less strongly mixed sediments [112]. The decreased importance of macrofauna in these environments may be due to the disturbance regime [112, 113].

A recent study examined the effects of physical mixing, although perhaps not to the extremes of the Amazon and Fly River deltas, on the fate of organic matter in sediment cores obtained from sites representing contrasting mixing conditions in the York River (USA) estuary [55, 114]. The sites differed in the extent and mechanism by which they were mixed: one station was characterized by the confluence of tidal and fluvial energy leading to resuspension, erosion, and episodic disturbances to the upper 50–100 cm

(Dellapenna et al. [115]); in contrast, the second site experienced lower bottom currents, but was dominated by bioturbation in the upper 10–15 cm. Calculated residence times for organic carbon and total nitrogen were two to four times higher at the physically mixed site than the bioturbated site (Table 2). Consistent with the Lehmann et al. [101] and Hulthe et al. [106] studies, apparent rate constants for labile compounds (e.g., diatom-derived fatty acids) were similar under the two mixing regimes, while rate constants for more stable compounds (n-alcohols, sterols, and long-chain fatty acids) were higher in the physically mixed sediments. Arzayus and coworkers [114] also found evidence for the degradation of quite stable compounds such as polycyclic aromatic hydrocarbons in sediments at the physically mixed site through differences in isomer ratios.

The influence of physical mixing of sediments on diagenesis reported by Arzayus et al. [55, 114] contrasts with a microcosm experiment conducted by Sun et al. [109]. Three different mixing regimes were simulated in the study by Sun and coworkers: bioturbated, episodically physically mixed, and no mixing. Algal lipids degraded at different rates under the different mixing conditions, with slow degradation under episodic physical mixing. The interpretation put forward by Sun et al. [109] was that the mechanical stirring that constituted the physical mixing moved otherwise labile substrates into the subsurface anoxic zone where anaerobic metabolism was slow. Degradation rates of lipids subjected to oscillating redox conditions via bioturbation were similar to those of unmixed cores in which aerobic decomposition dominated at the sediment surface. As suggested further by Arzayus and Canuel [114], part of the difference between the field and microcosm findings could derive from the very different timescales relevant to the two investigations, and to the fact that field conditions represent "open" systems while laboratory experiments are "closed".

6
Concluding Remarks

Despite the progress we have made in recent years, there are still considerable gaps in our understanding of the mechanisms by which physical and biogeochemical processes control organic matter degradation in marine sediments. In addition to further developments in the areas highlighted in this review, there are several new areas in which studies should be developed. The first of these is investigation of the role of suboxic processes (e.g., denitrification, reduction of iron and manganese oxides) on organic matter diagenesis. Recent studies [55, 114, 116] have illustrated the potential importance of suboxic processes, but additional studies are needed to explore the specific role metal oxyhydroxides play in enhancing rates of degradation of bulk carbon as well as specific biomolecules. Efforts in this area should be directed to deltaic and coastal regions where the delivery of manganese and iron oxide species is greatest. In addition, future work should attempt to tease apart the roles of sulfate reduction and methanogenesis in anoxic systems [100].

A second area for future investigations involves the study of co-metabolism. Co-metabolism is a process whereby the mixing of labile organic matter may enhance the degradation of refractory organic matter (i.e., terrestrial organic matter, or anthropogenic compounds such as polycyclic aromatic hydrocarbons). Coastal and deltaic regions are characterized by high rates of primary production due to human-induced nutrient loading. These regions are also affected by terrestrial and anthropogenic carbon inputs. Future studies should examine the role of co-metabolism in carbon diagenesis in these regions.

A third area in which additional studies are needed is in bridging the gap between benthic ecology and organic geochemistry. To date, studies in this area have involved laboratory experiments in which usually a single macrofaunal species has been manipulated. In the future, there is a need for field-based experiments in which complex benthic communities are manipulated to better understand the role of the microbial and macrofaunal communities in diagenesis. Additional areas of focus should include studies of benthic diversity and trophic processes on sediment organic carbon dynamics [117].

Finally, organic geochemists should pursue studies bridging the fields of organic geochemistry and molecular ecology. As tools in molecular ecology develop, biomarker information coupled with molecular (genetic) data will provide new insights about specific sediment microbial communities and their effects on sediment organic matter. An excellent example of this linkage comes from recent work on the anaerobic oxidation of methane in sediments (AOM; reviewed by Hinrichs and Boetius [123]). A variety of lipid biomarkers constructed of isoprenoid backbones derived from ether lipids similar to those of cultured methanogenic archaea have been identified in methane-rich sediments that are characterized by high rates of AOM. In the absence of culturable methanotrophic archaea, the presence of these biomarkers and their extremely depleted δ^{13}C values (often $-100‰$ or less) have been taken as biosynthetic products of anaerobic microorganisms using isotopically depleted methane as a carbon source. Parallel studies of the phylogeny of microorganisms in sediments with high rates of AOM and biomarkers associated with AOM reveal two groups of archaea, designated ANME-1 and ANME-2, involved in AOM. Phylogenetic analyses of archaeal ribosomal rRNA sequences place the ANME-1 and ANME-2 groups near the methanogenic *Methanosarcinales*. Fluorescent in situ hybridization (FISH) further shows that ANME-2 archaea occur in a syntrophic association with sulfate-reducing bacteria of the *Desulfosarcina/Desulfococcus* lineages, suggesting that a consortium of archaea–sulfate-reducing bacteria is involved in AOM. Subsequent biomarker analyses have now shown that some biomarkers derived from sulfate-reducing bacteria in methane-rich sediments are indeed strongly depleted in ^{13}C [123, 124]. Similar cross-disciplinary work should be quite fruitful in future studies of the impact that microorganisms have on organic geochemistry.

Acknowledgements This chapter was prepared with the support of National Science Foundation grants OCE-0223226 and OCE-0136318 to SGW and OCE-0223295 to EAC. Anna Boyette at Skidaway Institute of Oceanography prepared the figures. This is VIMS contribution number 2701.

References

1. Engel MH, Macko SA (1993) Organic geochemistry: principles and applications. Plenum, New York
2. Berner RA (1989) Palaeogeogr Palaeoclimatol Palaeoecol 73:97
3. Berner RA, Canfield DE (1989) Am J Sci 289:333
4. Hedges JI, Keil RG (1995) Mar Chem 49:81
5. Demaison GJ, Moore GT (1980) Am Assoc Pet Geol Bull 64(8):1179
6. Emerson S, Hedges JI (1988) Paleoceanography 3:621
7. Calvert SE, Pedersen TF (1992) Organic carbon accumulation and preservation in marine sediments: how important is anoxia? In: Whelan J, Farrington JW (eds) Organic matter. University Press, New York, p 231
8. van der Weijden CH, Reichart GJ, Visser HJ (1999) Deep-Sea Res I 46:807
9. Cowie GL, Calvert SE, Pederson TF, Schulz H, von Rad U (1999) Mar Geol 161:23
10. Schulte S, Mangelsdorf K, Rullkötter J (2000) Org Geochem 31:1005
11. Sinninghe Damsté JS, Rijpstra WIC, Reicharet G-J (2002) Geochim Cosmochim Acta 66:2737
12. Hartnett E, Devol AH (2003) Geochim Cosmochim Acta 67:247
13. Tegelaar EW, Derenne S, Largeau C, de Leeuw JW (1989) Geochim Cosmochim Acta 53:3102
14. de Leeuw JW, Largeau C (1993) A review of macromolecular organic compounds that comprise living organisms and their role in kerogen, coal, and petroleum formation. In: Engel MH, Macko SA (eds) Organic geochemistry: principles and applications. Plenum, New York, p 23
15. Hedges JI, Eglinton G, Hatcher PG, Kirchman DL, Arnosti C, Derenne S, Evershed RP, Kögel-Knabner I, de Leeuw JW, Littke R, Michaelis W, Rullkötter J (2000) Org Geochem 31:945
16. Henrichs SM (1993) Early diagenesis of organic matter: the dynamics (rates) of cycling of organic compounds. In: Engel MH, Macko SA (eds) Organic geochemistry: principles and applications. Plenum, New York, p 101
17. Aller RC (1982) The effects of macrobenthos on chemical properties of marine sediment and overlying water. In: McCall PL, Tevesa MJS (eds) Animal–sediment relations. Plenum, New York, p 53
18. Aller RC (1994) Chem Geol 114:331
19. Sun M-Y, Aller RC, Lee C, Wakeham SG (2002) Geochim Cosmochim Acta 66:2003
20. Hartnett HE, Keil RG, Hedges JI, Devol AH (1998) Nature 391:572
21. Hedges JI, Hu FS, Devol AH, Hartnett HE, Tsamakis E, Keil RG (1999) Am J Sci 299:529

22. Lee C (1992) Geochim Cosmochim Acta 56:3323
23. Smith CR, Walsh ID, Jahnke RA (1992) Adding biology to one-dimensional models of sediment-carbon degradation: the multi-B approach. In: Rowe T, Pariente V (eds) Deep-sea food chains and the global carbon cycle. Kluwer, Dordrecht, p 395
24. Deming JW, Barros JA (1993) The early diagenesis of organic matter: bacterial activity. In: Engel MH, Macko SA (eds) Organic geochemistry: principles and applications. Plenum, New York, p 119
25. Mayer LM (1994) Geochim Cosmochim Acta 58:1271
26. Mayer LM (1994) Chem Geol 114:347
27. Knicker H, Hatcher PG (1997) Naturwissenschaften 84:231
28. Nguyen RT, Harvey HR (1997) Org Geochem 27:115
29. Jørgensen BB (1978) Geomicrobiol J 1:29
30. Berner RA (1980) Early diagenesis: a theoretical approach. Princeton University Press, Princeton, NJ
31. Westrich JT, Berner RA (1984) Limnol Oceanogr 29:236
32. Middelburg JJ (1989) Geochim Cosmochim Acta 53:1577
33. Boudreau BP, Ruddick BR (1991) Am J Sci 291:507
34. Canfield DE (1994) Chem Geol 114:315
35. Canfield DE, Van Cappellen P (1992) Geol Soc Am Abst Prog 24:822
36. Canuel EA, Martens CS (1996) Geochim Cosmochim Acta 60:1793
37. Benner R, Fogel ML, Sprague EK, Hodson RE (1987) Nature 329:708
38. Hamilton SE, Hedges JI (1988) Geochim Cosmochim Acta 52:129
39. Cranwell PA (1981) Org Geochem 3:79
40. Haddad RL, Martens CS, Farrington JW (1992) Org Geochem 19:205
41. Harvey HR, Macko SA (1997) Org Geochem 27:129
42. Sun M-Y, Wakeham SG (1994) Geochim Cosmochim Acta 58:3395
43. Grossi V, Caradec S, Gilbert F (2003) Mar Chem 81:57
44. Camacho-Ibar VF, Aveytua-Alcázar L, Carriquiry JD (2003) Org Geochem 34:425
45. Arnosti C (2000) Limnol Oceanogr 45:1112
46. Arnosti C (2004) Mar Chem 92:263
47. Arnosti C, Holmer M (2003) Estuar Coast Shelf Sci 58:197
48. Harvey HR, Tuttle JH, Bell JT (1995) Geochim Cosmochim Acta 59:3367
49. Sarkanen KV, Ludwig CH (1971) Lignins: occurrences, formation, structure, and reactions. Wiley, New York
50. Kirk KT (1980) Studies of the physiology of lignin metabolism by white-rot fungi. In: Kirk KT, Takayoshi H, Hou-min C (eds) Lignin biodegradation: microbiology, chemistry, and potential applications, vol II. CRC, Boca Raton, p 51
51. Kawakami H (1980) Degradation of lignin-related aromatics and lignins by several pseudomonads. In: Kirk KT, Takayoshi H, Hou-min C (eds) Lignin biodegradation: microbiology, chemistry, and potential applications, vol II. CRC, Boca Raton, p 103
52. Gough MA, Mantoura RFC, Preston M (1993) Geochim Cosmochim Acta 57:945
53. Gong C, Hollander DJ (1997) Org Geochem 26:545
54. Hoefs MJL, Rijpstra WIC, Sinninghe Damsté JS (2002) Geochim Cosmochim Acta 66:2719
55. Arzayus KM, Canuel EA (2004) Geochim Cosmochim Acta 69:455
56. Reiley G, Raven AM, Lawson M, Evershed RP, Maxwell JR, Parkes RJ (1997) Abstract book, 18th international meeting on organic geochemistry. Forschungszentrum, Julich, p 9
57. Saddler JN, Wardlaw AC (1980) Antonie Van Leeuwenhoek J Microbiol 46:27

58. Harvey HR, Fallon RD, Patton JS (1986) Geochim Cosmochim Acta 50:795
59. Hernes PJ, Hedges JI, Peterson ML, Wakeham SG, Lee C (1996) Deep-Sea Res II, 43:1181
60. Tissot B, Welte D (1984) Petroleum occurrence and formation. Springer, Berlin Heidelberg New York
61. Zegouagh Y, Derenne S, Largeau C, Bertrand P, Sicre M-A, Saliot A, Rousseau B (1999) Org Geochem 30:83
62. Sinninghe Damsté JS, Rijpstra I, de Leeuw JW, Schenck PA (1988) Org Geochem 6:593
63. Werne JP, Lyons TW, Hollander DJ, Formolo MJ, Sinninghe Damsté JS (2003) Chem Geol 195:159
64. Gelin F, Volkman JK, Largeau C, Sinninghe Damsté JS, de Leeuw JW (1996) Org Geochem 30:147
65. Knicker H, Scaroni AW, Hatcher PG (1996) Org Geochem 24:661
66. Derenne S, Largeau C (1998) Mineral Mag 62A:372
67. Weiler RR, Mills AA (1965) Deep-Sea Res 12:511
68. Tanoue E, Handa N (1979) J Oceanogr Soc Jpn 35:109
69. Hedges JI (1977) Geochim Cosmochim Acta 41:1119
70. Christensen D, Blackburn TH (1982) Mar Biol 71:113
71. Gordon AS, Millero FJ (1985) Microbiol Ecol 11:289
72. Wang X-C, Lee C (1993) Mar Chem 44:1
73. Cranwell PA (1978) Geochim Cosmochim Acta 42:1523
74. Cranwell PA (1982) Prog Lipid Res 21:271
75. Kawamura K, Ishiwatari R (1984) Org Geochem 7:121
76. Suess E (1973) Geochim Cosmochim Acta 37:2435
77. Keil RG, Tsamakis E, Fuh CB, Giddings JC, Hedges JI (1994) Geochim Cosmochim Acta 57:879
78. Keil RG, Montluçon DB, Prahl FG, Hedges JI (1994) Nature 370:549
79. Keil RG, Tsamakis E, Giddings JC, Hedges JI (1998) Geochim Cosmochim Acta 62:1347
80. Bergamaschi BA, Tsamakis E, Keil RG, Eglinton T, Montluçon DB, Hedges JI (1997) Geochim Cosmochim Acta 61:1247
81. Ransom B, Bennett RH, Baerwald R, Shea K (1997) Mar Geol 138:1
82. Ransom B, Kim D, Kastner M, Wainwright S (1998) Geochim Cosmochim Acta 62:1329
83. Bock MJ, Mayer LM (2000) Mar Geol 163:65
84. Mayer LM (1999) Geochim Cosmochim Acta 63:207
85. Baldock JA, Skjemstad JO (2000) Org Geochem 31:697
86. King KJ, Hare PE (1972) Micropaleontology 23:180
87. Collins MJ, Muyzer G, Curry GB, Sandberg P, Westbroek P (1991) Lethaia 24:387
88. Ingalls AE, Lee C, Wakeham SG, Hedges JI (2003) Deep-Sea Res II, 50:713
89. Ensminger LE, Gieseking JE (1942) Soil Sci 50:205
90. Marshmann NA, Marshall KC (1981) Soil Biol Biochem 12:127
91. Zang X, Nguyen RT, Harvey HR, Knicker H, Hatcher PG (2001) Geochim Cosmochim Acta 65:329
92. Nguyen RT, Harvey HR (2001) Geochim Cosmochim Acta 65:1467
93. Nguyen RT, Harvey HR, Zang X, van Heemst JDH, Hetényi M, Hatcher PG (2003) Org Geochem 34:483
94. Collins MJ, Bishop AN, Farrimond P (1995) Geochim Cosmochim Acta 59:2387

95. Kristensen E, Ahmed SO, Devol AH (1995) Limnol Oceanogr 40:1430
96. Jørgensen BB, Bak F (1991) Appl Environ Microbiol 57:847
97. Ahmed SI, Williams BL, Johnson V (1992) Mar Microb Food Web 6:133
98. Sun M-Y, Wakeham SG, Lee C (1997) J Mar Res 61:341
99. Sun M-Y, Aller RC, Lee C, Wakeham SG (1999) J Mar Res 57:775
100. Teece MA, Getliff JM, Leftley JW, Parkes RJ, Maxwell JR (1998) Org Geochem 29:863
101. Lehmann MF, Bernasconi SM, Barbieri A, McKenzie JA (2002) Geochim Cosmochim Acta 66:3573
102. Henrichs SM, Doyle AP (1986) Limnol Oceanogr 31:765
103. Keil RG, Hu FS, Tsamakis EC, Hedges JI (1994) Nature 369:639
104. Cowie GL, Hedges JI, Prahl FG, de Lange GJ (1995) Geochim Cosmochim Acta 59:33
105. Prahl FG, de Lange GL, Scholten S, Cowie GL (1997) Org Geochem 27:141
106. Hulthe G, Hulthe S, Hall POJ (1998) Geochim Cosmochim Acta 62:1319
107. Ingalls AE, Aller RC, Lee C, Sun M-Y (2000) J Mar Res 58:631
108. Bianchi TS, Dawson R, Sawangwong P (1988) J Exp Mar Biol Ecol 122:243
109. Sun M-Y, Cai W-J, Joye SB, Ding H, Dai J, Hollibaugh JT (2002) Org Geochem 33:445
110. Keil RG, Mayer LM, Quay PD, Richey JE, Hedges JI (1997) Geochim Cosmochim Acta 61:1507
111. Hedges JI, Keil RG (1999) Mar Chem 65:55
112. Aller RC (1998) Mar Chem 61:143
113. Schaffner LC, Dellapenna TM, Hinchey EK, Neubauer MT, Smith ME, Kuehl SA (2001) Physical energy regimes, seabed dynamics and organism–sediment interactions along an estuarine gradient. In: Aller JY, Woodin SA, Aller RC (eds) Organism–sediment interactions. University of South Carolina Press, Columbia, p 161
114. Arzayus KM, Dickhut RM, Canuel EA (2002) Org Geochem 33:1759
115. Dellapenna TM, Kuehl SA, Schaffner LC (1998) Estuar Coast Shelf Sci 46:777
116. Van Mooy BAS, Keil RG, Devol AH (2002) Geochim Cosmochim Acta 66:457
117. Duffy JE, Richardson JP, Canuel EA (2003) Ecol Lett 6:637
118. Sun M-Y, Wakeham SG (1999) J Mar Res 57:357
119. Martens CS, Klump JV (1984) Geochim Cosmochim Acta 44:471
120. Klump JV, Martens CS (1987) Geochim Cosmochim Acta 51:1161
121. Zimmerman AR (2000) Organic matter composition of sediments and the history of eutrophication and anoxia in the mesohaline Chesapeake Bay. PhD dissertation. The College of William & Mary
122. McCaffrey MA (1990) Sedimentary lipids as indicators of depositional conditions in the coastal Peruvian upwelling regime. PhD dissertation, Woods Hole Oceanographic Institution and MIT
123. Hinrichs K-U, Boetius A (2002) The anaerobic oxidation of methane: new insights in microbial ecology and biogeochemistry. In: Wefer G, Billett D, Hebbeln D, Jørgensen BB, Schlüter M, van Weeging T (eds) Ocean margin systems. Springer, Berlin Heidelberg New York, p 457
124. Elvert M, Boetius A, Knittel K, Jørgensen BB (2003) Geomicrobiol J 20:403

Sources and Fate of Organic Contaminants in the Marine Environment

Josep Maria Bayona (✉) · Joan Albaigés

IIQAB-CSIC, Jordi Girona, 18, 08034 Barcelona, Spain
jbtqam@iiqab.csic.es

1	Introduction	325
2	Classes of Organic Contaminants of Major Concern in the Marine Environment	326
2.1	Organochlorine Pesticides	326
2.1.1	DDT	326
2.1.2	Lindane	328
2.2	Polychlorinated Biphenyls (PCBs)	329
2.3	Polychlorinated Dibenzodioxins (PCDDs) and Dibenzofurans (PCDFs) . .	330
2.4	Polycyclic Aromatic Hydrocarbons (PAHs)	331
2.5	Organotin Compounds	333
2.6	Emerging Contaminants	337
2.6.1	New Generations of Antifouling Agents	337
2.6.2	Brominated Flame Retardants (BFRs)	338
2.6.3	Perfluorinated Compounds (PFCs)	340
2.7	Endocrine Disruptors	340
3	Molecular Markers	341
3.1	Urban Molecular Markers	342
3.1.1	Fecal Steroids	342
3.1.2	Surfactant Derived Compounds	345
3.1.3	Other Molecular Markers of Urban Origin	347
3.2	Hydrocarbon Markers	348
4	The Transport and Transformation of Organic Contaminants	356
4.1	Major Transport Pathways	356
4.2	Biotic Processes	358
4.2.1	Assessment by Identification of Intermediates	359
4.2.2	Compositional Changes of Compound Assemblages	361
4.2.3	Enantioselective Biodegradation of Contaminants	362
4.3	Abiotic Processes	363
4.3.1	Gas-phase Reactions	363
4.3.2	Aqueous-phase Reactions	364
5	Concluding Remarks	365
	References	366

Abstract The sources, levels, and fate of organic contaminants of major concern in the marine environment are reviewed. Spatial distributions among the regional seas are discussed. Knowledge on the occurrence of emerging classes of contaminants (i.e., biocides, flame retardants, perfluorinated surfactants) and endocrine disruptors is also documented. The use of the molecular marker approach for tracing urban sewage inputs and oil contamination in the marine environment is illustrated and specific indices to improve the source specificity discussed. Furthermore, the major atmospheric and oceanic transport pathways and biotic and abiotic transformation processes of marine contaminants are summarized. Several indices based on the ratios of parent compounds and metabolites are defined to assess the biodegradation of contaminants (LABs, DDTs, TBT) and the age of their inputs in the marine environment. These include applications of enantioisomeric ratios of chiral contaminants (i.e., HCHs) to allow the discrimination between biotic and abiotic degradation processes. Finally, gaps in knowledge and research needs are identified.

Keywords POPs · Organic contaminants · Endocrine disruptors · Molecular markers · Transport · Transformation processes

Abbreviations
4,4′-DDE	1,1-Dichloro-2,2′-bis(4-chlorophenyl)ethylene (DDE)
4,4′-DDMU	1-Chloro-2,2′-bis(p-chlorophenyl)ethylene (DDMU)
4,4′-DDT	1,1,1-Trichloro-2,2′-bis(4-chlorophenyl)-ethane (DDT)
DBT	Dibutyltin
DOC	Dissolved organic carbon
dw	dry weight
HCH	Hexachlorocyclohexane
IRMS	Isotopic ratio mass spectrometry
K_{AW}	Partition coefficient air-water
K_{OA}	Partition coefficient octanol-air
LAB	Linear alkylbenzene
LAN	linear alkylnitrile
NO_2-PAH	Nitro-substituted polycyclic aromatic hydrocarbon
NOEL	No observable effect level
OCs	Organochlorinated compounds
PAH	Polycyclic aromatic hydrocarbon
PBDE	Polybromodiphenyl ether
PCBs	Polychlorobiphenyls
PCDDs	Polychlorodibenzo-*para*-dioxins
PCDFs	Polychlorodibenzofurans
PFCA	Polyfluorooctane carboxylate
PFOS	Polyfluorooctane sulfonate
POP	Persistent organic pollutant
RSD	Relative standard deviation
S-PAH	Sulfur heterocyclic aromatic hydrocarbon
SPM	Suspended particulate matter
TAB	Tetrapropylene-based alkyl benzene
TAM	Trialkylamine
TBBPA	Tetrabromobisphenol A
TBT	Tributyltin
TCDD	2,3,7,8-Tetrachlorodibenzodioxin (a dioxin)
TEQ	Toxicity equivalent
UCM	Unresolved complex mixture

1
Introduction

An increasing number of anthropogenic organic compounds have been detected in continental waters [1], although few of them are relevant in marine pollution studies. For compounds to be considered as target marine contaminants, several features need to be taken into account. They must belong to the high-production-volume class of chemicals or be produced in large amounts as by-products, or alternatively be continuously released into the marine environment. They must also be persistent against biotic and abiotic degradation processes and likely be bioaccumulated, thereby constituting a potential threat for the marine ecosystem or for human health.

A major group of these contaminants consists of organochlorinated compounds (OCs)—namely, PCBs, HCHs, DDTs, heptachlor, chlordane, toxaphene, mirex, aldrin, dieldrin, and endrin, which were banned in the 1960–1970s in most industrialized countries and recently included in the Stockholm Convention on Persistent Organic Pollutants (POPs) (May 2001) for regulation at a global scale. Other contaminants of marine concern are by-products of industrial or combustion processes such as PAHs, nitro-PAHs, PCDDs, PCDFs, etc. Despite the efforts to reduce their emissions, large amounts still enter the marine environment, mainly through atmospheric transport and deposition. Maritime transport is another important source of marine pollution due to intentional or accidental discharges, such as the release of deballasting waters or spills produced in oil tanker accidents. Although most hydrocarbons have a shorter lifetime than POPs, they are widely distributed in the marine environment as a result of multiple emission sources.

Emerging classes of contaminants receiving increasing attention include flame retardants, plasticizers, and surfactants, and personal care products such as antibacterial agents and pharmaceuticals that, due to their structure, can mimic natural hormones and disrupt endocrine functions (i.e., endocrinally active compounds, or EACs). They can be classified as estrogenic and androgenic, and those of longer lifetime have already been detected in the marine environment.

The levels and fate of those classes of organic contaminants of special concern in the marine environment are discussed in the following section. The molecular marker concept will be introduced in Sect. 3 and exemplified in the assessment of coastal urban sewage inputs and fossil fuel pollution. Finally, the transport and transformation pathways, including biotic and abiotic processes, will be presented in Sect. 4.

2
Classes of Organic Contaminants of Major Concern in the Marine Environment

2.1
Organochlorine Pesticides

Although they were phased out in the 1960s in most industrialized countries, OCs still constitute a class of widely distributed contaminants in the marine environment because of their conservative behavior and production or use over large areas. Environmental data collected over the last few decades, including sediment cores, show a steady decline in concentrations. Besides their conservative behavior, common properties of OCs are their hydrophobicity, bioconcentration, and trophic-chain biomagnification (Table 1). Due to their hydrophobicity, they become associated with suspended or sinking particles and accumulate in sediments. Taking into account that OCs represent a broad class of contaminants, attention will be focused on the most widely distributed and measured compounds in international network monitoring programs. Recently, a global assessment of the sources, levels, and transport pathways of these compounds in the environment has been published [2]. The most comprehensive regional study on the occurrence of the different OCs is probably the one carried out by AMAP [3] in the Arctic, although the largest data set relates to biota samples.

2.1.1
DDT

DDT was introduced during World War II to control insects that spread diseases like malaria, dengue, and typhus. More recently, it was widely used on a variety of agricultural crops. The technical product contains an isomeric mixture of approximately 85% 4,4'- and 15% 2,4'-isomers. The usual metabolites are DDE, DDD, and DDMU, but the proportions of these depends on the depositional environment. A predominance of DDT over its metabolites is an indication of recent inputs. DDT and its metabolites possess high bioconcentration factors (5×10^4 in fish and 5×10^5 in bivalves) [4].

Seawater DDT and its metabolites show a widespread distribution in oceanic waters worldwide. Hot spots at concentrations up to a few $\mu g\,L^{-1}$ have been detected in coastal areas (i.e., the Indian Ocean and the Caribbean), presumably indicating that it is still being used. In 1993, Iwata et al. [5] reported DDT values in surface seawaters representative of a large geographical area. The results shown in Fig. 1a clearly indicate that the highest levels were found in the coastal waters of the Indian Ocean. Elevated DDT concentrations have also been reported in suspended sediments of some Russian rivers (e.g., the Ob River), indicating that major inputs to the Arctic Ocean

Table 1 Organohalogen contaminants of concern in the marine environment

Name	Acronym	Formula	Isomers/congeners	Production Date	Log K_{ow}	Vapor pressure mm Hg (20 °C)
Dichlorodiphenyl-trichloroethane	DDT	$C_{14}H_9Cl_5$	2,4'- 4,4'-	1945	5.5–6.2	0.2×10^{-6}
Hexachlorocyclo-hexanes	HCHs	$C_6H_6Cl_6$	$\alpha, \beta, \gamma, \delta$	1940	3.8	3.3×10^{-5}
Hexaclorobenzene	HCB	C_6Cl_6	1	1945	3.9–6.4	1.1×10^{-5}
Toxaphene	TOX	$C_{10}Cl_{10}Cl_8$	300	1949	3.2–5.5	3.3×10^{-5}
Mirex	MIR	$C_{10}Cl_{12}$	1	1955	5.3	3×10^{-7}
Pentachlorophenol	PCP	C_6H_5Cl	1		3.3	16×10^{-5}
Endosulfan	EDN	$C_9H_6Cl_6O_3S$	α and β	1954	2.2–3.6	0.17×10^{-4}
Polychlorinated biphenyls	PCBs	$C_{12}H_xCl_y$ $x + y = 10$	209	1929	4.3–8.3	$1.6\text{–}0.003 \times 10^{-6}$
Polychlorinated dibenzodioxins	PCDDs	$C_{12}H_xCl_yO_2$ $x + y = 8$	75	n.a.	6.6–8.2	$2\text{–}0.0007 \times 10^{-6}$
Polychlorinated dibenzofurans	PCDFs	$C_{12}H_xCl_yO$ $x + y = 8$	135	n.a.	6.6–8.2	
Short-chain chlorinated paraffins	SCCPs	$C_xH_{(2x-y+2)}Cl_y$ $x = 10 - 13$	Not specified	n.a.	4.4–8.7	1.6×10^{-4}
Polybrominated diphenyl ethers	PBDEs	$C_{12}H_xBr_y$ $x + y = 10$	209	1960	4.3–9.9	3.8×10^{-3} $- < 10^{-7}$
Perfluorooctane sulfonate	PFOs	$C_8F_{17}SO_3^-$	1	1948	n.m.	2.5×10^{-6}
Perfluorooctane carboxylate	PCAs	$(C_8CO_2)^-$	1	n.a.	n.m.	n.a.

n.m. – not measurable
n.a. – not available

are from Russian sources [3]. Typical values of open oceanic waters not affected by direct discharges are in the low pg L^{-1}, DDE being the predominant metabolite.

Sediments DDT levels in Mediterranean coastal sediments are mainly within 47–227 ng g^{-1} dw whereas in the deep basin, levels are in the range of 0.5–1.2 ng g^{-1} dw. Levels in coastal sediments from South-east Asia and the South Pacific region are generally low, even though some hot spots have been detected, showing concentrations up to 0.1–1.9 μg g^{-1} dw [5].

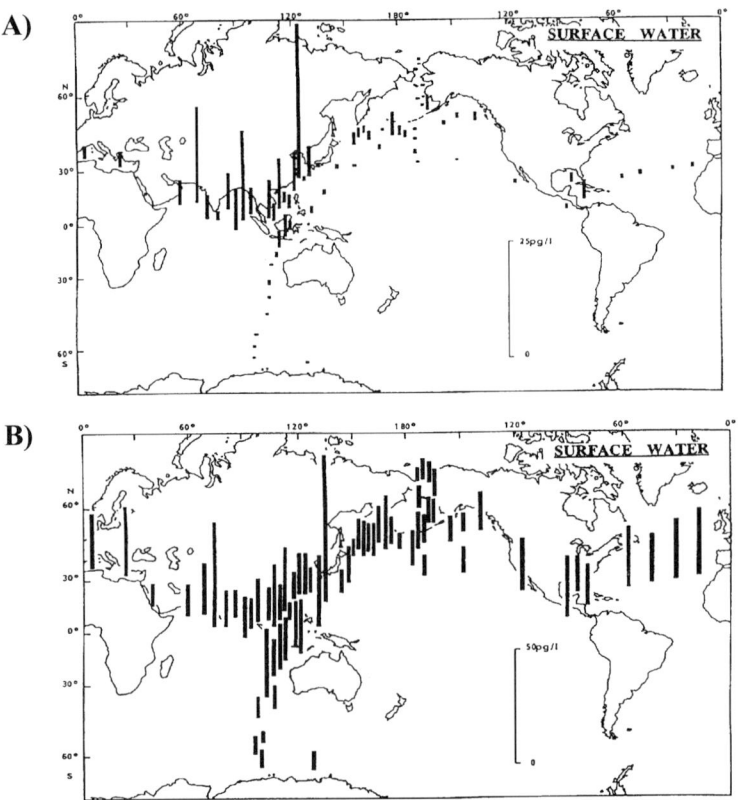

Fig. 1 Distribution of **A** total DDT and **B** PCB concentration in surface seawater. Reprinted with permission from [9]. Copyright (1993) American Chemical Society

2.1.2
Lindane

There are two main formulations of HCH—technical HCH, which is a mixture of various isomers including α-HCH (55–80%), β-HCH (5–14%), and γ-HCH (8–15%), and lindane, which is essentially the pure active product, γ-HCH. Historically, lindane was one of the most widely used insecticides in the world. Its insecticidal properties were discovered in the early 1940s for the control of a wide range of insects, for seed treatment and soil application, in household biocidal products, and as textile wood preservative.

HCHs are less bioaccumulative than other organochlorine compounds because of their high water solubility (Table 1). Moreover, their relatively high vapor pressures, particularly for the α-HCH isomer, contribute to their long-range atmospheric transport (see Sect. 4.1 and Table 7).

Seawater A major input to the sea is via rivers from areas where HCHs are applied. Some surveys have been carried out in estuaries and coastal waters.

Mean concentrations of α-HCH and γ-HCH of 0.28 and 1.1 ng L^{-1}, respectively, have been reported for the North Sea. Lindane levels in offshore waters in the Eastern Mediterranean ranged from 0.06 to 0.12 ng L^{-1}, whereas values one order of magnitude lower were found in the Western basin. Typical concentrations in oceanic waters are in the few to several hundreds of pg L^{-1}, increasing by one or two orders of magnitude in coastal waters of regions where lindane is presumably still used (e.g., India, Egypt). However, concentrations up to 1 µg L^{-1} have been detected in hot spots of Central America.

Surprisingly high levels of ΣHCH are found in the Arctic Ocean, especially in the Beaufort Sea and Canadian Arctic. ΣHCH levels measured in the late 1980s to early 1990s appear to increase in a smooth gradient with latitude from the tropical western Pacific Ocean to the Arctic Ocean, as a result of the combination of atmospheric mixing and gradient temperature that favors the gradual evaporation and deposition of volatile compounds in warm and cold regions, respectively, through the so called "cold-condensation" effect [6]. Data collected over the last decade confirm that the relative abundance in pelagic Arctic seawater is α – HCH > HCB > γ – HCH ≈ toxaphene > chlordanes ≈ PCBs > DDTs [7].

2.2
Polychlorinated Biphenyls (PCBs)

PCBs were introduced in 1929 and were manufactured in different countries until late 1980s under various trade names (e.g., Aroclor, Clophen, Phenoclor, Sovol, etc.) followed by a numeric code (i.e., 1242, 1254, 1260), the last two digits of which denote the percentage by weight of chlorine in the formulation. PCBs are chemically stable and heat resistant, and were used worldwide as transformer and capacitor oils, hydraulic and heat-exchange fluids, and lubricating and cutting oils. Theoretically, a total of 209 possible chlorinated biphenyl congeners exist, but only about 130 of these are likely to occur in commercial products.

Most PCB congeners, particularly those substituted on the 2,4,5-, 2,3,5-, and 2,3,6-positions are extremely persistent in the environment. Their half-lives are estimated from 3 weeks to 2 years in air, and with the exception of the mono- and dichlorobiphenyls, more than 6 years in aerobic sediments [8]. PCBs have also long half-lives in adult fish (more than 10 years for CB-153 in eel). The most toxic congeners are the so-called coplanar PCBs, not substituted at the two *ortho* positions, and the mono-*ortho*, when only one of these positions is substituted.

Seawater Any comparison of published results is generally hampered by the different calibration standards (e.g., number of individual congeners) used and the lack of distinction between the dissolved and particulate fractions. Predictably, the highest concentrations are reported in urban and industrial coastal wastewaters as well as in river discharge affected areas. Accordingly, decreasing concentration gradients have been found in transects offshore from these sources.

Concentrations of total PCBs in filtered ocean water are usually reported to be in the low to mid-pg L^{-1} range, and this makes reliable quantification difficult. A large survey conducted by Iwata et al. [9] showed a rather uniform distribution of PCBs in surface waters of the Atlantic, Pacific and Indian oceans, with relatively higher levels in the temperate Northern Hemisphere in the more industrialized areas (Fig. 1B). PCB concentrations in the suspended particulate matter (SPM) from coastal and open Western Mediterranean waters in the 1990s were in the range 1–35 pg L^{-1}. This is the same order of magnitude as those reported in other regions, e.g., the North Sea and North Atlantic. Higher concentrations were also detected in the Sea of Japan, where total PCBs analyzed in 1999 ranged from less than 0.01 ng L^{-1} for each congener to 150 ng L^{-1}, with a detection frequency of 131 among 171 samples [2].

Sediments PCBs are widely distributed in marine sediments including in remote regions (e.g., the Arctic). A mapping of PCBs in sediment samples (around 100) collected from different areas of the western Mediterranean basin revealed the widespread occurrence of these pollutants in the region and allowed a mass balance assessment [10]. While localized inputs or "hot spots" were identified near sewage outfalls from highly industrialized and populated cities (229 ng g^{-1} Σ9 PCB congeners) and river pro-deltas (34 ng g^{-1} Σ9 PCB congeners), the levels in the deep basin were 2–6 ng g^{-1} dw.

In New Zealand, South-east Asia, and the South Pacific region, the sum of 25 PCB congeners was in the range of 0.12–8.80 ng g^{-1} dw for sediment samples. Hot spots have been identified in the Antarctica where extremely high concentrations (250–4300 ng g^{-1}) were found in Winter Quarter Bay at the vicinity McMurdo Research Station (Ross Sea) [11]. A global accounting of PCBs based on individual congeners in the shelf sediments (4214 samples) suggested that the worldwide inventory of CB-153 was of the order of 1200 metric tons (95% confidence limit) [12]

2.3
Polychlorinated Dibenzodioxins (PCDDs) and Dibenzofurans (PCDFs)

PCDDs and PCDFs are produced as by-products in the combustion of organic matter and fuels at temperatures below 800 °C and in several industrial processes. They are characterized by their lipophilicity, low volatility, and refractory nature (half-life in soil: 10–12 years). They can also bioconcentrate and biomagnify through aquatic food webs. Their environmental fate and toxicity depends on the degree and position of chlorination. Higher toxicity is exhibited by those congeners substituted in the 2-, 3-, 7-, and 8-positions (17 congeners). The most toxic is the tetrachloro (2,3,7,8-TCDD) isomer, and hence this is used as the reference for reporting concentrations in toxic equivalents (TEQs).

Seawater Information on dioxin levels in seawater is scarce, but a few point-source emissions have been studied, such as in Frierfjorden (North Sea) and Japan. PCDD/PCDF concentrations are ten to twenty times higher in samples from the northern North Sea than in samples from the Barents Sea. In general, dioxin levels in coastal waters are an order of magnitude lower than those in rivers.

Sediments Levels of PCDDs/PCDFs have been determined in marine sediments from northern Norway, in the Mackenzie River Delta area, and in the Barents Sea [3]. PCDDs/PCDFs in the Barents Sea were 10 to 20 times lower than those in the northern North Sea. PCDD/PCDF isomer patterns were very similar for both the Barents Sea and North Sea samples and indicative of combustion sources. The Gulf of Finland, a sub-basin of the Baltic Sea, was found to be severely polluted, with concentrations as high as 101 ng g^{-1} dw and 479 pg TEQ g^{-1} dw attributable to their formation as by-products of chlorophenol manufacture. Sediments sampled in the Catalan coast (Spain) in the late 1980s to early 1990s showed concentrations in the range 0.4–8 pg TEQ g^{-1} dw [13]. In Japan, the levels of dioxins and furans in coastal sediments ranged, in 1997, from 0.012 to 49.3 pg – TEQ g^{-1} dw (av. 17.1).

All these data together point to a generalized pollution pattern in marine sediments by PCDDs/PCDFs, with higher contents in coastal shelf sediments close to the point sources (i.e., river mouths and outfalls) from industrialized areas. Some dated sediment cores provide a picture of large temporal variations in sedimentation of PCDDs/PCDFs, with a peak of pollution in the 1960s to 1970s followed by a stepwise decline attributable to better source control. The spatial and temporal dynamics of PCDD/PCDFs in the southern hemisphere are practically unknown, although the lower prevalence of industrial sources suggests that levels should be lower than in the northern hemisphere.

2.4
Polycyclic Aromatic Hydrocarbons (PAHs)

PAHs constitute a broad class of compounds containing from 2 to 7 aromatic rings with different numbers of alkyl substituents. They are present in fossil fuels and can be formed by combustion processes or from early diagenesis of organic matter, as will be described in more detail in Sect. 3.2. Hydrophobicity increases with the number of aromatic rings, and those containing more than three rings are typically found sorbed onto the organic matter of suspended particles or black carbon (e.g., soot). Recent studies have shown that sorption of PAHs in soot–water systems is exceptionally strong, and it could control their fate in the aquatic environment [14]. Both the atmospheric and aquatic pathways to the marine environment are important, although information on riverine inputs is very limited.

Half-life and bioconcentration potential increases with the number of aromatic rings. At higher trophic levels, where metabolization occurs, PAHs are of concern because most of the 5-ring and several alkylated derivatives are carcinogenic or mutagenic. Thus, the exposure to marine organisms is better assessed by the determination of hydroxy metabolites in the liver [15]. Moreover, the unresolved complex mixture (UCM) of monoaromatic hydrocarbons from crude oils elicits a sublethal toxic response in mussels [16].

Seawater Concentrations of PAHs in seawater vary considerably depending on the proximity of the source. The highest values are found in coastal waters as a result of terrestrial inputs (e.g., river outflows) and maritime transport (e.g., harbors). However, a global assessment is prevented by the different analytes and phases measured. Atlantic seawater concentrations range from 0.3 ng L^{-1} for individual low-molecular-weight PAHs (two- and three-ring compounds) to less than 0.001 ng L^{-1} for the high-molecular-weight PAHs (five or more rings). Coastal and estuarine waters contain total PAH concentrations ranging from not detectable to 8500 ng L^{-1}. The vertical profiles of SPM usually exhibit a decreasing concentration trend with a relative increase of the more polycondensed compounds derived from pyrolytic sources.

In the Danube, Dnieper, and Dniester river estuaries and other Black Sea coastal point sources, the PAH concentrations in SPM are approximately 1.6 ng L^{-1}. Concentrations of PAHs ($\Sigma 16$) associated with SPM range from 200 to 750 ng L^1 at the Rhone and Ebro river plumes. In the East Asia Region (Republic of Korea) concentrations reported for PAHs in seawater range from 25.9 to $10\,197$ ng L^{-1}. In the Russian sector of the Arctic region (Pechora and Kara seas), the concentration of the sum of 23 PAH compounds range from 10 to 69 ng L^{-1} [3].

Sediments PAHs are commonly found in marine sediments at relatively high levels compared to POPs, and are derived from both petrogenic and pyrogenic sources. In Europe, PAH concentrations up to 35.2 µg g^{-1} dw have been found in the southern Baltic Sea. Somewhat elevated concentrations were observed in the eastern Gulf of Finland as well as in the northern Gulf of Bothnia (up to 17–20.9 µg g^{-1} dw, respectively). Down-cores showed peak concentrations in the 1970s, reflecting a higher anthropogenic influence during that time [17].

PAHs are widespread in coastal zones in the Mediterranean where they are clearly associated with urban and industrial inputs. An increasing proportion of pyrolytic PAHs in transects from coastal areas towards the open sea indicates the predominance of atmospheric inputs in the latter, which account for 80–90% of the total PAHs in the deep basins. Extensive studies were conducted soon after the Gulf war on the determination of PAHs in sediments from the Arabian Gulf. The levels found in 1991 in the region ranged from 4.5 ng g^{-1} dw in Oman to 761 ng g^{-1} dw in Saudi Arabia.

In South America, most reports relate to harbors and ports in heavily affected areas, and thus, there is a great variation in the amounts reported

(0.1 ng g^{-1} – 286 μg g^{-1}). The most affected areas correspond to regions of intensive traffic in the estuaries of the main rivers, especially close to heavily populated areas such as Buenos Aires and Montevideo, and the Argentine Patagonian coastal area, where crude oil extraction and transport are very active.

In the Antarctic region the information on environmental levels of PAHs generally refers to local sources, which are unrepresentative of the region as a whole. Generally, PAH concentrations in marine sediments vary from undetectable to about 10 ng g^{-1}. An irregular combination of global input, low-level and long-term natural and anthropogenic sources, and accidental spills can be responsible factors. In the Arctic region, the baseline values in the Norwegian and Russian sectors are in the range of 10–160 ng g^{-1} (Σ24 PAHs), but concentrations up to 8 μg g^{-1} have been found in contaminated sites (e.g., harbors [3] AMAP, 1998).

2.5
Organotin Compounds

The main organotin compounds are tributyltin and triphenyltin (Table 2) and their di- and mono- degradation products. These compounds are mainly used as antifouling paints for underwater structures, ships, and cooling systems. Other applications of dibutyltin (DBT) are as a polymer stabilizer and catalytic agent in soft foam production. Minor applications include antiseptic and disinfecting agents in textile and industrial waters and pesticides.

Because of electrostatic and hydrophobic interactions, TBT binds strongly to suspended material and is accumulated in sediments, where a chemical or microbiological methylation may occur, giving rise to the formation of volatile species that can be transferred to the atmosphere [18]. TBT is lipophilic and tends to accumulate in aquatic organisms even at high trophic levels. Due to adverse effects on non-target organisms, the use of antifouling paints based on TBT was banned worldwide in 2003 by the International Maritime Organisation [19].

Seawater Butyltin compounds have been reported in seawater worldwide although the highest concentrations were found in harbors, estuaries and other semi-enclosed systems, with concentrations decreasing seaward. A survey was carried out in late 1980s in the Mediterranean region, as part of the MEDPOL monitoring program, which included the French and Northern Tyrrhenian coasts, the Nile Delta area, and the Southern coast of Turkey. The results of this survey can be considered as representative of the various contaminated sites occurring in the Mediterranean region before TBT regulation [20]. The concentrations of TBT in seawater from harbors and marinas on average varied between 100 and 1000 ng L^{-1}. Mariculture areas exhibited less contamination by TBT with concentrations < 20 ng L^{-1}, which is considered the NOEL for mollusc larval development.

Table 2 Biocides used in antifouling paints

Name	Acronym	Formula	Aqueous solubility (mg L^{-1})	log K_{ow}	Vapor pressure* (Pa) (20 °C)	Half-life** (days)
Tributyltin	TBT	[Sn(C$_4$H$_9$)$_3$]$^+$	1	3.19–3.84	133.3	700
Irgarol 1051	IRG		7	2.38	1.5×10^{-5}	100–200
GS26575	M1		n.a.	n.a	n.a.	226
Chlorothalonil			0.6	2.64	1.3×10^{-3}	< 0.5
Dichlofluanid			0.006	3.7	1.3×10^{-4}	< 0.5
SeaNine 211			n.a.	n.a.	n.a.	< 0.4

* According to [33]
** Anaerobic half-lives of dissolved biocide [27]
n.a. not available

Table 2 continued

Name	Acronym	Formula	Aqueous solubility (mg L^{-1})	log K_{ow}	Vapor pressure* (Pa) (20 °C)	Half-life** (days)
SeaNine 211			n.a.	n.a.	n.a.	< 0.4
Diuron			36.4	2.85	4.1×10^{-4}	14
1-(3-chlorophenyl)-3,1-dimethylurea	CPDU		n.a.	n.a.	n.a.	35

* According to [33]
** Anaerobic half-lives of dissolved biocide [27]
n.a. not available

Table 2 continued

Name	Acronym	Formula	Aqueous solubility (mg L^{-1})	log K_{ow}	Vapor pressure* (Pa) (20 °C)	Half-life** (days)
1-(3,4-dichlorophenyl)-3-methylurea	DCPMU		n.a.	n.a.	n.a.	1.0
1-(3,4-dichlorophenyl)urea	DCPU		n.a.	n.a.	n.a.	3
2-(Thiocyanomethylthio)benzothiazole	TCMTB		n.a.	n.a.	n.a.	1.5

* According to [33]
** Anaerobic half-lives of dissolved biocide [27]
n.a. not available

A research campaign conducted by IFREMER (Institute Française de Recherche pour l'Exploitation de la Mer) in 1997 confirmed that contamination of French coast was still a problem. Seventy-five percent of the measurements were above the threshold of $1\,\text{ng}\,\text{L}^{-1}$, which is known to cause toxic effects to some marine species [21].

Sediments The MEDPOL pilot survey reported TBT levels in sediments from Mediterranean harbors in the range of $30–1375\,\text{ng}\,\text{g}^{-1}$ [20]. The monitoring network program in France determined concentrations of $2–197\,\text{ng}\,\text{g}^{-1}$ of TBT in the Atlantic Bays of Brest and Arcachon, and concentrations of $9–127\,\text{ng}\,\text{g}^{-1}$ in the Mediterranean ports [22]. Further studies have also been performed in many Mediterranean coastal sites (e.g., Egypt, Malta, France, Spain, Italy, etc.) [23–25] with reported TBT values of 1 to $2067\,\text{ng}\,\text{Sn}\,\text{g}^{-1}$ dw, indicating that antifouling paints are still of concern in marinas and harbors. However, the relative predominance of the organotin degradation products over the parent compounds suggests that there are almost no recent inputs in marinas but still significant inputs in commercial harbors and, particularly, at sites adjacent to vessel repair facilities, where temporal trends do not show any decline. An additional minor source of organotin compounds is sewage sludge disposal [25].

TBT levels in India, in the Kochi, Marmagao, and Mumbai areas, respectively increased within the range $33–2333\,\text{ng}\,\text{g}^{-1}$. High TBT concentrations ($328.7–377.7\,\text{ng}\,\text{g}^{-1}$) were also found in sediments from the Front Channel of the Pearl River (Zhujiang) where more than 30 shipyards and ship-repairers are located [2]. Phenyltin compounds were detected in the late 1980s in marinas [26], but due to a faster degradation rate and lower usage, they are rarely detected in recent marine sediments.

2.6
Emerging Contaminants

This category of contaminants includes a wide variety of compounds having different applications and different physico-chemical properties, ranging from highly hydrophobic to hydrophilic.

2.6.1
New Generations of Antifouling Agents

A variety of biocides have been used as antifouling agents since TBT was banned from use on small vessels (Table 2). A common feature of the new generation of antifouling agents is their lower lipophilicity and shorter half-life compared to TBT, although a high persistence has been reported for some compounds associated with paint chips released by hull cleaning activities [27]. Although their environmental behavior and effects are not completely understood, some information about their occurrence is already available for some of these biocides [28]. Irgarol and diuron (Table 2) are particularly widespread in the marine environment.

Irgarol (2-methylthio-4-tertiary-butylamino-6-cyclopropylamino-s-triazine) belongs to the s-triazine group of compounds inhibiting photosynthetic electron capture transport in chloroplasts [29]. It was the first biocide to gain prominence as an environmental contaminant and was reported in 1993 for the first time in Cote d'Azur marinas [30]. At present it is one of the most widely used booster biocides, and it has been frequently identified (> 70%) in seawater from harbors and marinas in many European countries (UK, France, Spain, Netherlands, Sweden, Greece, Germany), Japan, Singapore, the United States (Florida), and Bermuda. Concentrations in seawater ranged from a few µg L^{-1} to the detection limit (< 1–5 ng L^{-1}, depending on the analytical technique) and occur mostly in the dissolved phase. Although it is one of the most lipophilic triazines due to the *tert*-butyl and cyclopropyl groups, it exhibits a low apparent sediment–water partitioning coefficient (log K_d = 3.4). Therefore, concentrations in sediments are usually in the low ng g^{-1} range except when paint chips occur in the sediment, which can give rise to quite high values [31].

The most commonly found Irgarol metabolite is the product referred as GS26575 or M1 (Table 2), formed by elimination of the cyclopropyl substituent group (2-methylthio-4-*tert*-butylamino-6-amino-s-triazine) through a suspected photochemical pathway [32]. Its concentrations are usually lower than those of Irgarol, especially in sediments, but there are contradictory results about its environmental stability in comparison with the parent compound.

Diuron is one of the major urea-based herbicides that has been in use since the 1950s as a biocide booster. It has been detected at higher concentrations in coastal waters than in freshwaters, suggesting a source from antifoulants rather than from herbicides [33]. Highest levels of diuron in seawater were reported in the United Kingdom (up to 6742 ng L^{-1}), Japan (up to 3054 ng L^{-1}), and Spain (2000 ng L^{-1}), following the yachting season, and exceeding the Irgarol concentrations in the same regions [28]. In estuaries and coastal areas, additional inputs of diuron from agricultural runoff are likely.

The solubility of diuron in water is rather high, but its ability to sorb to suspended particulate matter is low, being almost unaffected by salinity, based on experiments performed in the laboratory with different suspended particulate matter concentrations and materials [34]. As a consequence, it is considered to be mainly associated with the dissolved phase and exhibits negligible concentrations in sediments underlying contaminated waters [35].

2.6.2
Brominated Flame Retardants (BFRs)

BFRs comprise a variety of chemical compounds considered to be significant marine environmental contaminants. They include polybrominated biphenyls (PBBs), polybrominated diphenyl ethers (PBDEs), tetrabromobisphenol A (TBBPA), hexabromocyclododecane (HBCCD), *bis*(2,4,6-tribromophenoxy)ethane (BTBPE), *tris*(2,3-dibromopropyl)phosphate (Tris) (Table 1). The commercial use of some BFRs, namely PBDEs and PBBs, are banned or under increased regulation.

BFRs are divided into three subgroups depending on the mode of incorporation into the polymers: (a) brominated monomers, (b) reactive and, (c) additive. A monomer such as brominated styrene or brominated butadiene is used in the production of brominated polymers, which are then blended with non-halogenated ones. Reactive flame retardants, such as tetrabromobisphenol A (TBBPA), are chemically bonded into the polymer. Additive flame retardants, which include the polybrominated diphenyl ether (PBDE) are blended with the polymers and are more likely to leach out of the products [36].

Tetrabromobisphenol A (TBBPA)

This is the highest volume BFR on the market (2.1×10^5 metric tons). Despite its primary use as a reactive flame retardant (covalently bound to the polymer), data on its occurrence are rather scarce, as it is detected in seawater, sediment and fish from Japan [37] at low concentrations.

Polybromodiphenyl Ethers (PBDEs)

Since the 1960s, three commercial PBDE formulations have been in production: penta-, octa- and deca-BDEs. The pentabrominated product is used to render flame-retardant polyurethane foams in furniture, carpet underlay and bedding. Commercial octa- is a mixture of hexa- (10–12%), hepta- (44–46%), octa- (33–35%) and nona-bromodiphenyl (10–11%) ethers. It is used to render flame-retardant a wide variety of thermoplastics and is recommended for injection molding applications. The deca-BDE (a single congener) is used predominantly for textiles and denser plastics. The global market demand for deca-, octa-, and penta-BDEs in 1999 was 54 800, 3825, and 8500 tons, respectively. Recently, the EU announced an immediate ban on the marketing of octa-BDE and penta-BDE. The latter was voluntarily withdrawn from the market in Japan but is still used in North America.

The bromination of diphenyl ether is rather specific due to the directing properties of the oxygen and steric hindrance and, consequently, results in a limited number of PBDE congeners [38]. Due to their structural similarity to PCBs, the same congener nomenclature was adopted for PBDEs.

The environmental behavior of PBDEs depends on the degree of bromination. The more highly brominated compounds are less mobile in the environment, because of their low volatility, water solubility, bioconcentration, and strong adsorption to sediments. Thus, high concentrations of the deca-brominated congener (BDE-209) have been found in river mouths (4–6000 ng g^{-1}) [37], and similar levels (ng g^{-1} to µg g^{-1} dw) were also found in coastal sediments from Europe and North America [39]. On the other hand, the less brominated congeners, including their degradation products, are predicted to be more volatile, water soluble, and bioaccumulative, and thus exhibit an environmental behavior closer to PCBs. Incubation studies of the BDE-209 in sediments for 32 weeks in an anaerobic microcosm shows that biodegradation is not an important degradation pathway, but photodegradation leading to the formation of less brominated BDEs may play a significant role. Less-brominated BDE congeners have already been identified in high concentrations in marine birds and mammals from remote areas and in mussels and fish from the coastal zones [40, 41].

2.6.3
Perfluorinated Compounds (PFCs)

PFCs include a wide range of compounds, such as carboxylic acids, alcohols, sulfonates, sulfonamides, etc., the most common being the perfluorooctyl sulfonate (PFOS) and the perfluorodecyl carboxylate (PFCA) (Table 1). PFCs have been used for more than 50 years and are involved in different industrial processes and applications such as surfactants, polishers, surface protectors, pesticides, paper, photographs, etc. The principal production of PFOS was reduced in 2000 with total elimination in 2002.

PFOS does not hydrolyze, photolyze, or biodegrade under most environmental conditions. It is persistent in the environment and has been shown to bioconcentrate in fish. It has been detected in a number of wildlife species including mammals collected worldwide [42] and, recently, PFCAs ranging in length from 9 to 15 carbons have also been found in polar bear livers [43]. Nevertheless, the mechanism for long-range transport is not understood because they are not volatile. Oxidation of the highly volatile alcohols to carboxylic acids has been suggested.

2.7
Endocrine Disruptors

Evidence is accumulating that many compounds present in the marine environment have the potential to interfere with the actions of endogenous hormones. These compounds are known as endocrinally active compounds (EACs), since they affect target receptors of sex hormones as well as other organs [44]. The EACs can be classified as natural sex hormone-like compounds (e.g., phytoestrogens and mycoestrogens) and anthropogenic sex-hormone-like compounds (e.g., pharmaceuticals, estrogens, androgens, phenolic compounds, polychlorinated compounds, and phthalates). In the marine environment, the most relevant in terms of concentration, persistence and bioconcentration are the polychlorinated compounds and organotins. Among them, organotin compounds, affecting more than 70 marine gastropod species through the development of male sex organs in female animals (imposex) with resulting sterility, are of general concern [45]. The sources, occurrence, and trends of organotins in different regions have already been presented in Sect. 2.5.

Estrogenic effects have been associated with organochlorine compounds (e.g., DDTs, metoxychlor, dieldrin, endosulfan, chlorecone, PCDDs, PCDFs, and PCBs). It is known that since the 1960s DDTs have caused adverse effects in the reproduction of birds, especially the thinning of eggshells, but other OCs have also shown estrogenic activity. Recently, new evidence based on monitoring hormone and vitellogenin levels, together with gonad histology, indicate that in the central Mediterranean, about 14% of male swordfish *(Xiphias gladius)* have undergone sex inversion [46].

On the other hand, the estrogenicity of fishes living downstream of sewage plants was attributed to the occurrence of alkylphenols (i.e., octyl- and nonylphenols), a refractory intermediate in the degradation of non-ionic surfactant precursor. Similar effects have not been reported in the marine environment, probably because alkylphenols occur at lower concentrations than in rivers. However, ascertaining the biological effects of low doses of multiple EACs that may occur in the marine environment needs further research.

3
Molecular Markers

In general, a molecular marker is a compound or a suite of compounds, which can be unambiguously associated with a specific biogenic source [47]. Molecular information can be carried at three principal levels: (1) the isotopic composition, (2) the stereo- or structural isomerism, and (3) the composition of assemblages of contemporaneously formed compounds. The concept has also been used in environmental chemistry for recognizing the source of a contaminant and its environmental fate or tracing its spatial and temporal trends [48].

Major candidates in the marine environment are compounds present at elevated source concentrations (i.e., $> \mu g \, L^{-1}$ up to $mg \, L^{-1}$) in order to circumvent the high dilution factors in off-shore waters, and which display source-specific molecular features. Obviously, synthetic organic compounds (i.e., PCBs, phthalates, polysiloxanes) can be used as generic molecular markers of present or past human activities depending on their production period. However, the multiple sources of these compounds in the marine environment make it difficult to use them as tracers. Nevertheless, in some specific regions where they can be found at extremely high concentrations, they can be useful for source apportionment.

Other important requirements are their conservative behavior once they are transferred to the receiving system and their hydrophobicity (log $K_{ow} > 4$), so that they can be sorbed onto suspended particles and sink to sediments or be advectively transported over long distances. Indeed, knowledge of the degradation rate and pathways is needed for more quantitative applications of the molecular marker concept. In some studies, the determination of two molecular markers of different stability can bring some information on the extent of the affected area.

To illustrate the application of this concept, molecular markers of both biogenic and anthropogenic origin as tracers of urban pollution in coastal waters and those used in oil-spill fingerprinting will be described in the following sections.

3.1
Urban Molecular Markers

Several organic compounds occurring at high concentrations in urban sewage and sewage sludges have been proposed as molecular markers of urban origin. They can be biogenic or anthropogenic (Table 3). While the first ones occur in human feces or urine, the second are related to household products (i.e., surfactants, whitening agents, and fabric softeners).

3.1.1
Fecal Steroids

Coprostanol (5β(H)-cholestan-3β-ol) is one of the most widely studied urban molecular marker. Its structure is characterized by the β-stereochemistry of the – CH_3 and – H in the 10 and 5 positions, respectively (ring A–B fusion in *cis* configuration). The coprostanol epimer, epicoprostanol (5β-cholestan-3α-ol), can be of fecal origin while the 5α(H)-cholestanol stereoisomer can be of diagenetic origin, because it is the most stable thermodynamically.

Origin

Two different pathways of coprostanol formation from the bacterial reduction of cholesterol have been reported. In addition to the stereospecific direct reduction of cholesterol to coprostanol by fecal bacteria, the indirect reduction via the formation of Δ^4-unsaturated ketones and coprostanone (5β(H)-cholestan-3-one) can occur [49]. Although coprostanone is also present in sewage waters, it cannot be used as a marker because it lacks source specificity and its concentration is much lower than that of coprostanol.

Table 3 Classical urban molecular markers used in marine pollution studies

Name	Acronym	Number of compounds	Origin	Specificity	Log K_{ow}	Conservative behavior	Applications
BIOGENIC							
Coprostanol	COP	1	Human and mammalian feces	Moderate*	6.5–7.5	Low–moderate	Point-source tracing, bacterial guideline
Aminopropanone	APR	1	Urine	High	n.a.	Low	Near-source tracing
Urobilin	URO	1	Urine and human feces	High	n.a.	Moderate	Near-source tracing
ANTHROPOGENIC							
Linear alkylbenzenes	LAB	26	Anionic surfactant intermediate	High	6.9 – 9.3	Intermediate in oxic conditions	Point-source tracing, biodegradation index
Branched alkylbenzenes	TAB	80 000	Anionic surfactant intermediate *used until* mid-60 s	High	n.a.	High	Geochronology
Linear alkylbenzene sulfonates	LAS	26	Anionic surfactant	High	Anionic	Low in oxic conditions	Near-source tracing
Fluorescence Whitening agents	FWAs		Surfactant additive	High	Anionic	High	Long-distance transport from point and non-point sources
Trialkylamines	TAM	9	Cationic surfactant intermediate	Low–moderate**	Neutral = 8 (estimated)	High	Near-source tracing
Linear alkyl nitriles	LAN	5	Cationic surfactant intermediate	Low–moderate	n.a.	Poor	Near-source tracing
Tocopheryl acetate	VE	1	Vitamin E derivative	High	n.a.	Poor	Near-source tracing
Benzothiazoles	BT	3	Crumb rubber	High	n.a.	Poor	Street runoff

* It can occur in the feces of animals [50] and can be produced in situ by cholesterol reduction [57]
** Industrial sources can be predominant in western Europe
n.a.: information not available

Occurrence

Approximately 30% of the coprostanol in human feces occurs in the esterified form [50], but in sewage it accounts for only the 8–15% due to the presence of microorganisms that hydrolyze the ester bond [49]. The high hydrophobicity of coprostanol (log K_{ow} = 6.5–7.5) leads to adsorption on suspended particles in wastewaters (82–96% of total coprostanol) and to a lesser extent to the colloidal phase. In the marine environment, biodegradation is rather fast in oxic waters and sediments (i.e., a half-life of weeks), whereas in anoxic sediments it takes place more slowly (several months).

Typical concentrations of coprostanol in raw sewage and sewage sludge span from 60 to 2700 $\mu g\,L^{-1}$ and from 0.26 to 7.8 $mg\,g^{-1}$ dw, respectively [51]. In treated effluents, coprostanol decreases in concentration by 1 or 2 orders of magnitude, depending on the type of treatment (i.e., primary or activated sludge). The highest concentrations in marine sediments are detected in urbanized harbors (10–935 $\mu g\,g^{-1}$ dw) and estuaries (1.6–176 $\mu g\,g^{-1}$ dw) where sewage waters are discharged [52, 53]. Contents in coastal sediments exhibit a large variability, but values generally decrease with distance from the source. Reasonably high concentrations of coprostanol and epicoprostanol have been detected in deep sea sediments (0.18–1.09 $\mu g\,g^{-1}$ dw) reflecting direct sewage sludge disposal (e.g., at Deep Water Dump Site 106, 185 km offshore New Jersey, at a depth of 2400–2900 m) [54] or by advective transport of fine sediments combined with its preservation in anoxic conditions (e/g., at southern California Bight at a depth of 908 m) [55].

Source Indices

One of the limitations of coprostanol as a sewage tracer is its limited source specificity, especially in highly reducing environments (i.e., anoxic sediments, sea marshes). Several authors reported the occurrence of coprostanol in sediments (i.e., coastal, remote sea, deep ocean) and waters (coastal and estuary) where there are no known sewage inputs [55, 56]. Therefore, a variety of indices have been reported to improve the specificity of coprostanol as sewage marker.

The limited source specificity of fecal steroids can be mainly attributable to first, the *novo* synthesis of 5β-stanols from stenols in strongly reducing conditions such as anoxic sediments [57] or second, the fact that marine mammals can also excrete coprostanol, although at lower concentrations than its epimer epicoprostanol [58]. Therefore, the ratio of coprostanol to epicoprostanol has been proposed as an index to ensure that coprostanol is derived from human waste [55].

Another approach to assess the fecal origin of 5β(H)-stanols is based on the fact that the reduction of cholesterol to coprostanol in the environment leads to the formation of significant amounts of 5α(H)-cholestanol, whereas in bacterial reduction coprostanol is the most abundant compound formed [49]. Based on the stereospecificity of cholesterol reduction, an index based on the isomeric stanol ratio ($5\beta/5\alpha + 5\beta$) was introduced to assess the coprostanol origin [56]. Although values of 0.7 to 1.0 were initially proposed for that index in sewage-polluted temperate seawaters, lower values have been reported in other areas [59].

Related indices to trace sewage pollution based on the ratio of coprostanol to cholestanol + cholesterol ($5\beta/(5\alpha + \text{cholesterol})$) have also been suggested [60]. However, the widespread occurrence of cholesterol (i.e., sewage and natural sources) makes that index questionable in eutrophic aquatic environments.

The proportion of coprostanol + epicoprostanol to total sterols (% COP) and the ratio between coprostanol + epicoprostanol to dinosterol have also been proposed [55, 61] as indices of sewage addition into marine sediments. Based on the $5\beta/(5\alpha + \text{cholesterol})$ ratio, a mass balance model was developed to correlate the observed fecal stanol-to-sterol ratio with the sewage sludge loading on sediments. This modeling revealed the nonlinear relationship between the stanol-to-sterol ratio and the sludge deposition flux [59]. Furthermore, a concurrent measurement of coprostanol and bacterial indicators has been used to discern fecal pollution entering a bay through stormwaters [62].

3.1.2
Surfactant Derived Compounds

Linear Alkylbenzenes (LABs) and Tetrapropylene-Based Alkyl Benzenes (TABs)

Linear alkyl benzenes (LABs) comprise 26 compounds with an n-C_{10}–C_{14} n-alkyl chain (Table 3). They are the raw material for the production of linear alkylbenzene sulfonates (LAS), one of the most widely used anionic surfactants. They have replaced the tetrapropylene-based alkylbenzenes (TABs) (Table 3) used until the 1960s for the production of the corresponding alkylbenzene sulfonates (ABS), which were rather refractory to biodegradation. Theoretically, 80 000 TAB compounds are possible, due to alkyl chain branching, but their resolution is limited by the GC column efficiency. The predominant alkyl chain length is C_{12} with minor proportions of C_{11} and C_{13}.

Small amounts of LABs escape the sulfonation process and are present in commercial formulations being released to the environment when LAS-type detergents are used. Typical concentrations of the LAB residues in the LAS-type detergents range from 20 to 800 μg g^{-1} [63]. Alternative sources of LABs, other than surfactant-related products, are fluid oils used in offshore drilling, some crude oils and coal pyrolysis oils [51]. However, the different distribution of homologs (C_3–C_{29}) depending upon whether they are of diagenetic or synthetic origin, allows the distinction of these alternatively sourced LABs from surfactant-related LABs. The highly hydrophobic nature of LABs leads to their adsorption onto suspended particles and sediments, and so are useful molecular markers to trace urban pollution in the marine environment.

The use of LABs as molecular markers was first reported by Ishiwatari et al. [64] in Tokyo Bay and Eganhouse et al. [63] in southern California. Since then, LABs and coprostanol have been widely used in many regional studies as urban molecular markers. Typical LAB values found in raw wastewater are in the range of 18–85 µg L^{-1}, which can be compared with 2.6–11.8 and < 0.08 to 32.6 µg L^{-1} in primary and primary-secondary effluents, respectively. Concentrations in sewage sludges range from 206 µg g^{-1} in primary sludges to 44–433 µg g^{-1} in digested samples. Concentrations in sediments exhibit a large variability depending on the source distance. Typical values range from 0.15 µg g^{-1} in coastal sediments up to 43 µg g^{-1} in harbor sediments. LABs have also been used to assess the fate of hydrophobic organic compounds released to a large urban harbor and adjoining offshore waters [65].

LABs have been analyzed in ^{210}Pb-dated sediment cores to assess the temporal changes in the waste water pollution of Santa Monica Bay (southern California), as a consequence of improved treatment processes and better source control [66]. Surface distribution showed maximum concentrations near sewage outfalls, but no temporal trend was observed in contrast with other organic contaminants. LABs have also been reported to bioconcentrate in zooplankton, polychaetes, and bivalves. However, due to metabolic degradation of the aromatic ring, no biomagnification is apparent at higher trophic levels. The highest concentrations have been found in bivalves (6.3 µg g^{-1}) [51].

Trialkylamines (TAMs) constitute a suite of tertiary aliphatic amines varying in the length of their alkyl substituents; the most abundant class contains a methyl group and two linear alkyl chains, ranging from n-C$_{14}$ to n-C$_{18}$ with an even carbon number predominance (Table 3).

TAMs and linear alkylnitriles (LANs) are intermediate products in the synthesis of quaternary amines (ditallowdimethylammonium chloride, DTDMAC), a cationic surfactant widely used as fabric softener. TAMs and LANs are present as impurities in the commercial products (450–500 µg g^{-1} and 300–350 µg g^{-1}, respectively) and are released into the environment after the products are discharged in wastewater [67]. The elevated concentrations of TAMs found in urban wastewaters (27–390 µg L^{-1}) and receiving ecosystems is an indication of their persistence in the environment, probably due to their high molecular weight (i.e., MW = 423–535) and the low reactivity to biotic and abiotic processes. In fact, TAMs have demonstrated a marked stability in direct photolysis experiments [68], and the conservative homolog distribution along seaward transects suggests no selective biodegradation or adsorption [69]. Therefore, TAMs have been proposed as a tracer of urban inputs in the marine environment [70].

TAMs have been used to trace the transport of suspended organic matter from river and point sources to the open western Mediterranean and Black

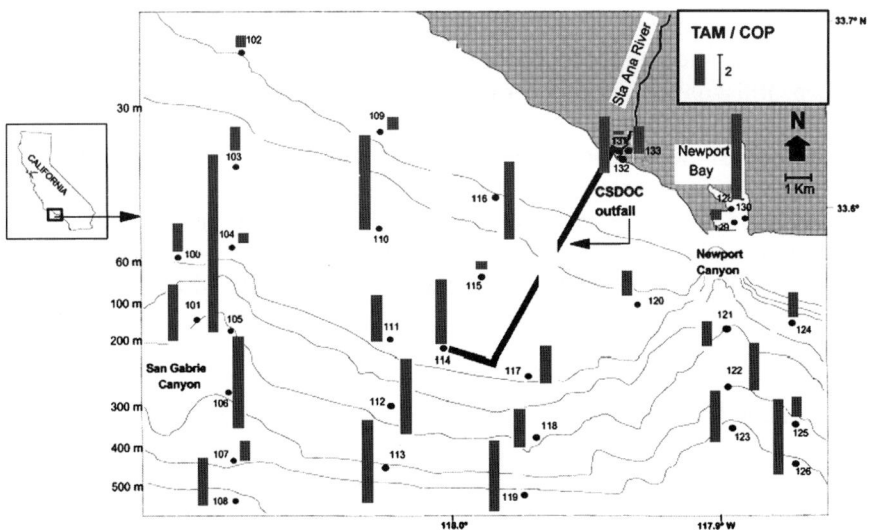

Fig. 2 Concentration ratios between TAMs and coprostanol in surface sediments from the San Pedro shelf in the vicinities of the Orange County outfall (southern California)

seas [71], of the sedimentary organic matter from enclosed bays [72], and sedimentary material from the continental shelf to depocenters in the open ocean [73] (Fig. 2).

On the other hand, LANs, exhibiting lower concentrations in waste waters and shorter half-lives in the aquatic environment, can only be used to trace urban pollution in continental waters [67]. The positive correlation between TAMs and coprostanol in urban wastewaters suggest that direct industrial sources of TAMs are negligible. The cationic form of TAMs is predominant in seawater giving rise to a strong adsorption onto negatively charged clays, and their long-chain alkyl substituents may be responsible for hydrophobic interactions with organic matter.

During the last decade, the quaternary alkyl amines were replaced in western Europe by other classes of more environmentally friendly cationic surfactants (i.e., esterquats), but they are still used in many technical applications (i.e., as anti-static agents, as asphalt additives, and in the textile industry) and consequently recent inputs into the marine environment still occur, and thus, it can be used as a marker of generic anthropogenic activity.

3.1.3
Other Molecular Markers of Urban Origin

A variety of compounds occurring in wastewaters of urban origin such as caffeine, α-tocopheryl acetate, urobilin, fluorescent whitening agents, and aminopropanone have been reported (Table 3). They have been used to trace urban sewage in continental waters, enclosed bays, and near-sewage outfalls [51], but their applicability to the open sea has not been demonstrated.

Benzothiazoles (BZs) and derivatives have been proposed to trace storm runoff contamination into the aquatic environment [74]. These compounds leach from crumb rubber material (CRM) and asphalt containing CRM. They enter the environment from a number of sources such as leaching rubber products, fine particles of rubber tires, and antifreeze products. They can be volatilized and biodegraded [75]. Recent studies have shown that some BZ can be rapidly photodegraded upon exposure to sunlight, which limits its use as runoff indicator in coastal regions [76].

3.2
Hydrocarbon Markers

Hydrocarbons are ubiquitous in the marine environment being of biogenic and anthropogenic nature, the former contributing to the background source of hydrocarbons and the latter mainly as a result of accidental or intentional oil spills, urban runoff, and atmospheric deposition. The molecular marker approach can be advantageously used for the characterization and differentiation of hydrocarbons from different sources, which is essential in any objective spill study in order to distinguish spilled from background hydrocarbons, to determine the pollutant source, and to evaluate the extent of the impact on the ecosystem.

Biogenic hydrocarbons exhibit specific distribution patterns, usually characterized by the large predominance of certain components, namely odd-carbon numbered n-alkanes, certain acyclic isoprenoids (e.g., pristane and squalene) or PAHs (e.g., perylene), or specific stereoisomers (e.g., 6R,10S-pristane). Those derived from shipping accidents or atmospheric deposition display other salient features that petroleum geochemists have been using since the 1970s for oil–oil and oil–source rock correlations [77]. These can be easily analyzed by gas chromatography-mass spectrometry (GC-MS) using selected ion monitoring. Particularly valuable are sterane derivatives and pentacyclic triterpanes, as shown in Fig. 3 and Table 4. As a consequence, GC-MS has emerged as the most powerful tool for oil-spill fingerprinting [78]. This technique has been used in assessing the fate of major oil spills [79–83] and the sources of tar balls chronically polluting coastal zones [84, 85].

Initially, the triterpane profiles (m/z 191) (Fig. 3 and Table 5) provide the T_m-to-T_s (C_{27}) ratio and the C_{29}-to-C_{30} ratio as useful initial parameters for oil source recognition. The occurrence of 28,30-bisnorhopane (C_{28}) is a distinctive feature of some North Sea (e.g., Brent) and Californian oils (e.g., Monterey). The presence of this triterpane was found of diagnostic value in the *Exxon Valdez* and the *Aegean Sea* oil spills—in the first case to differentiate between the oil cargo and the presence of former residues of Californian oils, and in the second to ascertain the occurrence of the spilled North Sea oil in the area [82, 83].

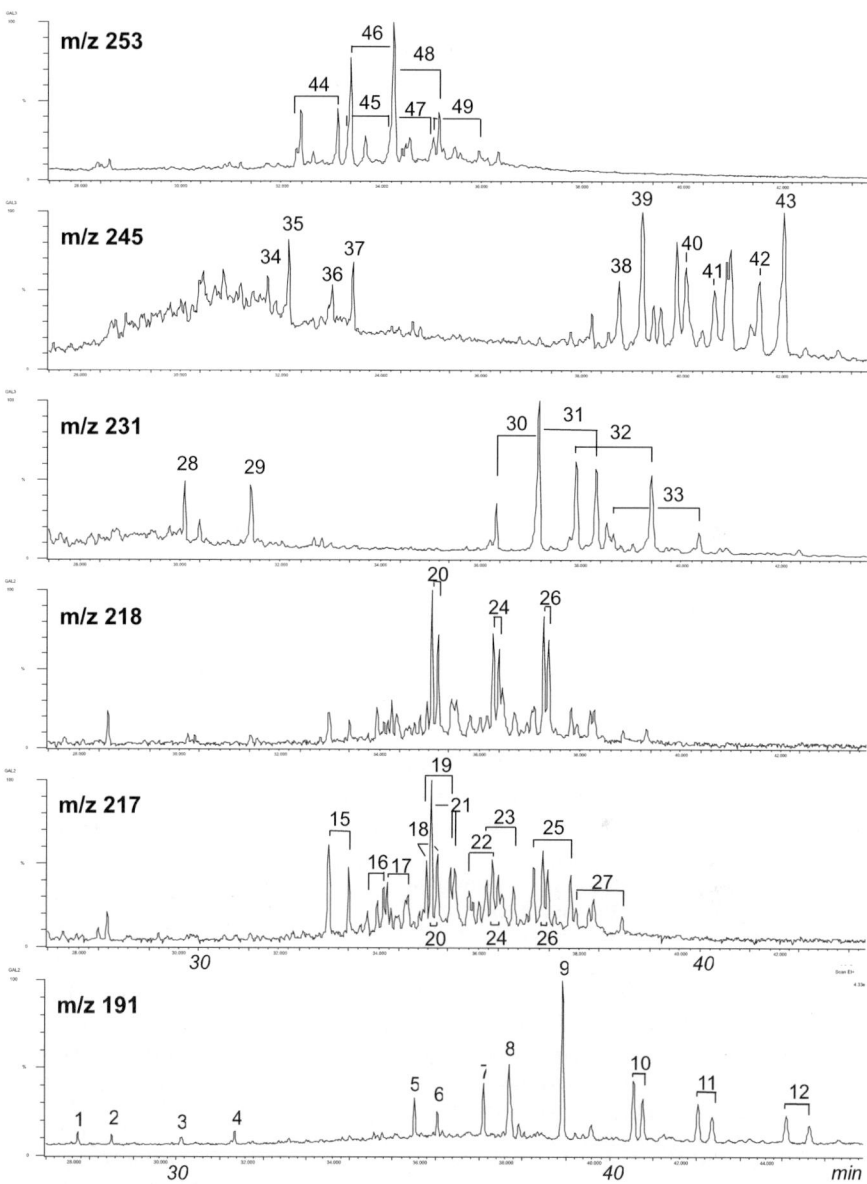

Fig. 3 Reconstructed ion traces of hopanes and steranes used for oil fingerprinting. Compound identification is indicated in Table 4

Within the triterpane profile, the abundance of tricyclic alkanes relative to hopanes is a valuable feature, which can also provide a convenient fingerprint in oil spill studies. For example, the differences in the abundance of two C_{26} tricyclic alkane isomers relative to the C_{24} tetracyclic

Table 4 Identification of individual components of ionic traces shown in Fig. 3

GC-MS	Peak label	Compound**	Carbon number	Structure*
m/z 191	1.	tricyclic terpane (13β,14α)	C-23	I, R = i-C$_4$H$_9$
	2.	tricyclic terpane (13β,14α)	C-24	I, R = i-C$_5$H$_{11}$
	3.	tricyclic terpane (13β,14α)	C-25	I, R = i-C$_6$H$_{13}$
	4.	17,21-secohopane	C-24	II
	5.	18α(H)-22,29,30-trisnorneo-hopane (T$_s$)	C-27	III
	6.	17α(H)-22,29,30-trisnor-hopane (T$_m$)	C-27	IV, R = H
	7.	17α(H), 18α(H),21β(H)-28,30-bisnorhopane	C-28	V
	8.	17α(H),21β(H)-30-norhopane	C-29	IV, R = C$_2$H$_5$
	9.	17α(H),21β(H)-hopane	C-30	IV, R = i-C$_3$H$_7$
	10. S, R	17α(H),21β(H)-homohopane (22S and 22R)	C-31	IV, R = i-C$_4$H$_9$
	11. S, R	17α(H),21β(H)-bishomohopane (22S and 22R)	C-32	IV, R = i-C$_5$H$_{11}$
	12. S, R	17α(H),21β(H)-trishomohopane (22S and 22R)	C-33	IV, R = i-C$_6$H$_{13}$
	13. S, R	17α(H),21β(H)-tetrakishomohopane (22S and 22R)	C-34	IV, R = i-C$_7$H$_{15}$
	14. S, R	17α(H),21β(H)-pentakishomohopane (22S and 22R)	C-35	IV, R = i-C$_8$H$_{13}$
m/z 217	15. S, R	13β(H),17α(H)-diacholestane (20S and 20R)	C-27	VI, R = H
	16. S, R	13α(H),17β(H)-cholestane (20S and 20R)	C-27	VII, R = H
	17. S, R	24-methyl-13β(H),17α(H)-diacholestane (20S and 20R)	C-28	VI, R = CH$_3$
	18. S, R	24-methyl-13α(H),17β(H)-diacholestane (20S and 20R)	C-28	VII, R = CH$_3$
	19. S, R	14α(H),17α(H)-cholestane (20S and 20R)	C-27	VIII, R = H
	20. R, S	14β(H),17β(H)-cholestane (20S and 20R)	C-27	IX, R = H
	21. S, R	24-ethyl-13β(H),17α(H)-diacholestane (20S and 20R)	C-29	VI, R = C$_2$H$_5$
	22. S, R	24-ethyl-13α(H),17β(H)-diacholestane (20S and 20R)	C-29	VII, R = C$_2$H$_5$

* The description of the structures is shown in Fig. 4
** Basic structures corresponding to aromatic steranes

Table 4 continued

GC-MS	Peak label	Compound**	Carbon number	Structure*
	23. S, R	24-methyl-14α(H),17α(H)-cholestane (20S and 20R)	C-28	VIII, R = CH$_3$
	24. R, S	24-methyl-14β(H),17β(H)-cholestane (20S and 20R)	C-28	IX, R = CH$_3$
	25. S, R	24-ethyl-14α(H),17α(H)-cholestane (20S and 20R)	C-29	VIII, R = C$_2$H$_5$
	26. R, S	24-ethyl-14β(H),17β(H)-cholestane (20S and 20R)	C-29	IX, R = C$_2$H$_5$
	27.	24-propyl-cholestanes	C-30	VIII-IX, R = C$_3$H$_7$
m/z 231	28.	pregnane	C-20	X, R = H
	29.	20-methylpregnane	C-21	X, R = CH$_3$
	30. S, R	cholestane (20S and 20R)	C-26	XI, R = H
	31. S, R	24-methylcholestane (20S and 20R)	C-27	XI, R = CH$_3$
	32. S, R	24-ethylcholestane (20S and 20R)	C-28	XI, R = C$_2$H$_5$
	33. S, R	24-propylcholestane	C-29	XI, R = C$_3$H$_7$
m/z 245	34.	1-methylpregnane	C-21	XII, R = H
	35.	4-methylpregnane	C-21	XIII, R = H
	36.	1,20-dimethylpregnane	C-22	XII, R = CH$_3$
	37.	4,20-dimethylpregnane	C-22	XIII, R = CH$_3$
	38.	1-methylcholestane (20R)	C-27	XIV, R = H
	39.	4-methylcholestane(20R)	C-27	XV, R = H
	40.	1,24-dimethylcholestane (20R)	C-28	XIV, R = CH$_3$
	41.	4,24-dimethylcholestane (20R)	C-28	XV, R = CH$_3$
	42.	1-methyl-24-ethylcholestane (20R)	C-29	XIV, R = C$_2$H$_5$
	43.	4-methyl-24-ethylcholestane (20R)	C-29	XV, R = C$_2$H$_5$
m/z 253	44. S, R	5β(H)-cholestane (20S and 20R)	C-27	XVI, R = H
	45. S, R	5α(H)-cholestane (20S and 20R)	C-27	XVI, R = H
	46. S, R	5β(H)-methylcholestane (20S and 20R)	C-28	XVI, R = CH$_3$
	47. S, R	5α(H)-methylcholestane (20S and 20R)	C-28	XVI, R = CH$_3$
	48. S, R	5β(H)-ethylcholestane (20S and 20R)	C-29	XVI, R = C$_2$H$_5$
	49. S, R	5α(H)-ethylcholestane (20S and 20R)	C-29	XVI, R = C$_2$H$_5$

* The description of the structures is shown in Fig. 4
** Basic structures corresponding to aromatic steranes

terpane were also used to identify oils derived from the *Exxon Valdez* spill [82].

Less common markers may provide unique interpretational advantages in fingerprinting oil sources. Bicadinanes are significant constituents in oils

Table 5 Characteristic oil source indicators

Parameter	Definition	Diagnostic ion (m/z)
Pr/Ph	pristane/phytane	113
%27 Ts	100*Ts/(Ts+Tm)	191
%29$\alpha\beta$	100*29$\alpha\beta$/(29$\alpha\beta$ + 30$\alpha\beta$)	191
%27d	100*27d(R + S)/[27d(R + S)+ 27$\beta\beta$(R + S)]	217
%29$\alpha\alpha$S	100*29$\alpha\alpha$S/(29$\alpha\alpha$S+29$\alpha\alpha$R)	217
%29$\beta\beta$(R + S)	100*29$\beta\beta$(R + S)/[29$\beta\beta$(R + S)+ 29$\alpha\alpha$R + S)]	217
%27$\beta\beta$	100*[27$\beta\beta$(R + S)]/[27$\beta\beta$(R + S)+ 28$\beta\beta$(R + S)+ 29$\beta\beta$(R + S)]	218
%28$\beta\beta$	100*[27$\beta\beta$(R + S)]/[27$\beta\beta$(R + S)+ 28$\beta\beta$(R + S)+ 29$\beta\beta$(R + S)]	218
%29$\beta\beta$	100*[27$\beta\beta$(R + S)]/[27$\beta\beta$(R + S)+ 28$\beta\beta$(R + S)+ 29$\beta\beta$(R + S)]	218
D2/P2	100*C2–DBT/(C2–DBT+C2–P)	206/212
D2/P2	100*C3–DBT/(C3–DBT+C3–P)	220/226

from Malaysia, Bangladesh and Indonesia [86]. On the other hand, pentacyclic triterpanes, such as oleanane, ursane, and lupane, are particularly abundant in some terrestrial oils (e.g., China and Nigeria) [87].

The sterane profiles (m/z 217 and 218) (Fig. 3) show a number of isomers having different stereochemistry at positions 5, 14, 17, 10, and 24, and therefore, they are quite complex. The distribution of the C_{27}, C_{28}, and C_{29} parent components, which is characteristic of each crude oil (Table 5), is easily determined from the m/z 218 fragmentogram, as are the epimeric 20S/R ratios. A predominance of the C_{27} steranes is typical of Middle East crude oils.

The sterane 20S/S + R epimeric ratios are basically maturity indicators, as are the hopane 22S/S + R ratios, and they usually reach the thermodynamic equilibrium values in mature crude oils. Therefore, they are of limited interest for fingerprinting oil spills. However, they may afford information on the extent of oil biodegradation because it is known that compounds with the biological 20R configuration are removed faster than 20S compounds. Moreover, 14α(H), 17α(H) isomers are removed faster than the 14β(H), 17β(H); and carbon number also affects the degradation rate with C_{27} > C_{28} > C_{29} [77, 88]. The effects of biodegradation in crude oils were summarized by Volkman et al. [89], who found that the rates of degradation usually follows the order *n*-alkanes > alkylcyclohexanes > acyclic isoprenoids > bicyclic alkanes > steranes > hopanes > diasteranes. These alterations cannot be ignored when comparing the profiles of environmental samples with the suspected sources.

Although n-alkanes cannot be considered useful markers for oil fingerprinting, except for an initial screening of the type of samples (boiling range and weathering state), the introduction of compound-specific isotope ratio monitoring techniques (GC-IRMS), which provide ^{13}C-to-^{12}C ratios for individual hydrocarbons, has been proposed. Several studies have already appeared that describe the use of this technique to elucidate the origin of various products found in the marine environment [90, 91].

Oil spill investigations can also take advantage of the distributions of polycyclic aromatic hydrocarbons (PAHs). Low-molecular-weight (2–3-aromatic-ring) compounds together with their alkylated derivatives are typical constituents of petroleum, whereas larger-molecular-weight (> 4-aromatic-ring) non-alkylated compounds, particularly peri-condensed, derive from high-temperature combustion processes. The distributions of alkylated PAH homologous series have been proposed for fingerprinting crude oils [92], although their composition can be significantly affected by evaporation/dissolution and degradation of the lower ring and less alkylated species. In this respect, alkyl naphthalenes, which are predominant in crude oils, are usually present in minor amounts in the weathered samples.

The C_1–C_3 alkylphenanthrenes (P), chrysenes (C), and dibenzothiophenes (D) are more persistent but show typical petrogenic distributions, which are common to most oils, and so are of limited diagnostic value. However, recently, the C_2 and C_3 alkyl homologs of dibenzothiophene and phenanthrene/anthracene and the same homologs of dibenzothiophene and chrysene/benz[a]anthracene, were found useful in, respectively, differentiating hydrocarbon sources and characterizing the weathering of the spilled oil in sediments (Tables 5 and 6) [93]. The corresponding ratios were shown to be stable at moderate degradation levels in different marine environments, al-

Table 6 Characteristic oil weathering indicators

Parameter	Definition	Diagnostic ion (m/z)
n-alkanes	$(C_{13} + C_{14})/(C_{25} + C_{26})$	85
C18/Ph	n-C_{18}/phytane	85
N2	$(N + N1)/N2$	128/142/156
2MP/1MP	100*2MP/(1MP + 2MP)	192
1MP/9MP	100*1MP/(1MP + 9MP)	192
4MD/1MD	100*4MD/(1MD + 4MD)	198
2 + 3MD/1MD	100*2 + 3MD/(2 + 3MD + 1MD)	198
D2/C2	100*C2–DBT/(C2–DBT + C2–C)	212/256
D3/C3	100*C3–DBT/(C3–DBT + C3–C)	226/270

though the study of samples collected in the Arabian Gulf after the war did not afford conclusive results [80].

Mono and tri-aromatic steroid hydrocarbons can also be used for source recognition since they are not biodegraded until steranes and triterpanes are severely affected; they can be readily detected by GC-MS using, respectively, m/z 253 and 231 mass fragmentograms [94]. Although these hydrocarbons are closely related to the source-dependent sterane distribution, the coelution of several components makes it difficult to choose adequate ratios for oil characterization. Moreover, in samples containing high contents of pyrolytic PAH, the m/z 253 profile is misrepresented by the coelution of the benzofluoranthenes and pyrenes (m/z 252). For this reason the triaromatic steranes (m/z 231) as well as their ring-A methylated derivatives (m/z 245) (Fig. 3 and Table 4) were selected in the assessment of the *Aegean Sea* oil spill, where most of the cargo was burned and large amounts of pyrogenic PAHs were deposited in the area [83]. However, those marker profiles often show small variations among crude oils, and they should therefore be used with caution in environmental studies.

Superimposed on the petrogenic signature, GC-MS analyses of the aromatic fractions from marine samples usually show the occurrence of a mixture of pyrogenic hydrocarbons, such as 3- to 6-ring PAHs considered to be typically combustion-derived compounds. These series range from the 3-ring aromatic anthracene to the 6-ring indeno[1,2,3-*cd*]pyrene and benzo[*ghi*]perylene, including the benzo[*b*] + [*k*]fluoranthenes. Elevated concentrations of pyrogenic PAHs formed during the burning of the cargo oil were found in the accidents of the tankers *Haven* (off Genova, Italy) [93] and *Aegean Sea* (Coruña, Spain) [83]. These types of profiles are common in coastal sediments of urban/industrial areas [95, 96] and are currently attributed to chronic atmospheric inputs or urban runoff.

By examining the relative abundances of isomeric parent PAHs, e.g., Ph-to-A, Fl-to-Py, BaA-to-chrysene, and IP-to-BPe ratios, it is also possible to identify sources of PAHs and broadly determine the relative contribution of each source to coastal sediments (e.g., used oils, coal soot, diesel exhausts, etc.) [95, 97–99]. However, in remote areas, the source apportionment is hampered by the depletion of more reactive components during long-range transport [96]. Several studies have shown that degradation of the PAH is strongly dependent on their structural features and is influenced by the atmospheric content of reactive gases (see Sect. 4.1).

From all the above it is apparent that a large number of indicators can be used for GC-MS fingerprinting of marine hydrocarbons. Qualitative comparison of molecular marker distributions of a spilled oil with the candidate sources and background materials is, in many instances, sufficient for a forensic investigation. However, when the similarity is not obvious, a quantitative analysis is required, which can be enforced with the application of multivariate statistical methods. Moreover, only those indices that can be measured

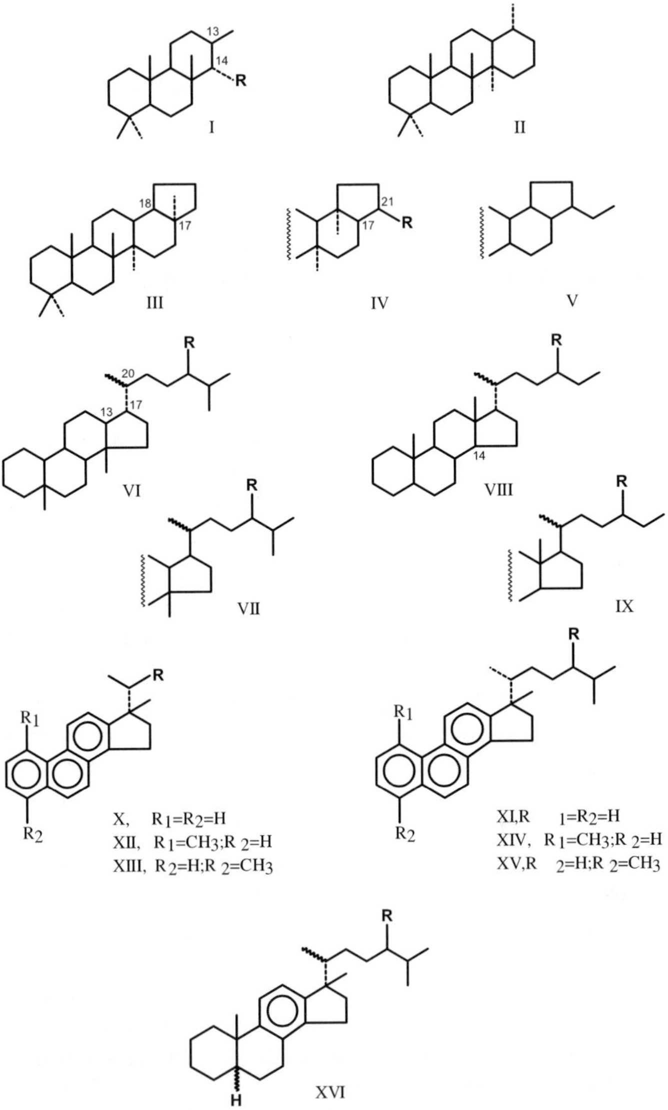

Fig. 4 Structure of selected molecular markers used in petroleum fingerprinting

precisely should be considered for comparing candidate sources to a spilled oil. The precision of the diagnostic indices in this type of oil spill correlation studies is fundamental to its defensibility. Stout et al. [100] have suggested a protocol by which candidate diagnostic ratios are evaluated in order to identify those that are more useful for correlation studies. Indices having relatively high % RSD (> 5%) values may be inappropriate for quantitative evaluation.

In summary, tiered analytical approaches including preliminary screening of the product (GC-FID), pattern recognition of target markers (GC-MS), determination of source diagnostic ratios, and quantitative assessment, are the most up-to-date methodology for oil-spill identification, which has been the basis of widely accepted standard methods [101, 102].

4
The Transport and Transformation of Organic Contaminants

4.1
Major Transport Pathways

The relative importance of transport relative to transformation processes in controlling the fate of the organic pollutants discussed so far is constrained by their persistence in the environment, a common feature among most of them. However, this is not sufficient for long-range transport in the atmosphere and the oceans. Another important characteristic is their distribution pattern in different environmental media (air, water, and sediments), which is mainly determined by their volatility (log K_{OA}) and water solubility (log K_{AW}). According to this method, three major groups of pollutants can be distinguished (Table 7). These are not sharply defined as these properties are temperature-dependent, and the particular categorization may differ according to the area or region where they are found.

Table 7 Categorization of organic pollutants in terms of their transport behavior [2]

Category	Characterization	Examples
A: single-hop	Involatile and water-insoluble contaminants that can undergo transport only by piggy-backing on suspended solids in air and water	Highly halogenated PCDD/Fs and PBDEs high-molecular-weight PAHs, mirex
B: multi-hop	Chemicals that readily shift their distribution between gas and condensed phase in response to changes in temperature and phase composition; can travel long distances in repeated cycles of evaporation and deposition	PCBs, HCB, lower-molecular-weight PCDD/Fs, toxaphene, dieldrin, chlordane
C: no hop required	Chemicals that are sufficiently water-soluble to undergo long-range transport by being dissolved in the water phase	HCHs, triazines, PFOS

The transport behavior of category A compounds (high K_{OA} values) is mainly controlled by the location of atmospheric contaminant sources relative to major atmospheric flow patterns. Atmospheric conditions at the time of release will have a strong impact on their transport behavior, and areas close to the sources are generally affected more strongly than those further away. Efficient atmospheric transport is restricted to episodes that are characterized by conditions that favor rapid horizontal air movement and lack of precipitation. Once deposited, these chemicals will only move toward the marine environment if the particles to which they sorb are remobilized (e.g., through heavy rainfalls, continental runoff, etc.) [103].

On the other hand, the behavior of category B substances (log K_{AW} : 0 to − 4; and log K_{OA} : 6 to 12) is controlled by the ease of transfer between the atmosphere and the sea surface. Persistent chemicals, which change from a gaseous state to a condensed state within the environmentally relevant temperature range, will undergo air–surface exchange more often and most likely travel long distances. Because cold temperatures favor deposition over evaporation and warm temperatures favor evaporation over deposition, temperature gradients in space in combination with atmospheric mixing will favor gradual transfer from warm to cold regions, as already observed on a global scale [6, 9].

When the organic contamination consists of a mixture of compounds of different volatility, such as the PCBs, they undergo air–surface exchange to a different extent and tend to experience shifts in their relative composition with distance from the source. The less volatile constituents of a mixture are retained more efficiently close to the source, whereas the more volatile travel further. This phenomenon, called "global fractionation" has been confirmed by measurements of compositional shifts of PCBs in marine sediments [104].

Category C chemicals (low K_{AW} and K_{OA} values) are so water soluble that they remain dissolved in the aqueous phase and can be dispersed by oceanic circulation. The ocean currents potentially play a role in transporting persistent organic pollutants, although very little experimental evidence exists for marine transport at lower latitudes. The best evidence comes from the extensive study of the distribution of HCHs in Northern waters [105]. However, because the same functional groups that impart water solubility often also increase the rate of degradation, chemicals in this category usually occur at only moderate distances from the emission sources.

The above contaminant features delineate several transport pathways of special relevance in marine pollution studies—namely, the land-to-ocean transfer, the role of the air–seawater exchange, ocean particle settling, and transoceanic transport. An assessment of the transfer processes at the land–ocean boundary is of special interest, particularly in arid or tropical regions affected by strong rainfall episodes. Rivers are major pollutant vectors to the marine environment. However, beyond the zone of influence of their discharges, concentrations drop rapidly, reflecting enhanced sedimentation processes, which take place at the fresh water–sea water interface many times in the river estuary. The continental shelf will thus likely constitute the final resting place of many persistent organic pollutants delivered by rivers to the oceans [12, 106]. Documented examples of regional importance are the transport of OCs from Japan to the Northern Pacific ocean, the input of land-derived pollutants to the Mediterranean Sea, and the transfer of organochlorine pesticides from agricultural areas to the Baltic Sea.

Once in the water column, hydrophobic semi-volatile compounds are partitioned between the dissolved, colloidal, and particulate phases depending on the POC and DOC contents of seawater. There is clear evidence now that gas exchange is a major contributor to atmospheric loadings of POP [107]. Air-water exchange of POPs has been the subject of intense research over the last decade [108, 109]. In this respect, the sea surface microlayer is a unique compartment where high enrichments of organic matter and POPs may occur [110, 111] and thus may play a key role in their transfer between the atmosphere and the marine environment [112].

On the other hand, particle-associated contaminants, depending on the particle size, can settle through the water column to sediments or be advectively transported. The extent of gravitational settling depends on the particle density, which in turn is affected by runoff and primary productivity and is thus likely highest in coastal and shelf areas, and marine regions of nutrient upwelling. Deposition from the surface ocean has been estimated to be highest in mid-high latitudes [113]. Along with the vertical transport of particles, biogeochemical processes can take place, and the more water-soluble contaminants are degraded or solubilized, leading to enrichment in the most hydrophobic compounds with depth, which can then be accumulated in sediment [114].

Interest in transoceanic transport arose from the evidence that the local sources of OCs in the Arctic and Antarctica do not account for the levels found, thus indicating the occurrence of transport from other world regions. Examples are transoceanic transport of DDT and other pesticides across the Pacific Ocean [115] and the deep sea outflow of Mediterranean DDTs, PCBs, and PAHs into the Northern Atlantic [116].

4.2
Biotic Processes

Microbial degradation is by far the most important route of cycling organic contaminants in the marine environment. It can occur in the water column, during the early stages of sedimentation, and in subsurface sediments. Aerobic processes tend to dominate in the water column, but anaerobic processes become more important in the sediments. The occurrence of biodegradation can be assessed either by the formation of intermediate compounds with the concurrent disappearance of the parent contaminant or by compositional changes of compound assemblages (e.g., PCBs, PAHs) or stereoisomer groupings, as will be illustrated below. In many cases, however, abiotic processes (i.e., photooxidation, hydrolysis, oxidation) can lead to the formation of the same intermediates, making it difficult to assess which is the prevailing process affecting the changes in the original composition. Furthermore, photooxidation may result in the production of biologically available photoproducts, thus leading to a more rapid oxidative turnover of materials [117]. The interactions between biological and photochemical turnover of organic pollutants in the marine environment, particularly in surface waters, has not been fully investigated.

4.2.1
Assessment by Identification of Intermediates

This approach is useful to assess the pathway of contaminant degradation, but can be limited in the marine environment by analytical constrains related to identification and quantification of the metabolites; these metabolites are usually more water soluble than the parent compound and occur at much lower concentrations.

Organochlorine Compounds

Hydrolytic and reductive dehalogenation are the most likely degradation pathways of OCs. The hydrolytic dehalogenation involves a nucleophilic substitution of a halogen by a hydroxyl group occurring in aerobic conditions. On the other hand, the reductive dehalogenation involves the chlorine substitution by hydrogen, and it can occur either aerobically or anaerobically [118]. In the aerobic route, the biodegradation is catalyzed by a monooxygenase leading to the formation of an oxidized intermediate that can be further degraded via reductive dehalogenation or dehydration.

The hydrolytic dehydrochlorination of DDT to DDE in oxic conditions, in the earliest stages of diagenesis, is probably the most widely reported process. However, the rate of the DDE degradation in field conditions is 10^2–10^3 slower than in microcosms [119], which highlights the extremely low degradation rate of DDTs in nature. The reductive dechlorination of DDT to DDD and of DDE to DDMU in anaerobic conditions has also been reported [120]. Therefore, the ratio between the main DDT metabolites referred to total DDTs, DDE-to-ΣDDTs and DDD-to-ΣDDTs can be used to assess the occurrence of recent DDT inputs in the marine environment as well as its stability in different depositional environments [10, 121].

Biocides

The biotic degradation pathway of TBT is a successive debutylation to DBT, MBT, and inorganic tin. Under aerobic conditions, the half-life of TBT is 1–3 months, but in anaerobic sediments it is over 2 years [122].

A degradation index for BTs has been calculated based on the ratio between TBT and the degradation products, DBT and MBT, expressed as:

$$\%BTs_{deg} = \left[1 - \left(\frac{TBT}{TBT + DBT + MBT}\right)\right] \times 100.$$

In recent harbor sediments, the ratio ranged from 28 to 54%, reflecting some variability in TBT inputs in the different harbor zones.

As for phenylureas, the aerobic and anaerobic degradation of the booster biocide diuron (Sect. 2.6.1) involves reductive dechlorination and demethylation producing the following biointermediates: CPDU, DCPMU, and DCPU (Fig. 5) [35]. The half-life of CPDU in vitro under anaerobic conditions is higher than diuron and the other intermediates, although no field data are available to confirm this.

For S-triazines, the main metabolites are the dealkylation of the N-alkyl substituents (ethyl, isopropyl, cyclopropyl, *tert*-butyl) leading to the formation of dealkylated derivatives. The decyclopropyl metabolite of Irgarol named M1 or GS26575 (Table 2) has been identified in seawater and sediments [32]. However, a photolytic process cannot be ruled out because it leads to the formation of the same intermediate.

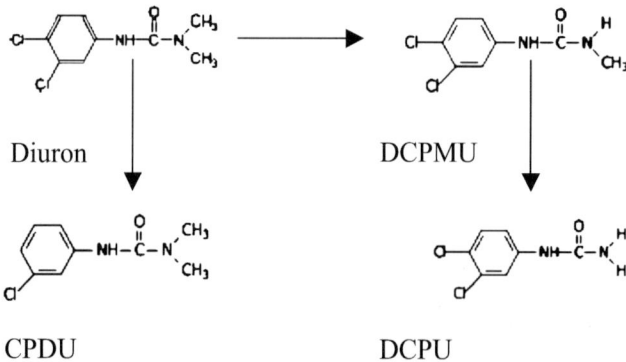

Fig. 5 Biodegradation pathways of diuron

4.2.2
Compositional Changes of Compound Assemblages

Aliphatic Hydrocarbons

Since normal alkanes are degraded faster than the isoprenoids, the n-C_{17}-to-pristane and n-C_{18}-to-phytane ratios have been widely used to assess the biodegradation of fossil fuel hydrocarbons in oil-polluted zones. As pristane also has a biogenic source, the latter ratio will usually be more informative. Other compound ratios have also been reported to assess oil weathering (see Table 6).

Linear Alkyl Benzenes (LABs)

The LABs include a mixture of isomers substituted at the 2, 3 and 4; and 5 and 6 positions of the alkyl chain, named as internal (I) and external (E) isomers, respectively. The different biodegradation rates of both groups of compounds—the external isomers (E) degrade faster than the internal ones (I)—can be used to assess the extent of degradation in the marine environment [123]. Takada and Ishiwatari [124] have proposed a degradation index based on the ratio between the different internal and external isomers of the linear alkylbenzene with a C_{12} alkyl chain:

$$I/E_{\text{ratio}} = \frac{(6 - C_{12}\text{LAB} + 5 - C_{12}\text{LAB})}{(4 - C_{12}\text{LAB} + 3 - C_{12}\text{LAB} + 2 - C_{12}\text{LAB})}$$

where the digit preceding C_{12} LAB denotes the alkyl isomer.

In commercial formulations, this index is about 0.7 and may increase to 5.7 depending on the type of wastewater treatment and the residence time in the environment. The highest values have been found in treated sewage (i.e., 3.5 ± 1.7) followed by aerobic sediments (i.e., 2.6 ± 1.2) [51]. Therefore, an I/E ratio in marine samples (e.g. suspended particles or sediments) lower than 2 suggests a recent input from raw wastewater or combined sewage overflow discharges [125].

Polycyclic Aromatic Hydrocarbons

Individual isomers of alkylated PAH and sulfur polycyclic aromatic compounds (S-PAC) are degraded at different rates. Therefore, a number of isomeric ratios can be used to assess biodegradation. In this regard, the 1- and 2-methylphenanthrenes (MP) and the 2- and 3-methyldibenzothiophenes (MD) are degraded faster than the 9-MP and 1-MD, respectively [126]; therefore, the 1 + 2MP-to-9MP and 2 + 3MD-to-1MD ratios can be used to trace early biodegradation of fossil PAHs in the marine environment (Table 6 in Sect. 3.2). In Sect. 3.2 other ratios, such as the C_2–DBT-to-C_2–C and

C_3–DBT-to-C_3–C ratios, where C_2 and C_3 correspond respectively to the di- and trimethyl substituted chrysenes (C), have been described to identify advanced stages in the degradation of oil spills.

4.2.3
Enantioselective Biodegradation of Contaminants

A number of organic pollutants consist of isomeric mixtures made up of the same atoms bonded in the same sequence, but possessing three-dimensional structures not superimposable (like two mirror images), a stereochemical feature known as chirality. The two stereoisomeric compounds are usually present in the same proportion in commercial formulations of chiral products. Taking into account that microbiological metabolism often reveals a distinction between enantiomers that is not evident in physico-chemical processes, by examining enantiomeric ratios (ER), it is possible, for example, to differentiate between new and old sources of organochlorine pesticides or to assess air–water exchange processes [127, 128].

In 1991, Faller et al. [129] were the first to report α-HCH enantiomers in seawater. α-HCH is the only chiral isomer of the eight conceivable HCH isomers, while γ-HCH is achiral, but is the only isomer exhibiting insecticidal activity. The transformation products of the two HCH parent compounds, β-pentachloro-cylcohexane (β-PCCH) and γ-PCCH, are both chiral. In turn, Ludwig et al. [130] were the first to prove the degradation pathway of chiral pollutants by enriched cultures of marine microorganisms, which was subsequently confirmed by field observations. They found that the enzymatic transformation of α-HCH and γ-HCH followed different pathways but led to ERs higher than unity. On the other hand, the microbial degradation of the chiral 2,4-dichlorophenoxyacetic acid (2,4-D), a polar herbicide that has attained considerable concentrations in the German Bight (SE North Sea), on the order of 10 ng L, showed preferential degradation of the R-enantiomer, consistent with enantiomer ratios found in marine ecosystems [130].

The enantioselective degradation of HCH isomers in waters of different oceanic regions (e.g., North, Baltic, and Bering seas) has shown comparable results—namely, a preferential degradation of (+)-α-HCH in all cases [7, 129]. However, a more complex situation was encountered in Arctic waters, where the ERs of dissolved α-HCH were generally > 1.0 in the Bering Sea and < 1.0 in the Greenland Sea [131], suggesting different microbial transformation pathways, as was found in the western and eastern parts of the North Sea [132].

These studies provided a better understanding of air–water exchange processes. The α-HCH in the atmosphere of the Arctic Ocean and the northern Atlantic, taken in 1993 and 1994, was depleted in the (+)-enantiomer (ER < 1.00), whereas air over the Bearing Sea was depleted in (–)-α-HCH (ER

> 1.00), following the same pattern shown the water [131]. The similarities in air and water enantiomeric profiles for the Bering and Greenland seas provide direct evidence of the "two-way street" of gas exchange, which in this case is an important source of α-HCH to the marine boundary layer. Air and water samples collected on another cruise in 1996 showed that HCHs are being lost from the Arctic Ocean through degradation, volatilization and advective outflow [133]. Organochlorine contamination in the Northern Hemisphere has been studied much more extensively than in the Southern Hemisphere. However, Jantunen et al. [134] have recently reported that transformation of α-HCH can also occur in the Antarctica.

4.3
Abiotic Processes

These involve a variety of chemical transformations usually taking place under sunlight radiation (photolysis). They can occur in the gas or in the aqueous phase, and their extent and rate largely depend on temperature, quantum yield of the analyte, analyte phase association, and solar irradiance (latitude). As already indicated, the combination of photolysis and biodegradation is suspected to favor the degradation of contaminants in aquatic systems because the photoproducts will increase their bioavailabitlity to microorganisms.

4.3.1
Gas-phase Reactions

Gas-phase reactions occur because of the presence of reactive species in the atmosphere, particularly the hydroxyl radical ($^{\cdot}$OH). The concentration of OH radicals, and, therefore, the rate of atmospheric reaction, vary greatly with season, time of day, latitude, and altitude from the sea surface. Highest OH radical concentrations, i.e., fastest degradation and thus reduced atmospheric long-range transport, occurs in low latitudes, at high altitudes, during daytime, and in summer. These reactions with the OH radical are also temperature-dependent with reactions occurring faster at higher temperatures. In the subtropical atmosphere, daytime depletions of PCB concentrations could be explained by efficient reaction with simultaneously measured OH radicals [135].

The fate of PAHs in the marine atmosphere and their subsequent long-range transport are also associated with the presence of reactive gases (e.g., NO_x, SO_x, O_3, free radicals), humidity, and temperature [136], as well as to their structural features. Considering the main PAH degradation pathways, the relative decay of atmospheric PAHs can be roughly rated as follows [137]: BaPy > BaA $\approx$ DBaA > BghiP $\approx$ Ipy > A $\approx$ Py > BePy $\approx$ Chr > Fl.

A consistent distribution of PAH in Mediterranean sediment transects going further off-shore has been observed [96], which can be attributed to the deposition of aerosols changing in composition during transport and aging. Simo et al. [138] have shown that the relative content of petrogenic PAHs with respect to the pyrogenic ones decreased from combustion sources to remote samples, suggesting that both present a decoupled association to soot particles. Whereas the soot–PAH interactions protect pyrogenic PAHs from major transformations, fossil PAHs are more exposed to vapor–particle partitioning.

The reaction products—namely, aromatic ketones, quinones, nitro-PAHs, etc.—have been identified in urban atmospheres [139], but only a few of them have been identified in the marine environment despite their ubiquity in the atmosphere [140, 141]. In fact, the 2-4 ring PAHs, which exist mainly in the gas phase, react faster with the hydroxy radical (calculated half-life less than 1 day). Further research is needed to establish the fate of these compounds in the marine environment.

4.3.2
Aqueous-phase Reactions

Aqueous-phase reactions can occur in the surface microlayer and in the photic layer of the water column via a direct or indirect photooxidation mechanism. Molecules that can absorb UV radiation are usually degraded directly; otherwise, a photosensitizer is needed. Marine humic and fulvic acids can act as photosensitizers or quenchers depending on their structure [142].

Hydrocarbons

Photooxidation of n-alkanes and alkylbenzenes in simulated marine environmental conditions was achieved with anthraquinone as a sensitizer [143, 144]. The main photoproducts identified were ketones and alcohols and a variety of minor components. Lower-molecular-weight carbonyl compounds (i.e., formaldehyde, acetaldehyde, and acetone) have also been identified from the anthraquinone-sensitized photochemical decomposition of aliphatic hydrocarbons in seawater [145].

Chlorinated Hydrocarbons

A variety of photoreactions at wavelengths > 290 nm have been reported for chlorinated aromatic contaminants, including photoreductive dechlorination and photoisomerization [146]. PCDDs and PCDFs can be photochemically produced from polychlorophenols and PCBs from chlorinated benzenes in irradiated solvent solutions. Some of these reactions may possibly occur in the sea surface microlayer.

Herbicides

The photodegradation of S-triazines, acetanilides, and thiocarbamates with sunlight at various temperatures and different waters has been extensively studied [147]. Their half-lives range from 26 to 73 days, but the presence of humic acids reduces the degradation rates in comparison with distilled water. The degradation products identified are hydroxy- and dealkyl- derivatives for the S-triazines, dechlorinated and hydroxy derivatives for the anilides, and the keto-derivative for the thiocarbamates. All of them are more polar and probably more available for biodegradation.

5
Concluding Remarks

Improvements during the last decade in the sensitivity and selectivity of analytical techniques have allowed the measurement of a variety of organic contaminants—namely, hydrocarbons and chlorinated compounds—in seawater, sediments, and biota worldwide. However, the accuracy in the determination of trace organics in seawater is still challenging. The highest concentrations have been found in "hot spots" located in the proximity of urban areas, dumping sites, harbors, and estuaries. However, atmospheric deposition is also relevant on a global scale and may affect remote areas, including the polar regions. As expected, the situation is very different from region to region and from chemical to chemical. There are regions with a tradition in gathering information on POPs since the 1970s, whereas in others there are significant data gaps or even entirely missing information. The levels of those chemicals widely used in the past but now subject to regulation (e.g., DDT) are declining in temperate countries, but not in the Arctic, where an increase in some environmental levels has been detected. This is of concern because environmental conditions favor a higher persistence and hence a higher probability for POPs entering into the marine food web.

Organic compounds that are still in use in some regions (e.g., PCBs or lindane) show detectable levels in practically all marine compartments and, in some cases, are quite high. Effective assessment of their environmental occurrence should be a priority.

New candidate chemicals for global concern are insufficiently covered to draw a complete picture, while there are clear evidences of ecotoxicological effects for some of them. Data are scarce in the developing world, representing a big data gap. This is the case of PCDD/PCDFs, brominated and perfluorinated compounds, alkylphenols, etc.

The molecular marker concept, successfully applied for tracing urban waste inputs into coastal waters or fingerprinting oil spills, can be extended and improved with a better understanding of the weathering processes of the organic contaminants.

Finally, limited information is available on the major transport pathways of special relevance in marine pollution studies—namely, the land-to-ocean transfer, the role of the air–seawater exchange, ocean particle settling and transoceanic transport. Most regional scale models were developed in Europe, North America and East Asia and therefore tend to describe POP transport and fate under temperate environmental conditions. Progress in quantitatively describing the regional and global scale transport pathways is limited by an incomplete understanding of fate processes under non-temperate conditions, incomplete or highly uncertain emission information, and the lack of measured data required to evaluate the simulation results.

References

1. Erickson BE (2002) Environ Sci Technol 36:140A
2. Whylie P, Albaigés J, Barra R, Bouwman H, Dyke P, Wania F, Wong M (2003) Global assessment of persistent toxic substances. UNEP/GEF, Geneva
3. AMAP (1998) AMAP Assessment Report: Arctic pollution issues. Arctic Monitoring and Assessment Programme, Oslo
4. van der Oost R, Beyer J, Vermeulen NPE (2003) Environ Tox Pharm 13:57
5. Iwata H, Tanabe S, Sakai N, Nishimura A, Tatsukawa R (1994) Environ Pollut 85:15
6. Wania F, Mackay D (1996) Environ Sci Technol 30:390A
7. Bidleman TF, Patton GW, Hinckley DA, Walla MD, Cotham WE, Hargrave BT (1990) Chlorinated pesticides and polychlorinated biphenyls in the atmosphere of the Canadian Arctic. In: Kurtz DA (ed) Long range transport of pesticides. Lewis, Chelsea, Michigan, p 347
8. Sinkkonen S, Paasvirta J (2000) Chemosphere 40:943
9. Iwata H, Tanabe S, Sakai N, Tatsukawa R (1993) Environ Sci Technol 27:1080
10. Tolosa I, Bayona JM, Albaigés J (1995) Environ Sci Technol 29:2519
11. Kennicutt MC II, McDonald SJ, Sericano JL, Boothe P, Oliver J, Safe S, Presley BJ, Liu H, Wolfe D, Wade TL, Crockett A, Bockus D (1995) Environ Sci Technol 29:1279
12. Jonsson A, Gustafsson O, Axelman J, Sundberg H (2003) Environ Sci Technol 37:245
13. Eljarrat E, Caixach J, Rivera J (2001) Chemosphere 44:1383
14. Jonker MTO, Koelmans AA (2002) Environ Sci Technol 36:3725
15. Collier TK, Anulacion BF, Stein JE, Goksyr A, Varanasi U (1995) Environ Toxicol Chem 14:143
16. Rowland S, Donkin P, Smith E, Wraige E (2001) Environ Sci Technol 35:2640
17. Witt G, Trost E (1999) Chemosphere 38:1603
18. Amoroux D, Tessier E, Donard OFX (2000) Environ Sci Technol 34:988
19. IMO (2003) International Convention on the Control of Harmful Anti-fouling Systems on Ships, 2001. International Maritime Organization, London
20. Gabrielides GP, Alzieu C, Readman JW, Bacci E, Aboul Dahab O, Salihoglu I (1990) Mar Pollut Bull 21:233
21. Michel P, Averty B (1999) Mar Pollut Bull 38:268
22. RNO, Surveillance du Milieu Marin (1999) IFREMER. In: www.ifremer.fr/delpc/rno.htm (accessed by 06/04/2004)

23. Bressa G, Sisti E, Cima F (1997) Mar Chem 58:261
24. Barakat AO, Kim M, Qian Y, Wade T (2002) Mar Poll Bull 44:1422
25. Díez S, Ábalos M, Bayona JM (2002) Wat Res 36:905
26. Tolosa I, Merlini L, de Bertrand N, Bayona JM, Albaigés J (1992) Environ Toxicol Chem 11:145
27. Thomas KV, McHugh M, Hilton M, Waldock M (2003) Environ Pollut 123:153
28. Konstantinou IK, Albanis TA (2004) Environ Intl 30:235
29. Dahl B, Blanck H (1996) Mar Pollut Bull 32:342
30. Readman JW, Wee Kwong LL, Grondin D, Bartocci J, Villeeneuve JP, Mee LD (1993) Environ Sci Technol 27:1940
31. Bou-Carrasco P, Díez S, Jiménez J, Marco M-P, Bayona JM (2003) Wat Res 37:3658
32. Liu D, Maguire RJ, Lau YL, Pacepavicious GJ, Okamura H, Aoyama I (1997) Wat Res 31:2363
33. Voulvoulis N, Scrimshaw MD, Lester JN (1999) Appl Organomet Chem 13:135
34. Voulvoulis N, Scrimshaw MD, Lester JN (2002) Mar Environ Res 53:1
35. Thomas KV, Fileman TW, Readman JW, Waldock M (2001) Mar Pollut Bull 42:677
36. Hutzinger O, Thoma H (1987) Chemosphere 16:1877
37. Watanabe I, Sakai S (2003) Environ Internl 29:665
38. Alaee M, Arias P, Sjödin A, Bergman A (2003) Environ Internl 29:683
39. de Wit CA (2002) Chemosphere 46:583
40. Voorspels S, Covaci A, Schepens P (2003) Environ Sci Technol 37:4348
41. Bayen S, Thomas GO, Lee HK, Obbard JP (2003) Environ Toxicol Chem 22:2432
42. Giesy JP, Kannan K (2001) Environ Sci Technol 35:1339
43. Martin JW, Smithwick MM, Braune BM, Hoekstra PF, Muir DG, Mabury SA (2004) Environ Sci Technol 38:373
44. Metzler M, Pfeiffer E (2001) Chemistry of natural and anthropogenic endocrine active compounds, In: Metzler M (ed) The handbook of environmental chemistry, vol 3, part L. Springer, Berlin Heidelberg New York, p 63
45. Stroben E, Oehlmann J, Fiorini P (1992) Mar Biol 113:625
46. Fossi MC, Casini S, Ancora S, Moscatelli A, Ausili A, Notarbartolo G (2001) Mar Environ Res 52:477
47. Johns RB (ed) (1986) Biological markers in the sedimentary record. Elsevier, Amsterdam
48. Eganhouse RP (ed) (1997) Molecular markers in environmental geochemistry. ACS Symposium Series 671. American Chemical Society, Washington
49. Walker RW, Wun ChK, Litsky W (1982) CRC Crit Rev Environ Control 12:91
50. Leeming R, Ball A, Ashbolt N, Nichols PD (1996) Wat Res 30:2893
51. Takada H, Eganhouse RP (1998) Molecular markers of anthropogenic waste. In: Meyers RA (ed) Environmental analysis and remediation. Wiley, New York, p 2883
52. Bachtiar T, Coakley JP, Risk MJ (1996) Sci Total Environ 179:3
53. Readman JW, Mantoura RFC, Llewellyn CA, Preston MR, Reeves AD (1986) Int J Environ Anal Chem 27:29
54. Takada H, Farrington JW, Bothner MH, Johnson CG, Tripp BW (1994) Environ Sci Technol 28:1062
55. Venkatesan MI, Kaplan I (1990) Environ Sci Technol 24:208
56. Grimalt JO, Fernández P, Bayona JM, Albaigés J (1990) Environ Sci Technol 24:357
57. Nishimura M (1982) Geochim Cosmochim Acta 46:423
58. Venkatesan MI, Santiago CA (1989) Mar Biol 102:431
59. Chan K-H, Lam MHW, Poon K-F, Yeung H-Y, Chiu KT (1998) Wat Res 32:225
60. Leenher JA, Writer JH, Barber LB, Amy GL, Chapra SC (1995) Wat Res 29:1427

61. Hatcher PG, McGillivary PhA (1979) Environ Sci Technol 13:1225
62. Leeming R, Bate N, Hewlett R, Nichols PD (1998) Wat Sci Technol 38:15
63. Eganhouse RP, Blumfield DL, Kaplan IR (1983) Environ Sci Technol 17:523
64. Ishiwatari R, Takada H, Yun S-J, Matsumoto E (1983) Nature 301:599
65. Gustaffson O, Long C M, Gschwend PM (2001) Environ Sci Technol 35:2040
66. Bay SM, Zeng EY, Lorenson TD, Tran K, Alexander C (2003) Mar Environ Res 56:255
67. Fernández P, Valls M, Bayona JM, Albaigés J (1991) Environ Sci Technol 25:547
68. Valls M, Bayona JM, Albaigés J, Mansour M (1990) Chemosphere 20:599
69. Chalaux N, Bayona JM, Venkatesan MI, Albaigés J (1992) Mar Pollut Bull 24:403
70. Valls M, Bayona JM, Albaigés J (1989) Nature 337:722
71. Maldonado C, Dachs J, Bayona JM (1999) Environ Sci Technol 33:3290
72. Chalaux N, Takada H, Bayona JM (1995) Mar Environ Res 40:77
73. Maldonado C, Venkatesan MI, Phillips CR, Bayona JM (2000) Mar Pollut Bull 40:680
74. Spies RB, Andresen BD, Rice DW (1987) Nature 327:697
75. Reddy CM, Quinn JG (1997) Environ Sci Technol 31:2847
76. Zeng E, Tran K, Young D (2004) Environ Monit Assess 90:23
77. Peters KE, Moldowan JM (1993) The biomarker guide. Prentice-Hall, New York
78. Albaigés J, Albrecht P (1979) Int J Environ Anal Chem 6:171
79. Page DS, Foster JC, Fickett PM, Gilfillan ES (1988) Mar Pollut Bull 19:107
80. Sauer TC, Brown JS, Boehm PD, Aurand DV, Michel J, Hayes MO (1993) Mar Pollut Bull 27:117
81. Hostettler FD, Kvenvolden KA (1994) Org Geochem 21:927
82. Bence AE, Kvenvolden KA, Kennicutt II MC (1996) Org Geochem 24:7
83. Pastor D, Sánchez J, Porte C, Albaigés J (2001) Mar Pollut Bull 42:895
84. Albaigés J (1980) Fingerprinting petroleum pollutants in the Mediterranean Sea. In: Albaigés J (ed) Analytical techniques in environmental chemistry. Pergamon, Oxford, p 69
85. Barakat AO, Mostafa AR, Rullkotter J, Hegazi AR (1999) Mar Pollut Bull 38:535
86. Currie TJ, Alexander R, Kagi RI (1992) Org Geochem 18:595
87. Ekweozor CM, Okogun JI, Ekong DEU, Maxwell JR (1979) Chem Geol 27:11
88. Chosson P, Lanau C, Connan J, Dessort D (1991) Nature 351:640
89. Volkman JK, Alexander R, Kagi RI, Rowland SJ, Sheppard PN (1984) Org Geochem 6:619
90. Dowling LM, Boreham CJ, Hope JM, Murray AP, Summons RE (1995) Org Geochem 23:729
91. O'Malley VP, Abrajano TA, Hellou J (1996) Environ Sci Technol 30:634
92. Wang Z, Fingas M, Page DS (1999) J Chromatogr 843:369
93. Douglas GS, Bence AE, Prince RC, McMillen SJ, Butler EL (1996) Environ Sci Technol 30:2332
94. Wardroper AMK, Hoffmann CF, Maxwell JR, Barwise AJG, Goodwin NS, Park PJD (1984) Org Geochem 6:605
95. Broman D, Colsmjo A, Ganning B, Naf C, Zebühr Y (1988) Environ Sci Technol 22:1219
96. Tolosa I, Bayona JM, Albaigés J (1996) Environ Sci Technol 30:2495
97. Yunker MB, Snowdon LR, Macdonald RB, Smith JN, Fowler MG, Skibo DN, McLaughlin FA, Danyushevskaya AI, Petrova VI, Ivanov GI (1996) Environ Sci Technol 30:1310
98. Sicre MA, Marty JC, Saliot A, Aparicio X, Grimalt J, Albaigés J (1987) Atmos Environ 21:2247
99. Burns WA, Mankiewicz PJ, Bence AE, Page DS, Parker KR (1997) Environ Toxicol Chem 16:1119

100. Stout SA, Uhler AD, McCarthy KJ (2001) Environ Forensics 2:87
101. ASTM (1990) Standard practice for oil spill identification by gas chromatography and positive ion electron impact low resolution mass spectrometry. D-5739-95, American Society for Testing and Materials, W. Conshohocken, PA, USA
102. NORDTEST (1991) Nordtest method for oil spill identification. NT CHEM 001, Nordic Innovation Center, Espoo, Finland
103. Raoux Ch, Bayona JM, Miquel J-C, Teyssie L-L, Fowler SW, Albaigés J (1999) Estuar Coast Shelf Sci 48:605
104. Muir DGC, Omelchenko A, Grift N, Savoie D, Lockhart W, Wilkinson P, Brunskill G (1996) Environ Sci Technol 30:3609
105. Li Yf, Macdonald RW, Jantunen LMM, Harner T, Bidleman TF, Strachan WMJ (2002) Sci Total Environ 291:229
106. Bayona JM, Fernández F, Porte C, Tolosa I, Valls M, Albaigés J (1991) Chemosphere 23:313
107. Bidleman TF (1999) Water Air Soil Pollut 115:115
108. Nelson ED, McConnell LL, Baker FE (1998) Envrion Sci Technol 32:912
109. Bamford HA, Offenberg JH, Larsen RK, Ko F, Baker FE (1999) Environ Sci Technol 33:2138
110. Hardy JT, Crecelius EA, Antrim LD, Kisser SL, Broadhurst VL, Boehm PD, Steinhauer WG, Cooghan TH (1990) Mar Chem 28:333
111. García-Flor N, Guitart C, Ábalos M, Dachs J, Bayona JM, Albaigés J (2005) Mar Chem 96:331
112. Booij K, Van Drooge B (2000) Chemosphere 44:91
113. Dachs J, Lohman R, Ockenden WA, Méjanelle L, Eisenreich SJ, Jones KC (2002) Environ Sci Technol 36:4229
114. Dachs J, Bayona JM, Fowler SW, Miquel JC, Albaigés J (1996) Environ Sci Technol 52:75
115. Bailey R, Barrie LA, Halsall CJ, Fellin P, Muir DCG (2000) J Geophys Res 105:11805
116. Martí S, Bayona JM, Albaigés J (2001) Environ Sci Technol 35:2682
117. Dutta TK, Harayama S (2000) Environ Sci Technol 34:1500
118. van Pee K-H, Unversucht S (2003) Chemosphere 52:299
119. Eganhouse RP, Pontillo J, Leiker TJ (2000) Mar Chem 70:289
120. Quensen JF, Mueller SA, Jain MK, Tiedje JM (1988) Science 280:722
121. Maldonado C, Bayona JM (2002) Estuar Coast Shelf Sci 54:527
122. De Mora SJ, Stewart, Phillips D (1995) Mar Pollut Bull 30:50
123. Bayona JM, Albaigés J, Solanas AM, Grifoll M (1984) Chemosphere 15:595
124. Takada H, Ishiwatari R (1990) Environ Sci Technol 24:86
125. Eganhouse RP, Sherblom PM (2001) Mar Environ Res 51:51
126. Bayona JM, Albaigés J, Solanas AM, Parés R, Garrigues P, Ewald M (1986) Int J Environ Anal Chem 23:289
127. Kallenborn R, Hühnerfuss H (2001) Chiral environmental pollutants. Trace analysis and ecotoxicology. Springer, Berlin Heidelberg New York
128. Bidleman TF, Falconer RL (1999) Environ Sci Technol 33:206A
129. Faller J, Hühnerfuss H, König WA, Ludwig P (1991) Mar Pollut Bull 22:82
130. Ludwig P, Hühnerfuss H, König WA, Gunkel W (1992) Mar Chem 38:13
131. Jantunen LMM, Bidleman TF (1996) J Geophys Res 102:19279
132. Hühnerfuss H, Faller J, König WA, Ludwig P (1992) Environ Sci Technol 26:2127
133. Harner T, Kylin H, Bidleman TF, Strachan WMJ (1998) Organohal Comp 35:355
134. Jantunen LMM, Kylin H, Bidleman TF (1998) Organohal Comp 35:347
135. Mandalakis M, Berresheim H, Stephanou EG (2003) Environ Sci Technol 37:542

136. Kamens RM, Guo Z, Fulcher JN, Bell DA (1988) Environ Sci Technol 22:103
137. Masclet P, Mouvier G, Nikolau K (1986) Atmos Environ 20:439
138. Simo R, Grimalt JO, Albaigés J (1997) Environ Sci Technol 31:2697
139. Fernández F, Grifoll M, Solanas AM, Bayona JM, Albaigés J (1992) Environ Sci Technol 26:817
140. Atkinson R, Arey J (1994) Environ Health Persp 102:117
141. Ramdahl T, Zielinska B, Arey J, Atkinson R, Winer AM, Pitts Jr JN (1986) Nature 321:425
142. Zepp RG, Schotzhauer PF, Sink RM (1985) Environ Sci Technol 19:74
143. Ehrhardt M, Petrick G (1984) Mar Chem 15:47
144. Guiliano M, El Anba-Lurot F, Doumenq P, Mille G, Rontani JF (1997) J Photochem Photob A: Chem 102:127
145. Ehrhardt M, Weber RR (1991) Fresenius J Anal Chem 339:772
146. Choudhry CG, Webster GRB, Hutzinger O (1988) Toxicol Environ Chem 17:267
147. Konstantinou IK, Zarkadis AK, Albanis TA (2001) J Environ Qual 30:2001

Subject Index

Ace Lake, carotenoids 139
Acetyl-CoA 230
Aerosols 166
Akylbenzene sulfonates (ABS) 345
Akylbenzenes, tetrapropylene-based 345
Alcohols, alicyclic 46
–, triterpenoid 57
Algaenan 305
Algal lipids 214
Alkanes, straight-chain 31
Alkanols 46
Alkan-2-ones 49, 176
Alkenes 33
–, HBI 34
Alkenones 50, 214, 233, 235
Alkyl diols 46
Alkylnitriles 346
Alkylphenanthrenes 353
Alkylphenols, endocrine disruptors 341
Amino acids, nitrogen cycle 266
Ammonia 266
Annamox 216, 260
Anoxia, photic zone 127
Antifouling agents, biocides 334
– –, butyltin 333
– –, new generations 337
Archaea 119
–, carbon isotope fractionation 230
Archaeol 217, 240
Autotrophic assimilation, nitrogen 263
Azelaic acid 40

Bacterial artificial chromosomes (BACs) 116
Bacteriochlorin 74
Bacteriochlorophylls 74, 132, 215
Betaine lipids 45
Biliproteins 73
Biocides, antifouling paints 334

Biomarkers, lipids 2
–, petroleum 36
–, steroids 179, 214
–, terpenoids 179
Biomass, pigment content 96
Biomass burning 186
Biphytanes 217
Black carbon 4
Black Sea, anoxic, chemocline 141
Brominated flame retardants 338
Butyltin compounds 333
– –, biotic degradation 360

C3/C4 plants, carbon isotope composition 226, 233
Caffeine 347
Calvin cycle 210, 221
Carbon, dissolved organic (DOC) 218
–, organic 301
Carbon aerosol 183
Carbon cycling, food web 19
Carbon flux 5
Carbon isotope analysis, compound-specific (CSIA) 210
Carbon reservoirs 1
Carbonate system 218
Carbonic acid 218
Carbonic anhydrase 236
Carotenoids 61, 71, 75, 133
Chemotaxonomy, pigments 78
CHEMTAX 93
Chlorobactene 132, 217
Chlorophyllide a 89
Chlorophylls 56, 61, 71, 74, 215, 313, 314
Cholesterols 52, 342
Chromatiaceae 130, 132
Chrysenes 353
CO_2, atmospheric isotope ratio 226
Coprostanol, marker 342

Crocetane 217, 240
CTAB 111

Dansgaard-Oeschger 278
DDT 327, 341, 359
Degradation, organic matter 295
Degradation/decay 12
Denitrification 251, 257, 258
–, water column 78
DGCC 45
Diacylglycerols 42
Dibenzothiophenes 353
Dibutyltin 333
Dicarboxylic fatty acids 40
4,4-Dimethyloxazoline 39
Dinosterol 214
Dinosterone 55
Dissolved organic carbon (DOC) 218
Diuron 338
–, biodegradation 360
DNA analysis 105
DOC 9, 358
DOM 1, 8
Dust, global 166

Ectothiorhodospiraceae 130
Endocrine disruptors 324, 340
Estrogens, endocrine disruptors 340

Fatty acid methyl esters 38
Fatty acids 29, 216, 298
– –, monocarboxylic 36
Flame retardants, brominated 338
Food web, microbial 19
Fossil fuels 3
– –, emissions 184
Fulvic acids 183

Galactolipids 44
Genetically modified organisms 116
Global carbon cycle 1
Glycerol bidiphytanyl glycerol tetraether 49
Glycerol tetraethers 217
Glycerophosphatides 43
Glycolipds 44
Green sulfur bacteria 128, 130, 217

HCHs 324, 328, 362
Herbicides, photodegradation 365

Hopanes 349
Hopanoids 57, 58, 216
Hopenes 59
Humic acids 183
Hydrocarbons 31
–, aerosol particulate matter 175, 360

Irgarol, antifouling agent 338
Isoprenoid alkanes 32
Isoprenoid ether lipids 48
Isoprenoids 214, 232
Isorenieratene 217
Isotope ratio mass spectrometry (IRMS) 213, 277

Kerogen 300
Ketones, mid-chain 50
Kyllaren Fjord, carotenoids 141

LABs 361
Lignin 299
Lincoln Creek Formation 241
Lipids, aerosol particulate matter 175
–, algal 214
–, analysis 30
–, biomarkers 27
–, carbon isotope record 209
–, decomposition/degradation 15, 300
–, extraction 29
–, isoprenoid ether 48
Long-chain wax esters 175
Lycopane 33

Mahoney Lake, carotenoids 139
Maleimides 215
Marine sediments, lipids, carbon isotopes 209
Marmorito Limestone 241
Metagenome 116
Methane, anaerobic oxidation (AOM) 49, 240
Methane cycling 239
Methanogens 230
Methylphenanthrenes 361
Microalgae, carbon isotope composition 232
Microbes, organic matter cycling 18
Microbial loop 18
Microbial phylogeny 117
Molecular fossils 128

Monogalactosyl diacylglycerols 42

N_2 fixation 251, 254
Nitrification 267
Nitrogen cycle, isotopic tracers 251
Nitrogen inventory balance 270
Nitrogenase 256
Nucleic acids 114

O_2 minimum zone (OMZ) 261
Oceanic anoxic effects 128
Oil weathering indicators 353
Okenone 132
Oleanane 57
Organic carbon 301
Organochlorinated compounds 325
– –, dehalogenation 359
Oxidations 306

PAHs 331, 353, 363
–, aerosol 181
Palmitic acids 36
Particle sizes 4
PBBs 339
PBDEs 339
PCBs 329
PCDDs 330
PCDFs 330
PCR, DNA 115
Pentamethylicosane 217
Perfluorodecyl carboxylate 340
Perfluorooctyl sulfonate 340
Peridinin 76
Pesticides, organochlorine 326
Petroleum biomarkers 36
Petroleum fingerprinting 355
PFCs 340
Phenylureas 360
3-Phosphoglycerate 221
Phospholipid fatty acids 43
Phospholipids 29, 43, 216
Photic zone, anoxia 127
Photoautotrophs 210
Phycobiliproteins 72, 77
Phytane 32, 215
Phytol 74
Phytoplankton production, pigment markers 71
Pigment distribution 210
Pigment retention times 86

Pigment suites 79
Pigments, extraction 82
Plant waxes, aerosol particulate matter 175
POC 1, 6, 233, 306, 358
Polysaccharide hydrolysis 298
POM 10, 267
POPs 324, 325, 358
Potash 174
Preservation 16
Pristane 32, 215
Prokaryotes, biomass 20
Proxies, anoxic deposits 129
PUFA 37
Purple sulfur bacteria 128, 130

Redox oscillation 295
Remineralization, organic N 266

Sargasso Sea, nirogen cycle 257
Sediment DNA 111
Sediments, lipids, carbon isotopes 209
Sitosterol 54
Size-reactivity continuum 10
Soil, carbon 3
Soot, aerosol 174
Sorptive preservation 303
Stanols 51
Steranes 349
Steroid biomarkers 179, 214
Steroid ketones 55
Steroids, fecal 342
Sterols 29, 51, 214, 298
Steryl chlorin esters 56
Steryl esters 56
Suboxia 259
Sulfate-reducing bacteria 240
Sulfoquinovosyldiacylglycerol 44
Sulfur bacteria 128
Sulfur compounds 62

TBBPA 339
TBT, debutylation 360
TCA cycle, reverse 230
TCDD 330
Terpenoid biomarkers 179
Thiocarbamates, photodegradation 365
Tin, organotin compounds 333, 360

Tracers, inorganic
Transport pathways, pollutants 356
Triacylglycerols 29, 43
Trialkylamines 346
Triazines 360, 365
Tributyltin 333
Trichodesmium spp. 255
Triterpenoid alcohols 57

Ursane 57

Vienna Pee Dee Belemnite 212
Violaxanthin 77

Wax (alkyl) esters 45
Waxes, aerosol particulate matter 175
Whiskey Creek deposits 241
Wind systems 166
Wood smoke, aerosol 174
Wüstenquarz 167

Xanthophylls 75

Printing: Krips bv, Meppel
Binding: Stürtz, Würzburg